2nd edition

histology and
cell biology

The National Medical Series for Independent Study

2nd edition
histology and cell biology

Kurt E. Johnson, Ph.D.

Professor of Anatomy
George Washington University
  Medical Center
Washington, D.C.

**NMS**

National Medical Series from Williams & Wilkins
Baltimore, Hong Kong, London, Sydney

Harwal Publishing Company, Malvern, Pennsylvania

Managing editor: Debra Dreger
Project editor: Judd Howard
Production: Keith LaSala, Laurie Forsyth, Judy Johnson
Art direction and illustration: Wieslawa B. Langenfeld
Composition and layout: Telecomposition, Inc.

Library of Congress Cataloging-in-Publication Data

Johnson, Kurt E.
   Histology and cell biology / Kurt E. Johnson.
      p.   cm. — (The National medical series for independent study)
  (A Williams & Wilkins medical publication)
    Includes index.
    ISBN 0-683-06210-7 (pbk. : alk. paper)
    1. Histology—Outlines, syllabi, etc.   2. Cytology—Outlines,
syllabi, etc. I. Title.   II. Series.   III. Series: A Williams & Wilkins medical
publication.
    [DNLM: 1. Cytology—examination questions.   2. Cytology— outlines.
3. Histology—examination questions.   4. Histology—outlines. QS
18 J67c]
  QM553.J624   1990
  611'.018'—dc20
  DNLM/DLC
  for Library of Congress   90-5226
                                                CIP

ISBN 0-683-06210-7

© 1991 by Williams & Wilkins, Baltimore, Maryland

10 9 8 7 6 5 4 3

Dedication

This book is dedicated to my squash and tennis partners, Steve Goldman, Ron Chander-bahn, Bryant Chomiak, Bob Walker, David Campbell, and Ralph Kramer. The many hours of fun and competition spent with them on the court have contributed immeasurably to my current psychological orientation. They get credit or are to blame, depending on your viewpoint. Whatever, I am grateful for their friendship. Thank you gentlemen!

This book also is dedicated to the memory of my late colleague, Ernest N. Albert, Ph.D. He was a good friend and will be missed.

Contents

Preface

NMS *Histology and Cell Biology,* 2nd edition, is an extensively revised version of NMS *Histology and Embryology.* There are four substantial changes in this new edition. First, material on embryology has been deleted and expanded into another NMS title, *Human Developmental Anatomy.* Second, the single chapter on cell biology contained in *Histology and Embryology* has been expanded into six new chapters that cover structural cell biology (Chapters 2–7). Third, there are numerous additions of material covering functional cell biology in Chapters 8–27. And finally, many new questions, answers, and explanations have been added. It is hoped that these changes will make the book more useful to students in the classroom and in preparing for licensure examinations.

Kurt E. Johnson

Acknowledgments

I would like to thank my colleagues, past and present, for their help with providing photographic material for this book: Dr. M.K. Reedy and Dr. J.D. Robertson (Department of Anatomy, Duke University); Dr. E.N. Albert, Dr. D.P. DeSimone, Dr. M.J. Keoring, Dr. J.M. Rosenstein, Dr. F.J. Slaby, Dr. R.J. Walsh, and Mr. Fred Lightfoot (Department of Anatomy, George Washington University); Dr. B. Gulyas (NICHD); Dr. J.A. Long (Department of Anatomy, UCLA); and Dr. H.A. Padykula (Department of Anatomy, University of Massachusetts).

The editorial assistance of Debra Dreger also is gratefully acknowledged. Her dedication to excellence and sense of humor are greatly appreciated. I am grateful for all of these contributions and acknowledge that any errors contained herein are my own.

To the Reader

Since 1984, the *National Medical Series for Independent Study* has been helping medical students meet the challenge of education and clinical training. In today's climate of burgeoning information and complex clinical issues, a medical career is more demanding than ever. Increasingly, medical training must prepare physicians to seek and synthesize necessary information and to apply that information successfully.

The *National Medical Series* is designed to provide a logical framework for organizing, learning, reviewing, and applying the conceptual and factual information covered in basic and clinical sciences. Each book includes a comprehensive outline of the essential content of a discipline, with up to 500 study questions. The combination of an outlined text and tools for self-evaluation allows easy retrieval of salient information.

All study questions are accompanied by the correct answer, a paragraph-length explanation, and specific reference to the text where the topic is discussed. Study questions that follow each chapter use the current National Board format to reinforce the chapter content. Study questions appearing at the end of the text in the Challenge Exam vary in format depending on the book. Wherever possible, Challenge Exam questions are presented as clinical cases or scenarios intended to simulate real-life application of medical knowledge. The goal of this exam is to challenge the student to draw from information presented throughout the book.

All of the books in the *National Medical Series* are constantly being updated and revised. The authors and editors devote considerable time and effort to ensure that the information required by all medical school curricula is included. Strict editorial attention is given to accuracy, organization, and consistency. Further shaping of the series occurs in response to biannual discussions held with a panel of medical student advisors drawn from schools throughout the United States. At these meetings, the editorial staff considers the needs of medical students to learn how the *National Medical Series* can better serve them. In this regard, the Harwal staff welcomes all comments and suggestions.

1
Methods Used in Histology and Cell Biology

I. INTRODUCTION. Histology and cell biology researchers use a variety of methods of extending visual perception to study the relationship between structure and function in human cells and tissues.

    A. Microscopes allow researchers to see cellular and macromolecular details that are not visible to the naked eye.

        1. Light microscopes employ transillumination and are used to examine living and prepared specimens and specimens with inherent or applied fluorescent properties.

        2. Electron microscopes illuminate specimens with an electron beam. They have 1000 times the resolving power of light microscopes and provide resolution to the threshold of atomic detail.

            a. Transmission electron microscopy (TEM) uses thin specimen sections and reveals details of the cell's interior.

            b. Scanning electron microscopy (SEM) provides a high resolution view of the cell surface and environment.

    B. Supplemental study methods. Often, microscopy is used in conjunction with one or more of the following techniques to study details of cellular anatomy and physiology.

        1. Staining is used in light and electron microscopy to increase the contrast of specimen structures. Some stains also convey information about chemical composition.

        2. Autoradiography uses radioactive isotopes to localize substances within cells.

        3. Differential centrifugation uses centrifugal force to separate cellular organelles and inclusions, thereby permitting biochemical analysis of subcellular fractions.

        4. Freeze-fracture-etch is used in conjunction with electron microscopy. This technique creates a metallic replica of a specimen and is used to study the details of membrane structure and intercellular junctions.

II. SPECIMEN PREPARATION varies for light and electron microscopy, although the steps are similar in both formats. Sections II A and B describe the basic steps in preparing specimens for light microscopy and TEM. Section II D describes specimen preparation for SEM.

    A. Fixation and sectioning

        1. Fixation is the first step in specimen preparation for light microscopy or TEM. Fixation preserves cell structure while introducing a minimal number of **artifacts** (artificial structures produced during fixation that may appear to be true histologic features).

            a. Small tissue or organ fragments are immersed in a buffer solution containing a fixative such as **glutaraldehyde**.

            b. Electron microscopy specimens typically are immersed in a buffered **osmium tetroxide (OsO_4)** solution after fixation with glutaraldehyde. Osmium tetroxide has a high affinity for cell components that contain lipids (e.g., plasma membrane, nuclear envelope, membranous components of the mitochondria and Golgi apparatus).

2. Dehydration. After fixation, specimens for light microscopy and TEM are dehydrated by immersion in a series of solutions containing increasing concentrations of ethanol. Dehydration is necessary because most embedding media (e.g., paraffin, plastic monomers) are immiscible in water.

3. Embedding. After dehydration, the specimen is treated with a liquid embedding medium that infiltrates and hardens the specimen so that it can be sliced into thin sections suitable for staining and microscopic examination.

   a. Light microscopy often uses paraffin, which hardens the specimen as it cools, as an embedding medium.

      (1) Paraffin is easy to work with, speeds specimen preparation, and stains reliably.

      (2) However, paraffin has low tensile strength and, therefore, cannot be cut into very thin sections.

   b. TEM uses plastic monomers as an embedding media.

      (1) Specimens are infiltrated with propylene oxide and then with unpolymerized mixtures of plastic monomers, cross-linking agents, and catalysts.

      (2) Specimens are then heated in an oven which polymerizes the plastic and hardens the specimen.

      (3) These plastics yield very hard specimens that can be cut into ultra-thin sections (20–100 nm thick).

4. Sectioning

   a. Light microscopic studies typically use 5–10 μm thick paraffin-embedded sections.

      (1) Sections are cut with a rotary microtome, a machine that moves the specimen across a sharp metal blade, advancing the specimen the desired thickness (e.g., 10 μm) after each pass.

      (2) Successive sections come off the microtome in a ribbon. Sections are cut from this ribbon and mounted on glass slides.

   b. TEM studies typically use 0.02–1.0 μm thick plastic-embedded sections.

      (1) A special microtome containing an extremely sharp glass or diamond blade is used to cut thin, smooth sections.

      (2) Sections are floated off the blade in a trough of water and then collected on small metal grids.

   c. Section thickness affects image resolution.

      (1) In thick sections, structures existing at different levels in the section overlap and interfere with each other in the image, thus reducing resolution.

      (2) Thin sections eliminate the overlapping problem and allow higher resolution views of specimens.

B. Staining. Light microscopy sections usually are stained with dyes or fluorescent tags. TEM staining involves treating the section with heavy metal salts.

1. Acidophilia and basophilia

   a. Acidophilia. Tissue components that bind acidic dyes (e.g., eosin, orange G) are **acidophilic**. Acidophilic substances bear a **net positive charge** and bind negatively charged (i.e., acidic, anionic) dyes. Acidophilic structures include:

      (1) Erythrocyte cytoplasm (due to its high concentration of hemoglobin)

      (2) Collagen fibers

      (3) Mitochondria

      (4) Lysosomes

   b. Basophilia. Tissue components that bind basic dyes (e.g., hematoxylin, methylene blue, toluidine blue) are **basophilic**. Basophilic substances bear a **net negative charge** and bind positively charged (i.e., basic, cationic) dyes. Basophilic structures include:

      (1) Nuclei

      (2) Aggregates of rough endoplasmic reticulum, such as those found in cells that secrete large amounts of protein (e.g., plasma cells, which secrete large amounts of immunoglobulins)

      (3) Extracellular matrix

2. Metachromasia

   a. Metachromasia, which literally means a change in color, is a property of certain dyes (usually basic dyes). Structures that stain with these dyes (e.g., mast cell cytoplasmic granules) are metachromatic structures.

(1) When a dye such as toluidine blue binds to tissue substances that occur in low concentrations, the substance stains **orthochromatically** (blue).

(2) When the same dye binds to tissue substances that occur in high concentrations, the substance stains **metachromatically** (purple).

   b. Cartilage staining is an example that clearly illustrates metachromasia.

(1) Areas of cartilage extracellular matrix containing high concentrations of chondroitin sulfate, which contains numerous sulfate (SO_4^{2-}) groups and consequently has a strong negative net charge, are **metachromatic**.

(2) Cartilage cell nuclei contain lower concentrations of negatively charged moieties (e.g., DNA) and are, therefore, **orthochromatic**.

3. Periodic acid-Schiff (PAS) reaction

   a. The PAS reaction is used to identify carbohydrates by exposing tissue sections to periodic acid oxidation and then reacting them with Schiff's reagent (essentially leukofuchsin in solution).

   b. For example, the PAS reaction can be used to identify glycogen (a glucose polymer stored in the cytoplasm of various cells) in liver cells.

(1) During periodate oxidation, hydroxyl groups on the glucose oxidize into aldehyde.

(2) The free aldehydes react strongly with bisulfite groups on leukofuchsin and yield a magenta condensation product.

   c. Other PAS-positive cell structures include the following.

(1) The **glycocalyx,** the carbohydrate-rich coat surrounding many cells, is strongly PAS-positive because it contains an abundance of glycoproteins.

(2) Most epithelia rest on a basement membrane that is strongly PAS-positive because it contains numerous glycoconjugate-rich macromolecules.

   d. Researchers use **histochemical reactions** in conjunction with PAS reaction to identify specific cellular constituents.

(1) Glycogen, for example, can be distinguished from other PAS-positive cell substances by preincubating slides with α-amylase, an enzyme that destroys only glycogen. (Cell structures that contain glycogen are PAS-positive and α-amylase sensitive.)

(2) Mild acid hydrolysis of DNA creates Schiff reagent reactive groups in the deoxyribose of DNA. This histochemical method of staining is a specific, sensitive test for DNA and is the basis for the Feulgen reaction.

C. Special stains

1. Staining elastic fibers and reticular fibers

   a. Elastic fibers contain elastin (a protein) and stain heavily with orcein dye. Also, Weigert's elastic stain selectively reveals elastic fibers.

   b. Reticular fibers (connective tissue fibers that exist in many organs) are a special kind of collagen fiber that has a rich coat of glycoproteins. They are especially prevalent in the liver, spleen, and lymph nodes.

   c. Reticular fibers are **argyrophilic** structures; their glycoproteins can reduce silver salts to silver metals. This reduction process deposits a black stain around these structures.

2. Fluorescent tags. Fluorescent molecules are used to tag antibodies to specific tissue antigens so they can be localized and examined in a fluorescence microscope.

   a. Fluorescent tags can be very specific for some intracellular or extracellular components. For example, antibodies can be prepared for specific proteins such as actin (intracellular) or collagen (extracellular).

   b. The antibodies are covalently coupled to **fluorochromes** (fluorescent chemicals), such as fluorescein isothiocyanate (FITC), and then bound to cell sections or whole cells where the tagged antibodies bind with specific antigens.

3. Histochemistry is used to localize enzymes within tissues.

   a. Typically, frozen sections are incubated in reaction mixtures that create reaction products from enzymatic activities in the specimen. Then, reaction products are precipitated in a way that minimizes diffusion away from the production site.

   b. For example, alkaline phosphatase activity in the renal brush border is demonstrated by incubating frozen sections with phosphorylated substrates in alkaline buffers and then precipitating the phosphates with lead salts, which appear black or brown in the sections.

4. Staining for transmission electron microscopy

   a. TEM specimens are stained by floating sections in solutions of heavy metal salts (e.g., lead citrate, uranyl acetate). In these solutions, heavy metals bind to the sections.

   b. The nuclei of heavy metal ions contain many protons and neutrons, which scatter electrons in the electron microscope's electron beam. This increases the specimen's electron density and, therefore, its contrast, resulting in a clearer image.

D. Specimen preparation for scanning electron microscopy

1. SEM specimens are fixed in buffered solutions of glutaraldehyde and osmium tetroxide.

2. The next step, **critical point drying,** is the most important stage of SEM specimen preparation.

   a. During critical point drying, specimens are infiltrated with ethanol or acetone, two solvents that are miscible with liquid carbon dioxide (CO_2), which dehydrates the specimen without subjecting it to the destructive surface tension forces of a liquid-gas interface.

   b. SEM examines surface details. Critical point drying preserves delicate surface features that would be destroyed by the surface tension forces of a liquid-gas interface.

3. After critical point drying, specimens receive a thin coating of gold and palladium, which scatters electrons in the microscope's electron beam, resulting in a clearer image.

E. Artifact interpretation. All specimen preparation techniques introduce artifacts. Scientists interpret their existence and significance by using more than one method to study a given organelle. For example, a cell nucleus present in a prepared specimen could be interpreted as an artifact of fixation, except that live, untreated cells studied with the phase contrast microscope also contain a nucleus.

III. LIGHT MICROSCOPY. Modern light microscopes are versatile scientific instruments used to study living and prepared specimens as well as specimens with inherent or applied fluorescent properties.

A. Light microscope components (Figure 1-1)

1. A light microscope has a built-in **light source** and an adjustable **substage condenser,** which project light into the objectives.

2. A mechanically operated **stage** carries the specimen.

3. **Objectives** magnify the specimen image, and **oculars** complete the image formation process.

B. Image formation

1. Modern light microscopes have a resolution limit of 0.2–0.4 μm, or approximately one-twentieth the diameter of a human erythrocyte.

   a. Many subcellular organelles exist in this size range. When specimens are carefully prepared, these organelles are visible in the light microscope.

   b. Scientists were not aware that organelles smaller than this existed until the electron microscope was developed.

2. The resolving power of a light microscope depends on three variables: objective magnification, objective **numerical aperture (N.A.),** and the wavelength of light used to illuminate the specimen.

   a. A typical compound microscope has a low-power ($4 \times$) objective, one or two intermediate-power ($10 \times$, $40 \times$) objectives, and a high-power ($100 \times$) oil-immersion objective. Oculars provide some magnification (usually $10 \times$) as well, so total magnification ranges from $40 \times$ to $1000 \times$.

   b. Every objective has an N.A., which is calculated by the formula

$$N.A. = n \text{ sine } \theta$$

   where n is the refractive index of the medium between the specimen and the objective lens, and θ is half the angle of the cone of light gathered by the objective.

   (1) A low-power objective has a low N.A.; a high-power objective has a high N.A. Thus, image resolution improves as the N.A. increases.

   (2) If air (refractive index of 1 by definition) is the medium between specimen and objective, no lens can gather light from a light cone when $\frac{1}{2}$ angle $\theta > 90°$. However,

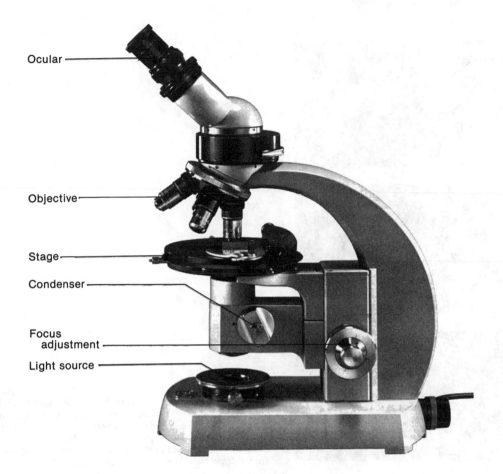

Figure 1-1. Components of the light microscope. (Photograph courtesy of Carl Zeiss, Inc., New York.)

if a high refractive index immersion oil is the intervening medium, the objective can gather light from a wider angle light cone, thereby increasing the objective N.A. and improving image resolution.

  c. The specimen produces a complex diffraction pattern in the back focal plane of the microscope, which is resynthesized into a recognizable **magnified** image by the oculars.

  d. Resolution improves as the wavelength of light used to illuminate the specimen shortens.

C. Types of light microscopy (Figure 1-2 shows representative micrographs)

  1. Brightfield microscopy uses standard lenses and condensers and has a limit of resolution of approximately 0.3 μm.

  2. Phase-contrast microscopy uses modified objective lenses and condensers to permit direct examination of living cells without fixation or staining.

  a. Phase-contrast microscopes have substage condensers and objectives that convert slight differences in the refractive indices of specimen structures and domains into distinct differences in light intensity.

  b. Each organelle and domain in a cell has a unique chemical composition or concentration of chemical constituents, which determines its refractive index. A phase-contrast microscope exaggerates these differences so that cell nuclei, cytoplasm ground substance, mitochondria, and other cytoplasmic structures appear in contrast to one another.

  c. This type of microscopy does not harm living cells and allows scientists to examine cell behavior under many artificial circumstances.

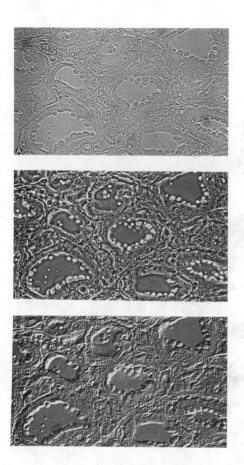

Figure 1-2. Light micrographs of thyroid follicular epithelium viewed in a brightfield microscope (*A*), a phase-contrast microscope (*B*), and a differential interference contrast microscope (*C*). (Photograph courtesy of Carl Zeiss, Inc., New York. Reprinted from Johnson KE: *Histology: Microscopic Anatomy and Embryology.* New York, John Wiley, 1982, p 3.)

3. **Differential interference contrast microscopy** (Nomarski light microscopy) uses special condensers and objective lenses to transform differences in refractive index into an image that appears to have a three-dimensional character. In this type of microscopy, the nucleus and various particulate cytoplasmic inclusions appear in low relief.

4. **Fluorescence light microscopy** is used to localize inherently fluorescent substances or substances labeled with fluorescent tags. Fluorescence microscopes have a high intensity light source and two special filters.
   a. The **excitor filter,** located between the light source and the specimen, blocks all light wavelengths except those that excite the fluorochrome.
   b. The **barrier filter,** located between the specimen and the ocular, blocks all light wavelengths except those emitted by the fluorochrome.

IV. ELECTRON MICROSCOPY is used to view fixed and sectioned or metal-coated specimens under magnification high enough to resolve fine details of the specimen.

A. Electron microscope components and image formation

1. In contrast to light microscopes, electron microscopes illuminate specimens with a short wavelength stream of electrons rather than photons, and they form images with magnetic lenses rather than glass lenses.
   a. The resolving power of modern electron microscopes is high enough to examine details of all types of cellular organelles and some macromolecules.
   b. Due to the short wavelength of electron radiation, electron microscopes have a high theoretical resolving power. However, the magnetic lens N.A. is so low that the theoretical limit is unattainable.

2. Specimens are illuminated in an evacuated column within the microscope, because the electron beam would be scattered by air.

3. The magnetic lenses form an image that is displayed on a video screen. Often, the image is photographed to generate a permanent record. These electron micrographs show the fine details of a specimen, often to the molecular level of resolution.

B. **Types of electron microscopy.** TEM and SEM are distinct types of electron microscopy that employ different methods of specimen preparation, irradiation, and image formation to study the details of cell structures and components.

   1. TEM uses thinly sliced, plastic-embedded sections that are stained with heavy metal salts. TEM is used to study the fine details of cell structure, such as the morphology of cell surfaces (in cross section) and the internal elements of cells; it can resolve features as small as 0.5 nm.

   2. SEM uses whole specimens that are subjected to critical point drying and then coated with a thin layer of gold and palladium. SEM is used to study three-dimensional features of cell surfaces; it can resolve features as small as 5 nm.

V. SUPPLEMENTAL HISTOLOGIC TECHNIQUES

A. **Autoradiography** is a valuable technique for localizing substances within cells using radioactive isotopes.

   1. **Utility**
      a. Autoradiography is used to localize DNA synthesis sites and hormone receptors and to identify the synthesis and distribution sites of metabolic products.
      b. Radioactive isotopes of cell components are introduced into cells and become involved in the metabolic events of the cell. For example, to detect cells involved in DNA synthesis, a labeled DNA precursor (e.g., tritiated thymidine) is introduced. The precursor is incorporated into the DNA, thus metabolically labeling cell nuclei.
      c. When autoradiography is used with electron microscopy, which provides much higher resolution, researchers can trace secretion products from their synthesis sites, through cellular organelles, to their eventual destinations.

   2. **Methodology**
      a. Labeled specimens are fixed and sectioned and then coated in the dark with a silver halide emulsion that is sensitive to beta particles released during the radioactive decay of tritium atoms.
      b. After an exposure period that can range from days to months, the exposed emulsion is developed. Silver grains appear in the emulsion directly above site of decaying isotopes.
      c. To follow the movement of labeled cells, researchers vary the time between the initial labeling period, called a **pulse,** and fixation to terminate the experiment, called a **chase**.
      d. During the chase, labeled cells may migrate from the site of mitosis (e.g., intestinal crypts) to another site (e.g., tops of the intestinal villi).

B. **Differential centrifugation** uses a centrifuge to isolate and collect large quantities of subcellular components.

   1. **Utility**
      a. Cell biologists use differential centrifugation to isolate large quantities of subcellular components (e.g., endoplasmic reticulum, lysosomes, Golgi vesicles) so they can study their anatomy, biochemistry, and physiology.
      b. Differential centrifugation is possible because most intracellular organelles are discrete particles that have unique densities and sizes.
      c. This process facilitated the discovery that enzymes of the Krebs cycle were localized in one intracellular organelle—the mitochondrion.

   2. **Methodology**
      a. Before centrifugation, the tissue sample is gently disrupted to release organelles without destroying them. Disruption occurs in buffer solutions so that the pH and ionic strength of the disruption medium can be controlled.

b. Next, the mixture of organelles is placed in a centrifuge tube containing a medium that varies in density at specific levels in the tube.

  (1) To obtain variations in medium density, researchers construct a concentration (density) gradient of a high molecular weight polymer, which permits the formation of a dense solution without significantly increasing its osmotic pressure.

  (2) The concentration gradient can be continuous or discontinuous. Often, discontinuous gradients are formed by varying the solute concentration in successive layers.

c. Then, the entire preparation is spun at high speed in a centrifuge. During centrifugation, organelles sediment through the medium until they reach the portion of the density gradient that matches their density, effectively separating the organelles according to their density.

d. After centrifugation, organelles are collected from the tube and analyzed by electron microscopy or by biochemical methods.

C. Freeze-fracture-etch

1. Utility

a. Freeze-fracture-etch is a technique that overcomes the inherent limitation of thin specimen sections to reveal details of the internal structure of membranes and intercellular junctions.

b. This technique allows researchers to study structural details of membranes in the hydrophobic portion of the membrane between the inner and outer membrane leaflets (see Chapter 2 III for more information about membrane structure and freeze-fracture-etch).

2. Methodology

a. Specimen modification by fixatives is minimized, thus reducing the number of artifacts introduced during this step.

b. Unfixed or lightly fixed specimens are infiltrated with glycerol, mounted on metal blocks, and then rapidly cooled in liquid nitrogen. Rapidly freezing the water-glycerol solution in the cells prevents ice crystal formation and the destructive effects of these macroscopic structures.

c. The frozen tissue specimen is placed in a special apparatus and fractured with a sharp metal blade in a vacuum of approximately 10^{-8} mm Hg.

  (1) The fracture plane passes between the inner and outer leaflets of the cell membranes and along cell surfaces, retaining the fine detail of membrane internal structures and cell surfaces.

  (2) The fracture faces are coated with a layer of metal that is evaporated from a heated filament.

d. After the specimen is removed from the freeze-fracture-etch apparatus, the tissue is thawed and then chemically digested, leaving a metal replica that is examined using TEM. Figure 1-3 is a micrograph of a specimen prepared by the freeze-fracture-etch technique.

Figure 1-3. Electron micrograph of a rat ileum specimen prepared by the freeze-fracture-etch technique. The ileal lumen (*upper left*) contains water. Numerous microvilli cover the epithelial cell apical surface. Many apical membrane delimited vesicles are visible as well. (Micrograph courtesy of Dr. J.D. Robertson, Department of Anatomy, Duke University. Reprinted from Johnson KE: *Histology: Microscopic Anatomy and Embryology.* New York, John Wiley, 1982, p 10.)

STUDY QUESTIONS

Directions: Each question below contains five suggested answers. Choose the **one best** response to each question.

1. Each of the following statements concerning the periodic acid-Schiff reaction (PAS) is true EXCEPT

(A) tissue sections are oxidized and then treated with Schiff's reagent
(B) it stains many glycoconjugates magenta
(C) it is used with α-amylase to identify DNA
(D) it stains the glycocalyx
(E) it stains many basement membranes

2. Each of the following statements concerning phase-contrast microscopy is true EXCEPT

(A) it can be used to observe live specimens
(B) it converts differences in the density of specimens into differences in light intensity in the image
(C) it uses special objectives and condensers
(D) it uses conventional incandescent illumination
(E) it requires stained specimens for proper visualization

3. Each of the following statements concerning image resolution in brightfield microscopy is true EXCEPT

(A) immersion oil reduces resolution
(B) the limit of resolution is less than 1 μm
(C) oculars are crucial for image formation
(D) high N.A. lenses yield high resolution
(E) N.A. is related to the refractive index of the medium between the specimen and the objective

4. Each of the following statements concerning autoradiography is true EXCEPT

(A) beta-particle disintegration causes exposure of emulsion grains
(B) it is used with transmission electron microscopy
(C) it can be performed in room light
(D) sections are coated with silver halide emulsions
(E) it can be used to trace the movement of cells or cell substances

5. Each of the following statements concerning scanning electron microscopy is true EXCEPT

(A) it is used to study details of the cell surface
(B) specimens are coated with metal
(C) critical point drying evaporates water from the specimen
(D) specimen fixation involves immersion in buffered glutaraldehyde and osmium tetroxide solutions
(E) its resolution is not as powerful as transmission electron microscopy resolution

6. Each of the following statements concerning freeze-fracture-etch is true EXCEPT

(A) it allows scientists to study membrane hydrophobic domains
(B) specimens are infiltrated with glycerol to induce ice crystal formation
(C) it is used with transmission electron microscopy
(D) frozen specimens are fractured in a vacuum
(E) a replica of the specimen is observed in the microscope

7. Each of the following statements concerning stains or staining is true EXCEPT

(A) staining reduces the contrast of specimen components
(B) acidophilic dyes bind to positively charged components
(C) antigens can be stained with fluorescent antibodies
(D) stains absorb light, scatter electrons, or add fluorescence
(E) eosin stains acidophilic structures

8. Each of the following statements concerning metachromasia is true EXCEPT

(A) metachromatic staining occurs in substances that bind low concentrations of dye
(B) toluidine blue is a metachromatic dye
(C) sulfated proteoglycans exhibit metachromasia
(D) many basophilic substances are metachromatic
(E) structures that stain orthochromatically with toluidine blue appear blue

Directions: The groups of questions below consist of lettered choices followed by several numbered items. For each numbered item select the **one** lettered choice with which it is **most** closely associated. Each lettered choice may be used once, more than once, or not at all.

Questions 9–13

For each light microscopic function listed below, select the lettered structure in the photograph that performs the function.

Courtesy of Carl Zeiss, Inc., New York.

9. Used to focus the image

10. Involved in image formation; gathers light rays emanating from the specimen

11. The movable structure that holds the specimen

12. Forms an image from the diffracted rays in the back focal plane

13. Supplies the light that illuminates the specimen

Questions 14–18

Match each structure listed below with the appropriate description of its staining characteristics.

(A) Acidophilic
(B) Basophilic
(C) Both
(D) Neither

14. A cell nucleus

15. Collagen fibers

16. Plasma cell cytoplasm

17. Erythrocyte cytoplasm

18. Fat vacuoles

ANSWERS AND EXPLANATIONS

1. The answer is C. [*II B 3*] The periodic acid-Schiff (PAS) reaction is a specific histochemical test used to identify glycoconjugates. It involves exposing a tissue section to oxidation and then staining it with Schiff's reagent. A magenta condensation product forms when the test is positive. The glycocalyx and many basement membranes that contain glycoconjugate-rich macromolecules are PAS-positive structures. Combining the PAS reaction and α-amylase treatment is a specific histochemical test for glycogen. The Feulgen reaction identifies DNA using Schiff's reagent and acid hydrolysis.

2. The answer is E. [*III C 2*] The invention of phase-contrast light microscopy expanded the utility of the brightfield microscope. Phase-contrast microscopy uses special objectives and condensers and conventional incandescent illumination. In phase-contrast microscopy, minute variations in the density or, more accurately, the refractive index of specimen structures and domains are converted into distinct differences in light intensity during image formation. Using this technique, living specimens can be observed without fixation or staining. It is especially useful for studying the movement of living cells in tissue cultures. The thin locomotory organelles on these cells, which are not visible under brightfield optics, are clearly visible.

3. The answer is A. [*III B, C 1*] The limit of resolution in brightfield light microscopy is 0.3 μm. Numerical aperture (N.A.) is defined by the formula: N.A. = n sine θ (n is the refractive index of the medium between the objective and the specimen; θ is half of the angle of the cone of light gathered by the lens). Lenses with a high N.A. provide high resolution; those with a low N.A. provide low resolution. Using immersion oil, which has a refractive index greater than 1.0, effectively increases the lens N.A., allowing it to gather more light from a given light cone, thereby increasing resolution. Image formation is impossible without oculars.

4. The answer is C. [*V A*] Autoradiography can be used in light and transmission electron microscopes. Specimen sections are coated with photographic emulsions of silver halides. The disintegration of radioactive molecules releases energetic beta particles, which expose emulsion grains. The photographic emulsions are sensitive to many kinds of electromagnetic energy. Thus, during preparation and exposure, specimens must be kept in the dark and protected from x-rays and other ionizing radiation. Cosmic rays contribute to the background level of grains detected in all autoradiograms. Autoradiography can be used to locate radioactive precursors during pulse experiments or to trace the movement of cells or organelles during pulse-chase experiments.

5. The answer is C. [*II D; IV B*] Typically, specimens for scanning electron microscopy are fixed using the same methods employed in transmission electron microscopy (i.e., immersion in buffered solutions of glutaraldehyde and then osmium tetroxide). Then, specimens are infiltrated with ethanol or acetone and critical point dried in liquid CO_2. Critical point drying dehydrates the specimen without passing it through an air-water interface, and thus minimizes specimen distortion due to surface tension forces. After critical point drying, specimens are coated with gold and palladium, which scatter the microscope's electron beam during image formation. Scanning electron microscopes have a resolution limit of 5 nm; transmission electron microscopes have a resolution limit of 0.5 nm.

6. The answer is B. [*V C*] In freeze-fracture-etch, specimens undergo minimal fixation and are infiltrated with aqueous solutions of glycerol to minimize ice crystal formation, which would otherwise damage the specimen and introduce artifacts. Next, specimens are placed in a vacuum and fractured. Then, the specimen receives a thin metal coating, and the tissue is thawed and chemically digested, leaving a metal replica of the specimen. This replica is viewed in the transmission electron microscope to study details of membrane structure, including the central hydrophobic domain.

7. The answer is A. [*I B 1; II B, C*] Specimens are stained to add contrast to their structures and domains. Stains may be colored dyes, heavy metals, or fluorescent probes. Dyes increase contrast in light microscope specimens by binding to specimen structures and absorbing light. Usually, dyes are selected to achieve a specific result. For example, acidic dyes (e.g., eosin) bind to molecules that have a net positive charge; basic dyes (e.g., alcian blue) bind to molecules that have a net negative charge; other dyes selectively reveal certain tissues (e.g., Weigert's stain reveals elastic tissue). Heavy metals scatter electrons in the electron microscope to improve contrast. Fluorescent probes highlight specific specimen structures, which are then studied in a fluorescence light microscope. For example, an antigen's location in a specimen can be revealed by binding a fluorescent probe to an antibody to tag the antigen.

8. The answer is A. [*II B 2*] Many basic dyes (e.g., toluidine blue) exhibit metachromasia. Most metachromatic substances contain numerous negatively charged groups such as the sulfate (SO_4^{2-}) groups in sulfated proteoglycans. Structures that bind low concentrations of dye stain orthochromatically (blue); structures that bind high concentrations of dye stain metachromatically (purple). For example, the extracellular matrix of hyaline cartilage is metachromatic because its proteoglycans contain a high concentration of chondroitin sulfate.

9–13. The answers are: 9-B, 10-D, 11-C, 12-E, 13-A. [*III A, B*] In a light microscope, the substage condenser gathers light rays from the light source (*A*) and projects them through the specimen, which rests on the microscope's movable stage (*C*). Diffracted light rays emanating from the specimen are collected by one of the objectives (*D*) and projected into the back focal plane of the microscope. The ocular (*E*), located above the back focal plane, synthesizes the complex diffraction pattern into a clear, magnified image. A knob (*B*) in the base of the microscope is used to focus the image by adjusting the distance between the stage and the objective.

14–18. The answers are: 14-B, 15-A, 16-B, 17-A, 18-D. [*II B 1*] Acidophilic structures have a net positive charge; basophilic structures have a net negative charge. Collagen fibers are acidophilic because they contain a high concentration of collagen, a protein that has a net positive charge. Erythrocytes are acidophilic because they contain a high concentration of hemoglobin, another protein with a net positive charge. Most cell nuclei are basophilic, because they have a high concentration of negatively charged DNA. Plasma cell cytoplasm is basophilic because it has abundant rough endoplasmic reticulum that contains negatively charged ribosomal RNA. The lipids in fat vacuoles often are extracted during specimen preparation. Consequently, fat vacuoles do not stain.

The Plasma Membrane and Cell Surface

I. INTRODUCTION

A. General characteristics

1. All human cells are surrounded by a plasma membrane (also called the cell membrane). Even the most complex cell surface irregularities, indentations, and projections are covered by a plasma membrane.

2. The plasma membrane is a **semi-permeable** boundary between the cell and its external environment that allows certain substances to pass through from the outside while keeping other constituents from escaping from the cell.
   a. The semi-permeable plasma membrane has an essential role in controlling the composition of cytoplasm.
   b. Water, ions, and small molecular weight metabolites such as glucose pass through the plasma membrane into the cytoplasm in a controlled fashion.

3. Cells use another mechanism to internalize extracellular macromolecules. The cell membrane surrounds and engulfs the macromolecule, and then transports it into the cell.

B. Cell membrane functions (Figure 2-1).

1. The cell membrane is the interface between a cell and its environment.
   a. Some cells exist in an aqueous environment with no close neighboring cells (e.g., free living unicellular organisms such as bacteria and protozoa).
   b. In multicellular organisms, the cell membrane fronts on an extracellular matrix or abuts on another cell of the organism.

2. Depending on the cell's environment, the cell membrane has receptors that bind extracellular matrix constituents or cell surface constituents.

3. The cell membrane also contains hormone receptors. Hormones signal changes in cell behavior.

4. Cell surfaces also contain enzymatic activities that allow cells to recognize other cells.

II. PLASMA MEMBRANE FUNCTIONS

A. Boundary functions

1. The cell membrane keeps nutrients in the cell.
   a. Animal cells accumulate organic molecules from their environment and retain them for the metabolic purposes of providing energy for life and building blocks for cytoplasm.
   b. Without a selective barrier, cells could not accumulate the materials necessary to sustain life.

2. Many important constituents are accumulated by the expenditure of metabolic energy. Membranes help retain transported substrates.

3. Useful substrates are gathered from the environment or synthesized by the cell and kept within the cell for the cell's own uses.

4. Cell waste products are expelled across the plasma membrane.

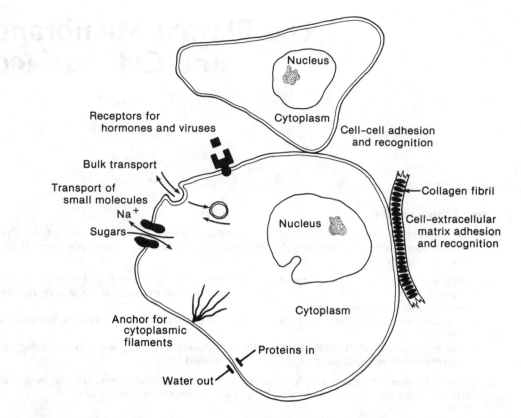

Figure 2-1. Diagram of the major plasma membrane functions. The plasma membrane performs barrier functions; anchors cytoplasmic filaments; transports small molecules; provides for bulk transports of water, ions, and macromolecules; contains receptors; and is involved in cell-to-cell interaction and cell-matrix interaction.

B. Interface functions

1. Cells respond to changes in their environment.
   a. A bacterium moves toward a source of nutrients (e.g., monosaccharides) by **chemotaxis,** because the bacterium membrane has receptors that bind nutrients, transport them into the cell, and then alter the cell's behavior so it remains near the nutrient source.
   b. In multicellular organisms, coordinated arrays of cells signal and respond using membrane receptors. An organism has a great selective advantage when equipped with these membrane receptors by mutation.

2. As multicellular organisms evolved from primitive single cell organisms, the plasma membrane developed receptors for recognizing other cells.

3. Organisms composed of many cells developed specialized subpopulations of cells for moving, feeding, and reproduction.

4. As multicellular aggregates evolved, their specialized subpopulations of cells required mechanisms to integrate their diverse functions. Cells developed the ability to elaborate secretion products called **hormones** or **neurotransmitters,** which allowed communication between different parts of the organism.

5. Hormone and neurotransmitter actions begin at special receptor sites in the plasma membrane.

III. MEMBRANE STUDIES IN THE ELECTRON MICROSCOPE

A. **Membrane structure.** Membranes are phospholipid bilayers with embedded protein molecules. This bilayer is about 10 nm thick. When cell membranes were first clearly viewed in the electron microscope, they appeared as two thin electron-dense lines separated by an electron-lucid gap.

B. Freeze-fracture-etch

1. The freeze-fracture-etch technique completed the modern view of membrane structure (Figure 2-2). In this technique, cells are frozen very rapidly in liquid nitrogen. Then the cells and their membranes are fractured in a vacuum device.

2. The fracturing process cleaves the cell membranes down the middle of the lipid bilayer, through the hydrophobic domain.

3. After fracturing, specimens are etched to reveal edges. The fractured membrane faces are then coated with a thin film of metal.

4. Finally, the specimen is chemically digested, leaving a metallic replica of the fractured membranes that can be studied in the electron microscope.

5. Freeze-fracture-etch images show that **intramembranous particles** are embedded within the hydrophobic domain of the membrane.

C. Membrane faces (see Figure 2-2)

1. The external half of the lipid bilayer has an **E-surface,** which faces the cell's external environment, and an **E-face,** which faces the hydrophobic compartment of the bilayer.

2. The internal half of the lipid bilayer has a **P-surface,** which rests directly on the protoplasm, and a **P-face,** which faces the hydrophobic compartment of the bilayer.

D. Intramembranous particles

1. The P-face and E-face of many membranes contain intramembranous particles, which consist of proteins dissolved in the hydrophobic compartment of the membrane. The particles are anchored to the cytoplasm by cytoplasmic fibrillar elements. The P-face is particularly rich in intramembranous particles.

2. The discovery of intramembranous particles and the observation that labeled probes diffuse rapidly in membranes led researchers to propose the current theory of membrane structure, the **fluid mosaic model,** in the early 1970s (Figure 2-3). This model proposes that membranous proteins resemble icebergs floating and moving in a sea of lipid.

3. Many membrane proteins are highly mobile in the plane of the membrane.

4. Many membrane proteins are **glycoproteins**. The exposed carbohydrate residues on the membrane E-surface are hydrophilic and thus interact with the fluid environment around the cell.

5. The hydrophobic portion of membrane proteins spans the hydrophobic phospholipid domain. This portion may have hydrophilic ends dissolved in cytoplasmic water or anchoring sites, binding them to cytoplasmic filamentous proteins (e.g., **actin**) that help move membrane proteins in the plane of the membrane.

6. Molecules with hydrophilic and hydrophobic portions are **amphipathic**.

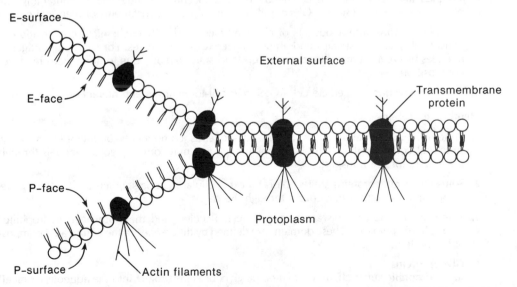

Figure 2-2. Diagram of plasma membrane structure, with the faces and surfaces revealed by the freeze-fracture-etch technique.

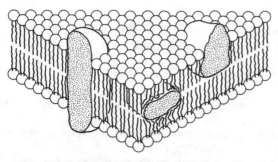

Figure 2-3. The fluid mosaic model of the plasma membrane. (Reprinted from Thorpe NO: *Cell Biology.* New York, John Wiley, 1984, p 116.)

IV. BIOCHEMICAL COMPOSITION OF THE PLASMA MEMBRANE

A. Membrane composition

1. The typical cell membrane is composed of lipids and proteins in approximately equal amounts.

2. The three major types of membrane lipids are phospholipids, cholesterol, and glycolipids. These lipids are amphipathic.

3. Cell membrane structure and chemical composition varies. For example, the red cell membrane contains comparatively few glycolipids and is rich in sphingomyelin, a variety of phospholipid. In contrast, the myelin sheath cell membrane is rich in glycolipids and contains little sphingomyelin.

B. **Phospholipids** are amphipathic molecules and therefore spontaneously form planar bilayered sheets or spherical **micelles** in an aqueous environment because these configurations represent minimum free energy conditions. The hydrophilic portions interact strongly with water and the hydrophobic portions interact strongly with one another.

C. **Cholesterol** is amphipathic and becomes intercalated between phospholipids in membranes. It increases the stability of the bilayers and prevents the loss of membrane liquidity at low temperatures. The concentration of phospholipids and cholesterol varies in the membranes of organisms that live at temperature extremes. Presumably, this variation maintains membrane fluidity above crucial threshold levels. If membrane fluidity falls below these hypothetical thresholds, vital functions such as selective membrane transport may cease and the cell will die.

D. **Glycolipids** are lipids that are covalently bonded to complex side chains containing various combinations of sugar residues. Glycolipids are amphipathic constituents of membranes.

1. Glycolipids constitute about 5% of all membrane lipids. Their chemical composition varies considerably among species and among tissues within a species. For example, antigenic differences between human blood group substances are partially due to differences in glycolipid composition.

2. Complex glycolipids mediate cell-to-cell and cell-to-environment interaction.

E. Membrane proteins

1. Proteins comprise about 50% of the non-aqueous components of membranes. They are structural components of the plasma membrane and may form aqueous channels through the membrane for ion transport.

2. Some membrane proteins are receptors for extracellular materials, others are anchoring sites for cytoplasmic structures that course through the cell.

3. Many membrane proteins are amphipathic molecules, and therefore have hydrophilic and hydrophobic domains. These domains are defined by differences in the amino acid composition of the protein.

4. **Glycoproteins**
   a. Hydrophilic sugar chains of membrane glycoproteins project into the aqueous extracellular compartment.

    b. These complex sugar chains contribute to the negative charge of the cell surface and comprise the glycocalyx.

    c. The glycocalyx is a thin layer of material outside the plasma membrane outer leaflet that is involved in cell-to-cell recognition.

    d. The glycocalyx is especially well developed in cells of the small intestines. Here, it prevents damage to cell membranes by digestive enzymes.

5. Membrane proteins that extend across the lipid bilayer are called **transmembrane proteins**. Most membrane proteins are transmembrane proteins.

6. Erythrocyte membrane proteins. The erythrocyte membrane is the best understood membrane. A dozen or so protein constituents exist. Three major proteins, **spectrin, glycophorin,** and **band III,** comprise more than 50% of the protein content of the membrane.

    a. Spectrin is a pair of 240 kD and 220 kD proteins loosely bound to the inner leaflet of the plasma membrane in a complex near the contractile protein **actin**. Although spectrin is present in highly purified membrane preparations, it is more appropriately considered a cytoplasmic protein involved in the maintenance of the biconcave-disk shape of the erythrocyte.

    b. Glycophorin is a transmembrane protein with a molecular weight of 30 kD.

       (1) A complex group of sugar molecules are attached to its N-terminus, which projects into the pericellular domain.

       (2) It also has a short segment composed of hydrophobic amino acids and a second hydrophilic region on the C-terminus facing the cytoplasmic watery domain.

V. MEMBRANE TRANSPORT MECHANISMS

A. Band III erythrocyte transport protein

1. Band III protein is a 100 kD dimer. It is a globular protein embedded in the hydrophobic domain. Because of the peculiarities of the folding of the globular portion of band III protein, a hydrophilic pore exists inside the molecule, providing an aqueous ionic channel.

2. The erythrocyte's main function is oxygen and carbon dioxide transport.

3. Hemoglobin, the dominant cytoplasmic protein of the erythrocyte, is an oxygen and carbon dioxide binding protein.

    a. Carbon dioxide is released from hemoglobin where it dissolves in cytoplasmic water. The resulting bicarbonate ions (HCO_3^-) are expelled from cells in exchange for chloride ions (Cl^-).

    b. Cl^- and HCO_3^- ions are exchanged across membranes through the aqueous channels formed by the hydrophilic pores of band III proteins.

B. Ion transport proteins

1. The electrochemical potential difference between the inside and the outside of a cell is maintained by differences in ion concentration, primarily due to a high extracellular concentration of Na^+ and a low extracellular concentration of K^+.

2. Cells must expend metabolic energy in the form of **adenosine triphosphate (ATP)** hydrolysis to **adenosine diphosphate (ADP)** to establish and maintain ionic concentration gradients. This process involves an important membrane protein complex called **Na^+-K^+-ATPase**.

3. The Na^+-K^+-ATPase consists of two transmembrane proteins, the **transmembrane catalytic subunit** (100 kD) and a glycoprotein (45 kD) of unknown function that is associated with the catalytic subunit.

4. The catalytic subunit has a Na^+-binding site on its cytoplasmic face and a K^+-binding site on its extracellular face. Na^+ ions bind to the catalytic subunit, which is then phosphorylated by ATP.

5. The phosphorylated protein undergoes a **conformational change,** expelling the Na^+ from the cell. Next, extracellular K^+ binds to the catalytic subunit, and the protein becomes dephosphorylated, returning to its original conformation. At this point the protein is ready to repeat the process.

    6. Conformational changes in the catalytic subunit and Na$^+$ expulsion use energy stored in the ATP molecule (Figure 2-4).

C. Bulk transmembrane transport

    1. Cells have mechanisms for moving large substances across their membranes.
       a. Proteins synthesized on the rough endoplasmic reticulum within cells are often expelled from cells by **exocytosis**.
       b. Macromolecules are brought into cells by **endocytosis**.
       c. Cells ingest large quantities of fluid by **pinocytosis,** a process similar to endocytosis.
       d. Cells engulf bacteria or degenerate cellular debris in a process called **phagocytosis**.

    2. These related processes use membrane-bound vesicles of various sizes to move materials from one compartment to another.

    3. **Exocytosis** is commonly used to secrete substances from cells.
       a. In the pancreas, hydrolytic enzymes are produced as **zymogen granules,** which digest the polysaccharides, proteins, lipids, and nucleic acids in food. Several mechanisms prevent these enzymes from degrading the body's macromolecules as well. For example, many hydrolytic enzymes are secreted as inactive enzymes, which become active only in the gastrointestinal tract.
       b. Proteins destined for secretion assemble on the ribosomes attached to the rough endoplasmic reticulum. Nascent polypeptide chains grow on ribosomes and then pass through ribosomes into the rough endoplasmic reticulum lumen.
       c. Membrane-bound vesicles bud from the rough endoplasmic reticulum, pass through the Golgi apparatus where the final stages of protein synthesis are completed, and then pass on to zymogen granules, which are membrane-bound packets of stored hydrolytic enzymes.

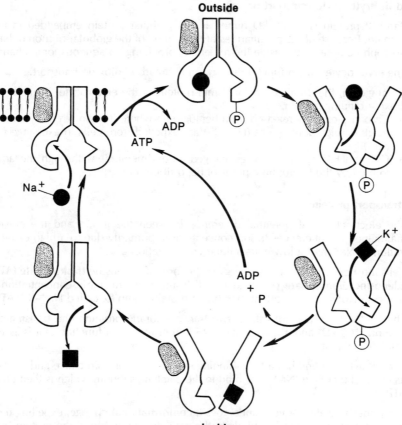

Figure 2-4. Diagram illustrating the Na$^+$-K$^+$-ATPase activity of the plasma membrane.

     d. Under appropriate stimulation, zymogen granules approach the apical surface of pancreatic acinar cells where their limiting membranes first adhere to, and then fuse with, the cell surface membrane.

     e. In this manner, potentially noxious enzymes are synthesized, carried through the cell, and then expelled from the cell while sequestered inside membrane-bound vesicles.

4. Endocytosis. Living cells are constantly engulfing their fluid environment by endocytosis. During this process, cells continuously replace old surface membrane with new membrane.

     a. The replacement process is so rapid that it is unlikely that plasma membrane constituents are synthesized de novo. It is more likely that the plasma membrane is recycled through endocytic vesicles and returned to the cell surface.

     b. Endocytic vesicles fuse with **primary lysosomes** to form secondary lysosomes.

     c. Ingested macromolecules in secondary lysosomes are degraded into small molecules, which are subsequently transported across lysosomal membranes for intracellular utilization (see Chapter 4 III). The lysosome membranes, however, appear to recycle through the Golgi apparatus and then return to the cell surface.

     d. Some endocytic vesicles don't fuse with lysosomes. Instead, they pass their contents through cells without degrading them.

5. Phagocytosis is the process by which cell cytoplasm engulfs large particles and whole organisms.

     a. Macrophages destroy bacteria by surrounding them with their cytoplasm and then engulfing them into membrane-delimited vesicles called **phagosomes**.

     b. Phagosomes fuse with lysosomes to form **phagolysosomes,** which contain enzymes that kill and digest the engulfed food.

VI. MEMBRANE RECEPTORS

A. Receptor functions

1. Most organisms have the ability to respond to changes in their environment. For example, viruses attach to cells and bacteria move toward nutrients.

2. Higher forms of animal life are vast arrays of cells, tissues, and organs whose activities are integrated by nervous system neurotransmitters and endocrine system hormones.

3. Integrated vital functions require cell membrane receptors. A hormone must be bound to a cell by a receptor to be perceived by the cell. Hormones set off a series of chemical reactions that modify cell behavior.

B. Insulin receptors

1. Insulin is a hormone produced by **beta cells** in the pancreatic **islets of Langerhans,** isolated groups of endocrine cells adjacent to capillaries and dispersed between the exocrine pancreatic acinar cells. **Endocrine cells** secrete hormones into the vascular system, which carries them to distant target organs.

2. Insulin binds to the **insulin receptor,** a membrane protein. Then, coated vesicles engulf the insulin and receptor.

3. Coated vesicles have an electron-dense layer of protein on the cell membrane P-surface. Coated vesicles can be isolated from cells and contain several membrane proteins including **clathrin** (180 kD) and other smaller molecular weight proteins.

4. Under appropriate conditions, proteins assemble spontaneously into three-legged macromolecules, which then assemble into a basket-like array around the coated vesicles. When the coating is removed from clathrin-coated vesicles, they fuse with lysosomes and the insulin receptors are degraded.

C. Receptors for extracellular macromolecules

1. Cell membranes also have receptors for extracellular matrix fibrous macromolecules such as **fibronectin** and **laminin,** two extracellular matrix proteins.

2. Many cells have **integrin,** a 140-kD glycoprotein complex, which is a fibronectin-laminin receptor. It is especially prevalent in the basal surface of epithelial cells, where these cells interface with the fibronectin and laminin in the basement membrane.

3. These receptors are crucially involved in cell adhesion to extracellular matrices containing fibronectin because they join cells to molecules in the surrounding extracellular matrix.

4. Integrin is a transmembrane protein that binds to the extracellular matrix on its extracellular face. It is anchored to the actin-containing cytoskeletal microfilaments by several intermediate proteins including **talin** and **vinculin**.

STUDY QUESTIONS

Directions: Each question below contains five suggested answers. Choose the **one best** response to each question.

1. The plasma membrane has all of the following functions EXCEPT

(A) it is a selectively permeable barrier surrounding the cell
(B) it is a site for hormone receptors
(C) it is a cell adhesion interface
(D) it facilitates the transport of sugars
(E) it is the primary protein synthesis site

2. Each of the following is an amphipathic constituent of the plasma membrane EXCEPT

(A) transmembrane proteins
(B) phospholipid fatty acids
(C) glycolipids
(D) cholesterol
(E) phospholipids

3. Each of the following statements concerning membrane transport proteins is true EXCEPT

(A) integrin is involved in fibronectin transport
(B) erythrocyte band III protein is involved in bicarbonate transport
(C) Na^+-K^+-ATPase is involved in cation transport
(D) most membrane transport proteins are transmembrane proteins
(E) most membrane transport proteins are amphipathic

4. Which of the following statements about the spectrin molecule is true?

(A) It is bound to the plasma membrane outer leaflet
(B) It helps maintain the shape of the erythrocyte
(C) It is closely associated with myosin
(D) It has a lower molecular weight than glycophorin
(E) It has a molecular weight of 200 kD

5. Each of the following statements concerning glycocalyx structure or function is true EXCEPT

(A) it contains glycoprotein constituents
(B) polysaccharides project from the glycoproteins into the aqueous environment around the cell
(C) it gives many cells a positive charge
(D) it is involved in cell-to-cell recognition
(E) it prevents digestive enzymes from degrading intestinal cell membranes

6. Freeze-fracture-etch is widely used to study membrane structure and function. This technique reveals intramembranous particles, which are most abundant in the

(A) E-surface
(B) E-face
(C) P-surface
(D) P-face
(E) none of the above

7. The fluid mosaic model of plasma membrane structure explains many structural and functional properties of membranes. The fluid mosaic model proposes all of the following concepts EXCEPT

(A) phospholipids exist in a bilayer
(B) intramembranous proteins resemble icebergs floating in the sea
(C) intramembranous particles have little mobility in the plane of the membrane
(D) the hydrophilic portion of membrane proteins often projects into the surrounding aqueous environment
(E) the hydrophobic amino acids in membrane proteins interact strongly with the hydrophobic portions of phospholipids

Directions: The groups of questions below consist of lettered choices followed by several numbered items. For each numbered item select the **one** lettered choice with which it is **most** closely associated. Each lettered choice may be used once, more than once, or not at all.

Questions 8–12

Match each protein function or characteristic listed below with the appropriate protein.

(A) Spectrin
(B) Glycophorin
(C) Na^+-K^+-ATPase
(D) Integrin
(E) Fibronectin

8. Binds cytoskeletal elements to the red blood cell membrane

9. Extracellular matrix constituent used in cell adhesion

10. Undergoes conformational changes during cation transport

11. A receptor for extracellular matrix glycoproteins

12. Interacts strongly with talin and vinculin to anchor cytoskeletal microfilaments to extracellular fibronectin fibrils

Questions 13–17

For each cell transport function described below, select the most appropriate cell transport process or processes.

(A) Endocytosis
(B) Exocytosis
(C) Both
(D) Neither

13. Uses membrane-bound vesicles

14. Used in pancreatic acinar cells to secrete zymogen granules

15. Used to bring large macromolecules into cells

16. Uses minute membrane-bound vesicles to bring large quantities of water into cells

17. Used to bring receptor-bound hormones into cells

ANSWERS AND EXPLANATIONS

1. The answer is E. [*II*] The plasma membrane's most important function is to form a barrier between the cell and its external environment. The barrier selectively allows materials to enter and leave the cell. The cell membrane contains proteins that facilitate ion and sugar transport and receptors for hormones, which regulate changes in cell behavior. It also contains receptors that bind the cell to adjoining cells. The plasma membrane contains a variety of proteins that perform many of these functions. These proteins are synthesized on the rough endoplasmic reticulum and not the plasma membrane per se.

2. The answer is B. [*IV B, C, D*] Amphipathic molecules have hydrophilic and hydrophobic portions. As membrane constituents, the hydrophobic portions of these molecules dissolve in the hydrophobic domain of the bilayer, and the hydrophilic portions dissolve in water inside or outside the cell. Transmembrane proteins, phospholipids, glycolipids, and cholesterol are amphipathic molecules. Fatty acids esterified to glycerol in phospholipids are extremely hydrophobic but are not amphipathic.

3. The answer is A. [*VI C*] Erythrocyte band III protein has a hydrophilic core for ion transport and is involved in bicarbonate transport. Na^+-K^+-ATPase uses energy stored in ATP to maintain the surface potential differences between the cell's inner and outer surfaces and is involved in cation transport. Most membrane proteins are amphipathic. Transmembrane proteins such as integrin span the entire membrane. They have hydrophilic domains associated with the water inside and outside the cell. They also have a hydrophobic domain associated with the hydrophobic portions of phospholipids. Integrin is a membrane receptor for extracellular matrix molecules such as fibronectin and laminin. It binds with extracellular fibronectin to anchor the cell to the extracellular matrix, but it does not transport fibronectin.

4. The answer is B. [*IV E 6 a*] Spectrin is one of the most important and abundant erythrocyte plasma membrane proteins. It is a pair of proteins with weights of 240 kD and 220 kD. The molecular weight of glycophorin, another erythrocyte membrane protein, is 30 kD. Spectrin is tightly bound to the P-surface of the membrane, where it binds to actin-containing microfilaments in the cytoplasm. Spectrin plays an important role in maintaining the biconcave disk shape of the erythrocyte.

5. The answer is C. [*IV E 4*] The glycocalyx is the layer of glycoconjugate-rich macromolecules that covers the external surface of the cell membrane. Its main constituent is a family of proteins rich in sialic acid. These sialic acid residues carry a negative charge and impart a net negative charge to most cells. Glycocalyx glycoproteins are important in many cell recognition phenomena. They also prevent digestive enzymes from reaching the membranes of columnar epithelial cells in the small intestine.

6. The answer is D. [*III B, C*] Freeze-fracture-etch involves several steps. First, cells are lightly fixed and frozen quickly in glycerol. Then, the frozen specimen is fractured and coated with a thin film of metal. Finally, the tissue is chemically digested, and the remaining metal replica is examined in the electron microscope. Intramembranous particles are revealed by this technique. Intramembranous particles are most abundant in the membrane P-face, although some exist in the E-face as well. The E-surface and P-surface do not contain intramembranous particles.

7. The answer is C. [*III A, D*] Membrane proteins are highly mobile in the plane of the membrane. According to the fluid mosaic model of membrane structure, membrane proteins move within the phospholipid bilayer like icebergs floating in the sea. Phospholipids and proteins are amphipathic molecules because they have hydrophilic and hydrophobic domains. The hydrophilic domains interact with water in the intracellular and extracellular compartments. The hydrophobic domains interact with the hydrophobic portions of phospholipids.

8–12. The answers are: 8-A, 9-E, 10-C, 11-D, 12-D. [*IV E 6; V B; VI C*] Spectrin is a dimer of a 240-kD and a 220-kD protein in red cell membrane proteins. It binds cytoskeletal elements to the red cell membrane and helps maintain the red blood cell shape.

   Glycophorin (30 kD), also present in red blood cell membranes, is a transmembrane protein whose function is poorly understood.

   Na^+-K^+-ATPase is used in cation transport. It undergoes conformational changes as it performs this function. The osmotic work of ion transport is driven by ATP hydrolysis.

   Integrin is an integral membrane protein and a receptor for extracellular molecules such as fibronectin. It is a transmembrane protein bound to the extracellular matrix and anchored to actin-containing cytoskeletal microfilaments by talin and vinculin.

13–17. The answers are: 13-C, 14-B, 15-A, 16-D, 17-A. [*V C 1, 2, 3, 4*] Both endocytosis and exocytosis use membrane-bound vesicles. Endocytic vesicles are formed by invagination of the surface membrane. Exocytic vesicles are added to the surface membrane during exocytosis by membrane fusion.

During endocytosis, cell-surface invaginations surround materials outside the cell (e.g., bacteria, hormones) and bring them into the cell for processing. Endocytic vesicles often fuse with lysosomes for processing contents. Endocytic vesicles are formed by invagination of surface membrane. Pinocytosis uses small vesicles to move large quantities of water into the cell. It is similar to endocytosis, but is distinct because it uses small vesicles and transports only water.

Exocytosis is the reverse process to endocytosis. Membrane-bound vesicles of enzymes, produced in the rough endoplasmic reticulum and Golgi apparatus, fuse with the plasma membrane, thereby releasing enzymes into the extracellular environment.

3

Cytoplasmic Organelles: Membranous Systems

I. INTRODUCTION

A. General concepts. Most cytoplasmic organelles are composed of membranes.

1. Endoplasmic reticulum and Golgi apparatus membranes are extensive interconnected saclike systems. Often, the boundaries of these structures are indistinct.

2. In contrast, intracellular organelles (see Chapter 4) are discrete, particulate structures bounded by membranes.

B. The plasma membrane and membranous organelles

1. The plasma membrane (see Chapter 2) is a lipoprotein bilayer that forms the boundary between a cell and the extracellular environment.

2. The endoplasmic reticulum and the Golgi apparatus are fluid-filled compartments within cells, which are bounded by lipoprotein bilayers with structural and biochemical similarities to the plasma membrane. Often, endoplasmic reticulum and Golgi apparatus membranes are in spatial and temporal continuity with the plasma membrane.

II. SMOOTH ENDOPLASMIC RETICULUM

A. Structure

1. **General features.** All endoplasmic reticular membranes are lipoprotein bilayers.
   a. Endoplasmic reticular membranes are thinner (approximately 5 nm thick) than the plasma membrane (8–10 nm thick), and they are more stable and less fluid than the plasma membrane.
   b. Endoplasmic reticular membranes have two surfaces: one facing the cytoplasmic domain surrounding the endoplasmic reticulum, and another facing the cavity (lumen) of the flattened sac, tubule, or vesicle.

2. **Specialized forms**
   a. **Three forms** of smooth endoplasmic reticulum exist.
      (1) **Lamellar form.** Often, smooth endoplasmic reticulum exists as extensive lamellae (sheets) of long flattened sacs bounded by membranes.
      (2) **Vesicular form.** In this form, smooth endoplasmic reticulum exists as small vesicles. The vesicular form is similar in composition to the lamellar form except that it exists as small, round structures bounded by membrane.
      (3) **Tubular form.** In this form, smooth endoplasmic reticulum exists as elongated tubules bounded by membranes. Frequently, tubular forms are dynamically related to lamellar and vesicular forms.
   b. The three forms of smooth endoplasmic reticulum probably are freely convertible, suggesting that endoplasmic reticulum is a highly dynamic, pleomorphic organelle. All three forms also are found in rough endoplasmic reticulum.

B. Functions. Smooth endoplasmic reticulum metabolizes small molecules, contains cellular detoxification mechanisms, and is involved in lipid and steroid synthesis.

III. ROUGH ENDOPLASMIC RETICULUM

A. Structure

1. Rough endoplasmic reticulum consists primarily of lamellar forms interconnected by short tubular segments.

2. Rough endoplasmic reticular membranes are studded with electron-dense particles called **ribosomes** (Figure 3-1).

3. Ribosomes are approximately 11 nm in diameter and consist of ribosomal RNA and specific ribosomal proteins.

4. Ribosomes are bound to rough endoplasmic reticular membranes by ribosome-binding proteins called **ribophorins,** which have molecular weights of 63 and 65 kD. Ribophorins add some structural rigidity to membranes.

B. Functions

1. The main function of rough endoplasmic reticulum is the **synthesis of proteins** that are eventually secreted from the cells. Ribosomes are protein synthesis sites. Nascent polypeptide chains grow on ribosomes. During synthesis, the protein folds into its natural tertiary structure as it passes into the rough endoplasmic reticular lamellae (Figure 3-2).

2. Rough endoplasmic reticulum can be **converted to smooth endoplasmic reticulum**.
   a. Injecting animals with phenobarbital causes an immediate proliferation of rough endoplasmic reticulum in liver parenchymal cells.
   b. After proliferation, ribosomes drop from the rough endoplasmic reticulum, converting it to smooth endoplasmic reticulum, which now has enzymes to detoxify the phenobarbital. The liver detoxifies many ingested poisons and, thus, has an abundance of rough endoplasmic reticulum (a source of smooth endoplasmic reticulum) and smooth endoplasmic reticulum (a source of detoxification enzymes).

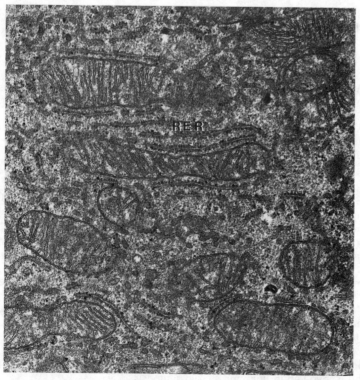

Figure 3-1. Electron micrograph of rough endoplasmic reticulum (*RER*).

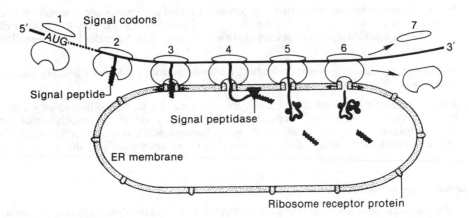

Figure 3-2. Diagram of protein synthesis in the rough endoplasmic reticulum illustrating the signal hypothesis. (Reprinted from Thorpe NO: *Cell Biology.* New York, John Wiley, 1984, p 324.)

IV. BIOCHEMICAL FUNCTIONS OF ENDOPLASMIC RETICULAR MEMBRANES

A. Chemical composition

1. Endoplasmic reticular membranes have a higher protein concentration (60%–70% protein by weight) than plasma membranes. Many of these proteins have important enzymatic functions.

2. Endoplasmic reticular membranes have a lower phospholipid concentration (30%–40% by weight) than plasma membranes.

3. The phospholipids of endoplasmic reticular membranes are mostly phosphatidylcholine (55%) and phosphatidylethanolamine (25%), and they have a very low concentration of sphingomyelin (5%) compared to plasma membranes.

B. Enzyme constituents include:

1. **Glucose-6-phosphatase** (see IV E 1)

2. **Cytochrome P-450 and NADPH cytochrome P-450 reductase.** With phosphatidylcholine, these enzymes are part of the endoplasmic reticular electron transport chain and are involved in hydroxylation reactions.

3. **Cytochrome b_5**

C. Endoplasmic reticular hydroxylation system. Enzymatic hydroxylation in the endoplasmic reticulum has important anabolic and catabolic functions within cells.

1. **Anabolic functions** include cholesterol biosynthesis, steroid synthesis, and bile acid synthesis.

2. **Catabolic functions** make drugs and toxic substances more hydrophilic and, therefore, more readily excreted. These functions include drug metabolism, excretion of insecticides, anesthetic metabolism, and conversion of carcinogens.

D. Steroid metabolism

1. **Cholesterol biosynthesis.** Mevalonate and squalene are key intermediates in cholesterol synthesis.
   a. The first step is formation of **3-hydroxy-3-methylglutaryl-coenzyme A (HMG-CoA)** from acetoacetyl CoA and acetyl CoA. HMG-CoA is present in the cytosol and in mitochondria.
      (1) **HMG-CoA reductase** in the endoplasmic reticulum converts HMG-CoA to mevalonic acid.
      (2) Mevalonic acid synthesis is the committed step in cholesterol synthesis.

    b. Squalene synthetase (an enzyme that converts farnesyl pyrophosphate to squalene) and other enzymes involved in the multistep process of squalene conversion to cholesterol (e.g., **squalene epoxidase, sterol cyclase**) also are constituents of the endoplasmic reticular membranes.

  2. Bile acid synthesis. Enzymes that synthesize bile acids from cholesterol also are bound to the endoplasmic reticular membranes.

  3. Steroid synthesis. Cholesterol precursors are synthesized into steroid hormones on the membranes of the endoplasmic reticulum. Steroid biosynthesis also involves mitochondrial enzymes, which explains the abundance of smooth endoplasmic reticulum and mitochondria in the liver and in steroidogenic tissues (e.g., the adrenal cortex, Leydig cells in the testis, thecal and granulosa cells in the ovary).

E. Carbohydrate metabolism

  1. Glucose-6-phosphatase is an endoplasmic reticular enzyme involved in the regulation of blood glucose levels.

    a. This enzyme cleaves phosphate from glucose-6-phosphate, thus liberating glucose into the bloodstream.

    b. This enzyme is particularly abundant in the liver, which stores and metabolizes glycogen.

  2. Proximal sugar residues of many glycoproteins destined for secretion from cells are added to protein backbones in the endoplasmic reticulum. More distal sugar residues are added in the Golgi apparatus.

V. GOLGI APPARATUS

A. Structure (Figure 3-3)

  1. The Golgi apparatus is an array of flattened discoid lamellae.

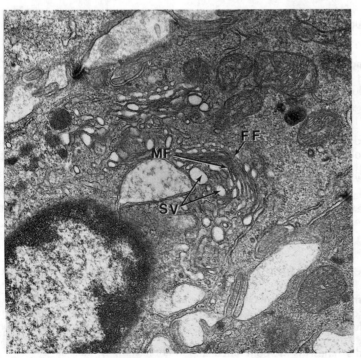

Figure 3-3. Electron micrograph of the Golgi apparatus forming face (*FF*) and maturing face (*MF*) with associated secretory vesicles (*SV*).

2. Golgi apparatus membranes are about 7.5 nm thick, and the lumen of each sac is about 25 nm wide.

3. Each Golgi apparatus contains 5–10 lamellae, which are separated by gaps of 20 nm.

4. Each lamella has numerous fenestrations and contains many small vesicles and anastomosing tubes of membranes associated with the stacks of lamellae.

5. The Golgi apparatus has a convex outer face, or **forming face,** and a concave inner face, or **maturing face** (see Figure 3-3).
   a. New membranes are added to the forming face from the endoplasmic reticulum. In many secretory cells (e.g., pancreatic acinar cells), the maturing face produces large numbers of secretory (zymogen) granules.
   b. Old membranes and vesicles bud from the maturing face.
   c. The membranes around secretory granules fuse with, and become an integral part of, the plasma membrane.

B. Functions

1. **Protein synthesis**
   a. Proteins initially synthesized in the endoplasmic reticulum pass to the Golgi apparatus forming face in **transition vesicles**.
   b. Many polypeptides have complex sugar chains added in the Golgi apparatus due to the action of Golgi membrane enzymes called **glycosyl transferases**.
   c. Newly synthesized glycoproteins leave the Golgi apparatus maturing face in secretory granules.

2. **Membrane synthesis** begins in the endoplasmic reticulum and continues in the Golgi apparatus. New membranes move to, and fuse with, the plasma membrane.

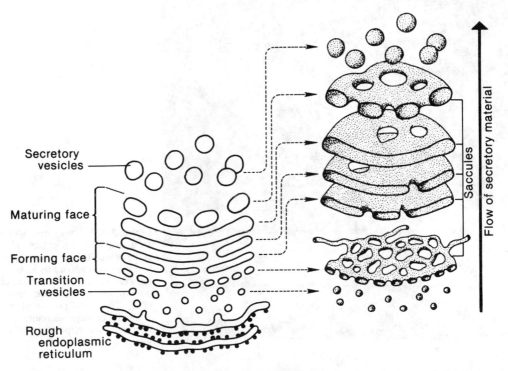

Figure 3-4. Diagram of the secretory pathway through the rough endoplasmic reticulum and Golgi apparatus. (Reprinted from Thorpe NO: *Cell Biology*. New York, John Wiley, 1984, p 360.)

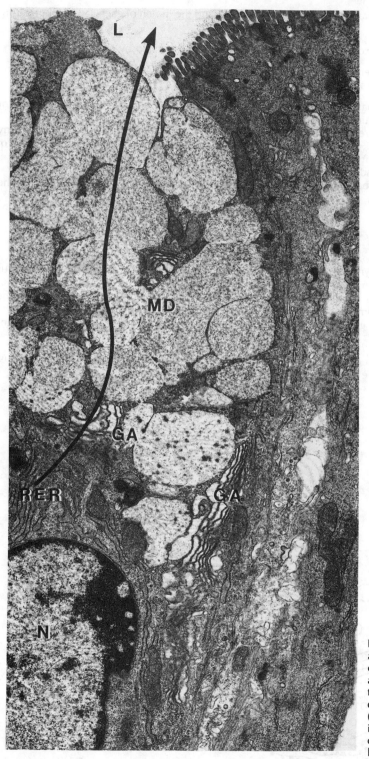

Figure 3-5. Electron micrograph of an active intestinal goblet cell. The cell nucleus (*N*) is in the lower left and the Golgi apparatus (*GA*) is near the center. This cell contains abundant rough endoplasmic reticulum (*RER*). The apical portion of the cell is filled with large secretory droplets of mucus (*MD*) moving toward the lumen (*L*).

VI. THE SECRETORY PATHWAY

A. Endoplasmic reticulum

1. Synthesis of proteins destined for secretion occurs in the rough endoplasmic reticulum (see III B 1).

2. Cells discriminate between proteins destined for secretion and proteins destined to be used internally.

3. Proteins destined for secretion are synthesized and transported in one direction. **The signal hypothesis** accounts for unidirectional secretion.
   a. According to the signal hypothesis, all secretory proteins are encoded in mRNA with a specific **signal codon** on the 3' end of the initiation codon (AUG).
   b. Signal codons are translated into **signal peptides** near the N-terminus. Signal peptides interact strongly with specific receptor proteins in rough endoplasmic reticular membranes.
   c. Signal peptides are cleaved by **signal peptidases**, allowing the nascent polypeptide to float freely in the rough endoplasmic reticular cisternal space and begin to fold into its normal secondary and tertiary structure.
   d. As the newly synthesized polypeptide undergoes this conformational change, it becomes too large to leave the rough endoplasmic reticular lumen.
   e. Then, the ribosome releases the polypeptide. A **detachment factor** subsequently releases the polysomes from the rough endoplasmic reticular membranes, sequestering the polypeptide in the rough endoplasmic reticular lumen (see Figure 3-2).

B. The Golgi apparatus

1. Newly synthesized polypeptides are sequestered in small vesicles that bud from the rough endoplasmic reticulum. These vesicles merge with the forming face of the Golgi apparatus.

2. Polypeptides are processed further in the Golgi apparatus as they pass from the forming face to the maturing face (e.g., distal sugar residues are added to glycoproteins).

3. Proteins pass from the maturing face into **condensing vacuoles,** where water is removed and **zymogen granules** are formed. Upon appropriate stimulation, zymogen granules fuse with the plasma membrane and are released from the cell by exocytosis (Figure 3-4).

C. Goblet cells (Figure 3-5) are abundant in the intestines and the upper respiratory tract. Their apices are distended by many mucus granules—another kind of secretory granule.

STUDY QUESTIONS

Directions: The groups of questions below consist of lettered choices followed by several numbered items. For each numbered item select the **one** lettered choice with which it is **most** closely associated. Each lettered choice may be used once, more than once, or not at all.

Questions 1–6

For each description of an organelle function, select the appropriate lettered organelle in the micrograph below.

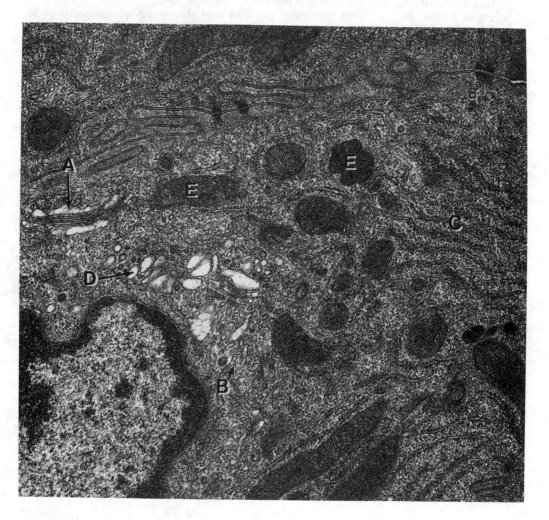

1. A ribosome attachment site

2. Involved in steroid biosynthesis and ATP production

3. Consists of flattened membrane lamellae, has forming and maturing faces, and forms secretory vesicles

4. Contains HMG-CoA reductase activity

5. Transition vesicles that move from endoplasmic reticulum to the Golgi apparatus

6. Initial site of protein synthesis including polypeptide chain assembly and proximal glycosylation

Questions 7–9

For each characteristic listed below, select the most appropriate organelle or organelles.

(A) Smooth endoplasmic reticulum
(B) Rough endoplasmic reticulum
(C) Both
(D) Neither

7. Abundant in liver parenchymal cells

8. Contains high concentrations of ribophorin

9. Consists of sheets, tubes, and vesicles

Questions 10–13

Match each characteristic or function listed below with the most appropriate enzyme.

(A) Sialyl transferase
(B) HMG-CoA reductase
(C) Glucose-6-phosphatase
(D) HMG-CoA
(E) Cytochrome P-450

10. Synthesized by mitochondrial enzymes

11. Abundant in the Golgi apparatus

12. Involved in hydroxylation that detoxifies drugs and anesthetics

13. Involved in cholesterol biosynthesis

ANSWERS AND EXPLANATIONS

1–6. The answers are: 1-C, 2-E, 3-A, 4-B, 5-D, 6-C. [*II B; III A, B; IV D; V A, B*] This is an electron micrograph of a liver parenchymal cell. The Golgi apparatus (A) consists of many stacks of flattened lamellae and has a newer forming face and an older maturing face. The Golgi apparatus adds distal sugar residues to glycoproteins during glycoprotein synthesis and then transfers the glycoproteins to secretory vesicles, which secrete them by fusing with the plasma membrane.

Smooth endoplasmic reticulum (B) contains HMG-CoA reductase and, with mitochondria, has an important role in cholesterol and steroid biosynthesis.

Rough endoplasmic reticulum (C) contains ribophorin, a protein involved in ribosome binding, and is the site of the initial phases of protein synthesis, including the assembly of nascent polypeptide chains and the addition of proximal carbohydrate residues to glycoproteins.

Small vesicles (D) may be components of smooth endoplasmic reticulum or the Golgi apparatus and may be transition vesicles that move from the endoplasmic reticulum to the Golgi apparatus.

Mitochondria (E) are involved in cholesterol and steroid biosynthesis and the production of ATP, an energy-rich compound.

7–9. The answers are: 7-C, 8-B, 9-C. [*II A; III A, B; IV D*] Smooth endoplasmic reticulum and rough endoplasmic reticulum both are abundant in liver parenchymal cells. Smooth endoplasmic reticulum is involved in cholesterol and bile acid biosynthesis, and rough endoplasmic reticulum is involved in protein synthesis. Both are pleomorphic, existing in lamellar, tubular, and vesicular forms. Ribophorin, an integral membrane protein that recognizes ribosomes and binds them to membranes, is present only in rough endoplasmic reticulum.

10–13. The answers are: 10-D, 11-A, 12-E, 13-B. [*IV C, D; V B*] Sialyl transferase is a Golgi apparatus glycosyl transferase involved in adding distal sialic acid residues to glycoproteins. HMG-CoA reductase is an endoplasmic reticular enzyme involved in cholesterol biosynthesis. Glucose-6-phosphatase is an endoplasmic reticular membrane protein involved in blood glucose homeostasis. HMG-CoA is a cholesterol precursor that is synthesized by mitochondrial enzymes. Cytochrome P-450 is an endoplasmic reticular membrane protein that is involved in hydroxylation and detoxification of drugs.

4
Cytoplasmic Organelles: Discrete Particulate Organelles

I. INTRODUCTION

A. General features. In contrast to the organelles discussed in Chapter 3, particulate organelles are structurally and functionally isolated from one another. However, this does not imply a lack of interaction between the particulate organelles and the membranous systems.

B. Membranous system and particulate organelle interaction

1. Often, a complex and dynamic interplay exists between membranous organelles (e.g., smooth endoplasmic reticulum) and particulate organelles (e.g., mitochondria). For example mitochondria and smooth endoplasmic reticulum work together during steroid biosynthesis (see Chapter 3 IV D 3).

2. Mitochondria supply the adenosine triphosphate (ATP) required for protein synthesis, consequently rough endoplasmic reticulum and mitochondria frequently exist close to one another in protein-secreting cells (Figure 4-1).

3. Lysosomes are discrete particulate organelles that are specialized secretory products of the endoplasmic reticulum and Golgi apparatus.
   a. Lysosomes contain hydrolytic enzymes that destroy engulfed materials.

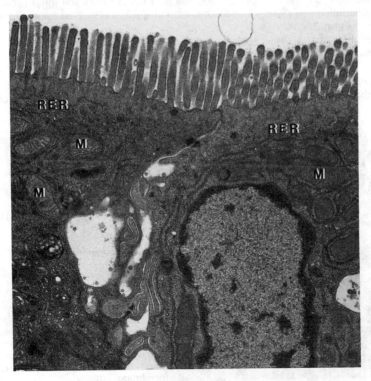

Figure 4-1. Electron micrograph illustrating the intimate relationship between mitochondria (*M*) and rough endoplasmic reticulum (*RER*) as an example of the complex interplay between the membranous and particulate organelles in active cells.

b. These enzymes are manufactured in the secretory pathway and are sequestered in membrane delimited vesicles.

II. MITOCHONDRIA

A. Structure

1. Mitochondria are about the size of bacteria, approximately 0.5–1.0 μm wide and 2.0–4.0 μm long. Often, mitochondria are distributed in cell areas that require substantial amounts of ATP.
   a. Mitochondria are located in the apical portion of ciliated cells and cells with many microvilli. In these cells, mitochondria produce the ATP required for ciliary beating and the transport of solutes.
   b. In striated muscle tissue, mitochondria are located between myofibrils and supply ATP for muscle contraction.
   c. Mitochondria exist near the endoplasmic reticulum, supplying ATP for protein synthesis.

2. Cells may have few mitochondria (e.g., few mitochondria surround the flagellum of a spermatozoon) or many mitochondria (e.g., liver parenchymal cells typically contain 500–1000 mitochondria).

3. Mitochondria move within cells and undergo dramatic changes in shape. Therefore, they have some characteristics of autonomous organisms.

B. Ultrastructure. The mitochondrion consists of two unit membrane systems, one surrounding the other.

1. The **outer membrane,** the outer boundary of the mitochondrion, is a continuous membrane with an ultrastructure similar to the plasma membrane ultrastructure. It has the characteristic trilaminar structure and consists of a phospholipid-protein bilayer.

2. The **inner membrane,** which is completely surrounded by the outer membrane, also consists of a phospholipid-protein bilayer. This membrane has numerous folds called **cristae,** which project into the internal compartment of the organelle (Figure 4-2).
   a. Mitochondrial cristae often are shelflike projections that fold inward from the inner membrane. They often interdigitate so closely that the mitochondrion has a striated appearance in the transmission electron microscope.
   b. Cristae also can assume other shapes, such as the vesicles and tubules found in many steroidogenic tissues or the fenestrated sheets in some muscular tissues.

C. Function. Mitochondria contain the **tricarboxylic (TCA) cycle** enzymes, which generate reduced nucleotides from the metabolism of citric acid, **electron transport chain** enzymes, and enzymes for **ATP synthesis.**

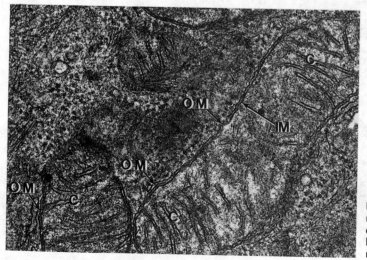

Figure 4-2. Electron micrograph of mitochondria with shelflike, interdigitating cristae (*C*). The outer membrane (*OM*) surrounds the inner membrane (*IM*).

    1. The primary function of mitochondria is the synthesis of ATP from citric acid.

    2. Mitochondria also assist in steroid biosynthesis, fatty acid oxidation, and nucleic acid synthesis.

D. Functional compartmentalization

    1. The **outer membrane** contains the enzymes:
        a. NADH-cytochrome c oxidoreductase
        b. Cytochrome b
        c. Acyl-coenzyme A synthetase
        d. Monoamine oxidase
        e. Kynurenine hydroxylase

    2. The **inner membrane** contains the enzymes:
        a. Cytochromes b, c, c_1, a, and a_3 (for electron transport)
        b. Succinate dehydrogenase (the only TCA cycle enzyme in the inner membrane)
        c. ATP-synthesizing enzymes

    3. **Inside the mitochondrion** is an amorphous matrix, bounded by the inner membrane.
        a. Within this matrix are found the rest of the TCA cycle enzymes, including:
            (1) Citrate synthetase
            (2) Aconitase
            (3) Isocitrate dehydrogenase
            (4) Fumarase
            (5) Malate dehydrogenase
            (6) Pyruvate dehydrogenase
        b. This matrix also contains enzymes for the synthesis of proteins and nucleic acids, most likely for endogenous utilization, and enzymes for fatty acid oxidation.

E. Mitochondrial nucleic acids

    1. Mitochondria contain DNA that is distinct from the cell's nuclear DNA. This DNA can self-replicate and directs the synthesis of mitochondrial mRNA, rRNA, and tRNA.

    2. Nuclear genes are required for mitochondrial nucleic acid metabolism. Products of these nuclear genes include DNA polymerase for the synthesis of mitochondrial DNA, and RNA polymerase for the synthesis of mitochondrial RNA. Most mitochondrial ribosomal proteins are made in the cytoplasm and then transported into mitochondria.

    3. Mitochondrial DNA genes make most of the enzymes for the electron transport system and some of the enzymes for ATP synthesis.

    4. Mitochondrial replication involves DNA synthesis and then division of a single mitochondrion into two daughter mitochondria. Division occurs as the inner membrane grows to form a partition across the mother mitochondrion.

    5. **Mitochondrial genome functions**
        a. The mitochondrial genome may have a mechanism for gene amplification, providing multiple copies of the genes for proteins needed in large quantities. Many mitochondrial proteins are extremely hydrophobic and are not easily transported across mitochondrial membranes after synthesis in the cytoplasmic domain. The mitochondrial genome may circumvent this difficulty.
        b. Mitochondria may be derived from endosymbiotic bacteria that have lost most of their nuclear functions. The cell's nuclear genome provides these lost functions, and the mitochondria respond by producing ATP efficiently.

III. LYSOSOMES

A. Composition and function

    1. **Enzyme composition**
        a. Lysosomes are particulate organelles, formed by the Golgi apparatus, which contain **acid hydrolases,** including:
            (1) Acid phosphatase
            (2) Deoxyribonuclease

        (3) Ribonuclease

        (4) Cathepsin D

        (5) β-Glucuronidase

    b. These enzymes help degrade resorbed protein in the kidney, destroy bacteria in phagocytic leukocytes, and degrade effete cytoplasmic organelles.

2. Types (Figure 4-3). **Primary lysosomes** are lysosomes that have not participated in any other metabolic events. **Secondary lysosomes** are lysosomes engaged in degradative activities.

    a. Heterolysosomes (digestive vacuoles) help degrade materials brought into the cell by endocytosis or phagocytosis.

    b. Autolysosomes (autophagic vacuoles) help degrade the effete components of the cell containing the autolysosome. Autolysosomes are formed when primary lysosomes fuse with membrane-bound vacuoles (cytosegresomes) containing worn cytoplasmic constituents.

    c. Telolysosomes are secondary lysosomes containing several degraded constituents. These may be resorbed into the cytoplasm, expelled from the cell (forming cytostools), or persist in the cell as **residual bodies**.

B. Theories of lysosome origin

1. Lysosomes as modified secretory vacuoles. One theory proposes that lysosomes are a variant of the secretory pathway. According to this theory, lysosomal enzymes are synthesized in the rough endoplasmic reticulum, finished in the Golgi apparatus, and then budded from the Golgi apparatus as finished lysosomes.

2. Golgi-associated endoplasmic reticulum–forming lysosomes (GERL) theory. This theory proposes that lysosomes are produced in a specialized portion of the smooth endoplasmic reticulum located near the Golgi apparatus. Lysosomal enzymes start in the rough endoplasmic reticulum and are finished in the Golgi-associated endoplasmic reticulum, rather than in the Golgi apparatus.

C. Lysosome ultrastructure

1. Lysosomes are surrounded by a unit membrane that is about 9 nm thick. They have a diameter of 0.25–0.50 μm and appear as rounded and unremarkable organelles when viewed in the electron microscope (Figure 4-4).

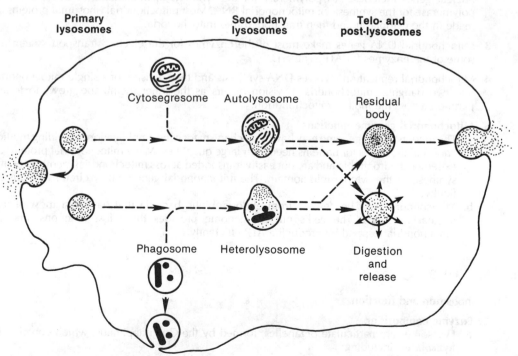

Figure 4-3. Lysosome terminology used to describe the various physiologic roles of each class of lysosome. (Reprinted from Thorpe NO: *Cell Biology.* New York, John Wiley, 1984, p 280.)

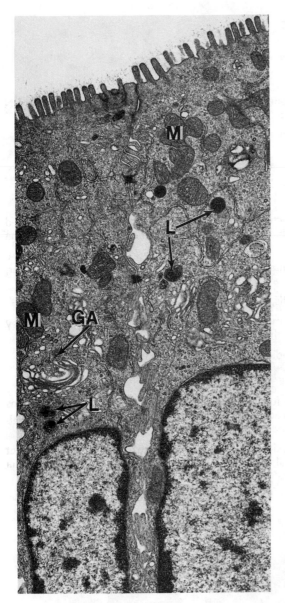

Figure 4-4. Electron micrograph of lysosomes (*L*). Notice that they are smaller than mitochondria (*M*) and often are located close to the Golgi apparatus (*GA*).

2. Lysosomes have granular contents of moderate electron density. Many modified lysosomes, such as those found in leukocytes, have complex crystalline arrays of hydrolytic enzymes.

3. Lysosomal membranes contain cholesterol and sphingomyelin, which are abundant in the plasma membrane but relatively rare in other cytoplasmic membrane systems. Their similarity to plasma membranes probably facilitates lysosome fusion with plasma membranes.

D. Lysosomal enzymes

1. Esterases
  a. Phospholipases and cholesterol esterases
  b. Phosphodiesterase
  c. Deoxyribonuclease II and ribonuclease II

2. Glycosidases
  a. Lysozyme
  b. β-Glucosidase and β-galactosidase
  c. α-*N*-Acetylgalactosaminidase and α-*N*-acetylglucosaminidase
  d. α-ʟ-Fucosidase

3. **Peptidases**
   a. Cathepsins B, D, E, and G
   b. Carboxypeptidases A, B, and C
   c. Collagenase and elastase
   d. Renin
   e. Plasminogen activator

4. Lysosomes also contain abundant supplies of other hydrolytic enzymes, which break C-N non-peptide bonds, as well as P-N and S-N bonds.

5. Many of these enzymes have a pH optima in the acidic range.

IV. MICROBODIES (PEROXISOMES)

A. Structure

1. Microbodies are spherical or oblong structures 0.3–1.5 μm in diameter, which are surrounded by a lipoprotein bilayer unit membrane that is 6–8 nm thick (somewhat thinner than the plasma membrane).

2. They have a moderately electron-dense **inner matrix**, which may appear amorphous, granular, or crystalline.

3. Microbodies are abundant in animal cells, often close to the endoplasmic reticulum.

4. Microbody membranes contain cytochrome b_5 and NADH-cytochrome b_5 reductase, and they have a lipid composition very similar to the endoplasmic reticulum.

5. These observations suggest that microbodies are derived from the endoplasmic reticulum.

B. Functions

1. Microbodies perform important oxidative functions. **Flavin oxidases** mediate oxidative reactions and use molecular oxygen as an electron acceptor, producing toxic hydrogen peroxide (H_2O_2). Then, **catalase** converts H_2O_2 to H_2O and O_2.

2. Microbodies also contain D-amino acid oxidases, which can metabolize D-amino acids brought into the cell by bacterial phagocytosis. All animal cell amino acids are L-amino acids. Bacteria cell walls, however, are rich in D-amino acids.

3. Microbodies in the liver can participate with mitochondria in β-oxidation of fatty acids.

STUDY QUESTIONS

Directions: Each question below contains five suggested answers. Choose the **one best** response to each question.

1. Which of the following enzymes is contained primarily in the outer mitochondrial membrane?

(A) Succinate dehydrogenase
(B) DNA polymerase
(C) Enzymes for fatty acid oxidation
(D) Monoamine oxidase
(E) Isocitrate dehydrogenase

2. Each of the following statements describing mitochondrial cristae is true EXCEPT

(A) they contain most of the TCA cycle enzymes
(B) they are folds in the inner mitochondrial membrane
(C) they contain succinate dehydrogenase
(D) they contain enzymes for electron transport
(E) they contain enzymes for ATP synthesis

3. All of the following TCA cycle enzymes are dissolved in the mitochondrial matrix EXCEPT

(A) aconitase
(B) fumarase
(C) malate dehydrogenase
(D) cytochrome c
(E) isocitrate dehydrogenase

4. Each of the following statements concerning mitochondrial genes is true EXCEPT

(A) they encode some mitochondrial ribosomal proteins
(B) they encode some electron transport proteins
(C) they encode DNA polymerase for mitochondrial DNA synthesis
(D) they encode some ATP synthesis proteins
(E) they encode mitochondrial mRNA

5. Of the following cell types, which would contain many mitochondria in the apical portion of the cell?

(A) Smooth muscle cells
(B) Ciliated cells
(C) Steroid-secreting cells
(D) Liver parenchymal cells
(E) Skeletal muscle cells

6. Each of the following statements concerning microbodies is true EXCEPT

(A) they contain catalase
(B) they are approximately the same diameter as mitochondria
(C) they are surrounded by a lipoprotein bilayer
(D) they are 0.3–1.5 μm in diameter
(E) they arise from mitochondria

Directions: The groups of questions below consist of lettered choices followed by several numbered items. For each numbered item select the **one** lettered choice with which it is **most** closely associated. Each lettered choice may be used once, more than once, or not at all.

Questions 7–11

Match each description below with the most appropriate organelle.

(A) Primary lysosome
(B) Phagolysosome
(C) Heterolysosome
(D) Autolysosome
(E) Residual body

7. A secondary lysosome containing effete components of the cell that produced it

8. A telolysosome that remains in a cell

9. GERL product before it engages in any other metabolic event

10. The product of a primary lysosome fusing with ingested bacteria

11. The product of a primary lysosome fusing with a substance from the extracellular environment

Questions 12–16

Match each description below with the most appropriate organelles.

(A) Mitochondria
(B) Lysosomes
(C) Both
(D) Neither

12. Membranous organelles

13. Contain enzymes for ATP synthesis

14. Functionally related to the rough endoplasmic reticulum

15. Contain many hydrolytic enzymes

16. Involved in steroid biosynthesis

ANSWERS AND EXPLANATIONS

1. The answer is D. [*II B 1, D 1*] The outer mitochondrial membrane is the outer boundary of this discrete cytoplasmic organelle. It is a lipoprotein bilayer that surrounds the cristae-laden inner mitochondrial membrane, which in turn surrounds the mitochondrial matrix. The outer membrane contains monoamine oxidase and several enzymes involved in electron transport, although most of these are located on the inner mitochondrial membrane, which also contains succinate dehydrogenase. The remaining tricarboxylic acid (TCA) cycle enzymes and enzymes for DNA synthesis and fatty acid oxidation are dissolved in the mitochondrial matrix.

2. The answer is A. [*II B 2, D 2*] Mitochondrial cristae are membranous folds that increase the surface area of the inner mitochondrial membranes. They provide an expanded surface for inner membrane functions, such as electron transport and ATP synthesis. The inner mitochondrial membranes contain most of the enzymes involved in oxidative phosphorylation and one TCA cycle enzyme: succinate dehydrogenase. Most TCA cycle enzymes are located in the matrix between cristae.

3. The answer is D. [*II D 3*] The mitochondrial matrix contains most of the TCA cycle enzymes, including aconitase, fumarase, malate dehydrogenase, and isocitrate dehydrogenase. The only TCA cycle enzyme not in the matrix is succinate dehydrogenase, which is contained in the inner mitochondrial membrane. Cytochrome c is an electron transport enzyme primarily localized in the inner mitochondrial membrane.

4. The answer is C. [*II E 1, 2, 3*] Mitochondria have a genome that is distinct from the nuclear genome. The DNA polymerase for synthesis of mitochondrial DNA is encoded in nuclear genes. Mitochondrial DNA encodes at least one ribosomal protein; nuclear genes produce the rest. The enzymes for oxidative phosphorylation, including most enzymes for electron transport and for ATP synthesis, are made in situ from mitochondrial genes.

5. The answer is B. [*II A 1 a*] Mitochondria typically exist in cell areas that use substantial amounts of ATP. They are abundant in the apices of ciliated cells because the beating action of cilia consumes ATP. They also exist in apices of cells that have a microvillous brush border (e.g., certain kidney cells), because solute transportation and pinocytosis of proteins in the glomerular filtrate consume energy and, therefore, require ATP. Mitochondria are distributed evenly throughout the cytoplasm of smooth muscle cells, steroid-secreting cells, skeletal muscle cells, and liver parenchymal cells rather than existing in apical concentrations.

6. The answer is E. [*IV A, B*] Microbodies (peroxisomes) contain catalase, which converts toxic hydrogen peroxide (H_2O_2) into H_2O and O_2. Microbodies also metabolize D-amino acids and fatty acids. Like many discrete particulate organelles, microbodies are surrounded by a lipoprotein bilayer. Evidence suggests that microbodies arise from the endoplasmic reticulum or de novo in the cytoplasm—not from mitochondria. They are 0.3–1.5 μm in diameter, about the same diameter as mitochondria.

7–11. The answers are: 7-D, 8-E, 9-A, 10-B, 11-C. [*III A 2, B 2*] The GERL produces lysosomes that are called primary lysosomes until they participate in another metabolic event. Primary lysosomes that fuse with phagosomes form phagolysosomes, one type of heterolysosome. Primary lysosomes that fuse with cytocogresomes form autolysosomes. Partially degraded autolysosomes and phagolysosomes may be reabsorbed by the cytoplasm, expelled from the cell as cytostools, or stored in the cell as residual bodies.

12–16. The answers are: 12-C, 13-A, 14-C, 15-B, 16-A. [*I B 2; II C; III A 1, B 1*] Mitochondria and lysosomes are particulate organelles bounded by membranes. Both are functionally related to other membranous organelles, including the rough endoplasmic reticulum. For example, lysosomes are modified secretory structures manufactured in the rough endoplasmic reticulum and Golgi apparatus. In cells that synthesize protein, mitochondria are located near the rough endoplasmic reticulum and produce the ATP required for protein synthesis. Lysosomes contain many hydrolytic enzymes that are used to degrade phagocytosed materials. Mitochondria contain enzymes that are involved in steroid biosynthesis.

5
The Cytoskeleton and Microtubule-Containing Organelles

I. INTRODUCTION

A. Cytoskeleton and cell motility. The cytoskeleton plays an important part in cell motility in all higher organisms. It consists of microtubules, microfilaments, and intermediate filaments.

1. **Microtubules** help maintain cell shape and have a dynamic role in chromosome movement during mitosis. They also are important components of cilia and flagella, motile organelles on the surface of many cells that propel secretions (e.g., mucus in the respiratory tract) or whole cells (e.g., spermatozoa).

2. **Microfilaments** also are abundant in many cells and have an important contractile function in cell motility. They are especially prominent in muscle cells (see Chapter 13 I) but are found in many other cells as well.

B. Cytoskeleton and organelle motility. Recent discoveries have revealed that most cells have a complex cytoarchitecture.

1. The cytoplasm has a complex cytoskeleton of interlocking microtubules, microfilaments, and intermediate filaments, which connects the plasma membrane to cytoplasmic organelles.

2. The cytoskeleton mediates cytoplasmic organelle motion and controls their distribution within the cell as the cell's metabolic requirements change during different physiologic states.

II. MICROTUBULES

A. Structure (Figure 5-1)

1. Microtubules are long, pipe-like structures. Direct observation of flagella indicates that microtubules can be 200 μm long or longer.

2. Microtubules are hollow and 24 nm in diameter. The hollow center is about 15 nm in diameter, yielding a wall thickness of about 4.5 nm.

3. Each microtubule consists of 13 **protofilaments,** which contain globular proteins called **tubulins**. Two classes of tubulin exist: α-tubulin and β-tubulin. Both weigh approximately 55 kD; however, there are subtle chemical differences between the two. The microtubular wall consists of tight spirals of intertwined α-tubulin and β-tubulin (Figure 5-2).

B. Function

1. Microtubules help move chromosomes during mitosis. Thus, they are a prominent feature of all dividing cells (e.g., cells in the stratum basale of the skin).
   a. Microtubules are attached to centrioles and to chromosomes.
   b. Some microtubules extend from one centriole but do not attach to chromosomes. Others extend from one centriole to a chromosome attaching to the kinetochore (centromere).
   c. As microtubules slide past one another, they move chromosomes from the metaphase plate to the centrioles at opposite poles of the dividing cell.

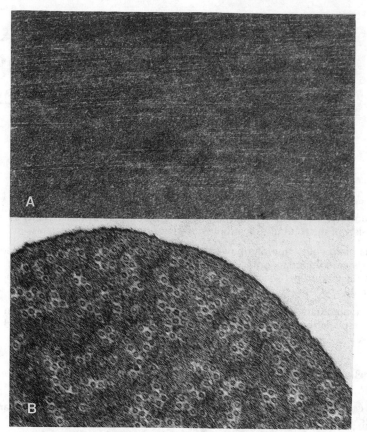

Figure 5-1. Electron micrographs of microtubules. In sections, microtubules appear parallel (*A*) or perpendicular (*B*) to their long axis.

2. Microtubules also help maintain the asymmetrical shape of cells.
   a. Many elongated cells (e.g., motor neurons) have an abundance of microtubules. The long axis of these microtubules is parallel to the long axis of the axon.
   b. Many tall columnar cells have an abundance of microtubules arranged parallel to the long axis of the cell.
   c. Some discoid cells (e.g., platelets) have a circular array of microtubules in their peripheral parts, which help maintain the cell's curved shape.

3. Microtubules are essential constituents of cilia, flagella, basal bodies, and centrioles.

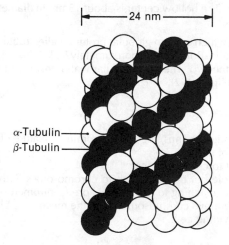

α-Tubulin
β-Tubulin

Figure 5-2. Assembly of α-tubulin and β-tubulin subunits into a tubular structure containing 13 protofilaments. (Adapted from Thorpe NO: *Cell Biology*. New York, John Wiley, 1984, p 616.)

III. CILIA AND FLAGELLA, BASAL BODIES, AND CENTRIOLES

A. Cilia and flagella

1. **Cilia** are movable organelles that are 5–10 μm long and 0.2 μm in diameter.
   a. **Locations.** Cilia are present on the apical epithelial surfaces of the:
      (1) **Respiratory tract** (nasal cavities, larynx, trachea, and bronchi), where they propel mucus and debris out of the system
      (2) **Female reproductive tract** (uterine tubes and uterus), where they propel ova and mucus through the system
      (3) **Sensory organs** (olfactory, auditory, and visual epithelia), where modified cilia help to form chemoreceptors, mechanoreceptors, and photoreceptors
   b. **Structure and function.** Cilia are visible in the light microscope. The electron microscope reveals the complex internal structure of cilia (Figure 5-3).
      (1) Each cilium is surrounded by the plasma membrane and contains a central doublet of microtubules surrounded by nine pairs of fused microtubules. This characteristic **9 + 2 arrangement of microtubules** and its associated structures constitute the **axoneme** (Figure 5-4).
         (a) The central pair is two complete microtubules connected by a **cm bridge** and surrounded by a **central sheath**.
         (b) Each of the **nine outer pairs** of doublets consists of a complete **subunit-A** and an incomplete microtubule called **subunit-B,** which abuts on subunit-A.
         (c) Projecting from subunit-A are a hooked **outer arm** and an angled **inner arm;** a **radial spoke** with a **spoke head** also projects from subunit-A toward the central sheath.
         (d) The subunit-A outer and inner arms have the adenosine triphosphatase (ATPase) protein **dynein.**
      (2) Cilia bend as microtubules in the axoneme slide past one another. This motion is driven by ATP hydrolysis caused by dynein.

2. **Flagella** are similar to cilia; however, they are much longer. For example, spermatozoan flagella are about 50 μm long. In spermatozoan flagella, each outer doublet has a large, electron-dense **outer fiber,** which is absent in cilia. Outer fibers provide the flagellum with rigidity.

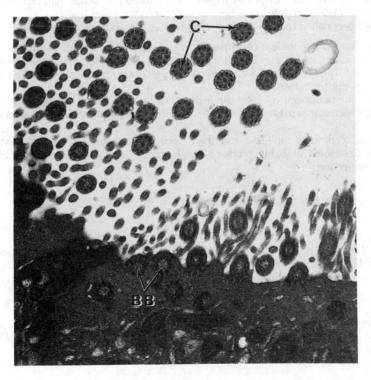

Figure 5-3. Electron micrograph of cilia (*C*) attached to the apical surface of tracheal epithelial cells. Basal bodies (*BB*) anchor the cilia to the apical cell surface.

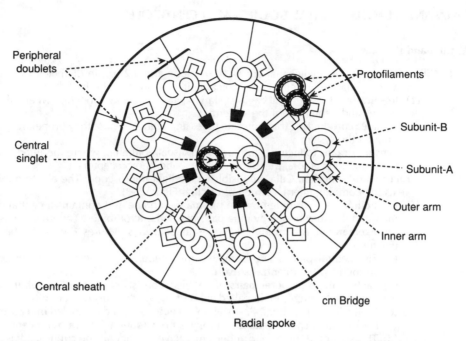

Peripheral doublets

Central singlet

Central sheath

Central spoke

Radial spoke

cm Bridge

Protofilaments

Subunit-B

Subunit-A

Outer arm

Inner arm

Figure 5-4. Diagram of the ciliary axoneme. (Adapted from Fawcett DW: *The Cell*, 2nd ed. Philadelphia, WB Saunders, 1981, p 589.)

B. Basal bodies and centrioles

1. **Basal bodies.** In ciliated epithelia, a basal body anchors each cilium to the cell's apical surface.
   a. The arrangement of microtubules in basal bodies is similar to the arrangement in cilia; however, basal bodies have nine peripheral **triplets** of microtubules rather than doublets, and the ciliary central doublet terminates where the cilia join the basal bodies.
   b. Striated bundles of fibers called **rootlets** anchor basal bodies to the surrounding cytoplasm.

2. **Centrioles.** Many cells have a perinuclear **centrosome,** which contains a pair of **centrioles**.
   a. Centrioles are paired during interphase. During cell division, centrioles separate and migrate to the poles of the cell. Then, centrioles form a microtubule-containing **mitotic spindle** that moves chromosomes during mitosis.
   b. Each centriole has nine peripheral microtubule triplets but lacks a central pair. It also has electron-dense **pericentriolar satellites** that radiate away from the triplets like the vanes of a pinwheel.
   c. In ciliated epithelia, centrioles replicate and then become concentrated in the apical epithelial surface. Centrioles probably produce basal bodies, which, in turn, probably produce a ciliary projection.

IV. INTERMEDIATE FILAMENTS

A. **Structure and function.** Intermediate filaments are less well understood, both structurally and functionally, than microtubules and microfilaments. They are 10 nm in diameter, consisting of four or five protofilaments. Intermediate filaments bind cytoskeletal elements such as microfilaments to the plasma membrane.

B. **Types of intermediate filaments.** Although all intermediate filaments appear similar in the electron microscope, five types exist and each has a distinct biochemical composition. Often, specific antisera are used to distinguish types of intermediate filaments.

1. **Desmin filaments** contain a protein called **desmin,** which has a molecular weight of 50–55 kD. Desmin filaments are abundant in muscular tissue, where they bind microfilaments to the plasma membrane.

2. **Tonofilaments** contain a protein family called **keratins,** which have molecular weights ranging from 40 to 65 kD. Tonofilaments are abundant in many epithelial cells, especially radiating from desmosomes (Figure 5-5). Keratin is the primary protein in epidermal epithelial cells. When extensively cross-linked, keratin produces a hydrophobic barrier in the skin.

3. **Vimetin filaments** are comprised of **vimetin,** a protein with a molecular weight of 50–54 kD. Vimetin is immunologically distinct from desmin and keratin. Vimetin filaments are a prominent component of many mesenchymally derived cells (e.g., fibroblasts) and are abundant in many types of differentiated cells.

4. **Neurofilaments** are intermediate filaments in neurons.

5. **Glial filaments** are intermediate filaments in glial cells.

V. MICROFILAMENTS

A. **Structure.** Microfilaments are very long filaments with a diameter of approximately 7 nm. They are composed of **G-actin** subunits arranged end-to-end to produce **F-actin**. Actin is abundant in non-muscle cells, typically accounting for 5%–10% of all cellular protein. Microfilaments interact strongly with heavy **meromyosin.**

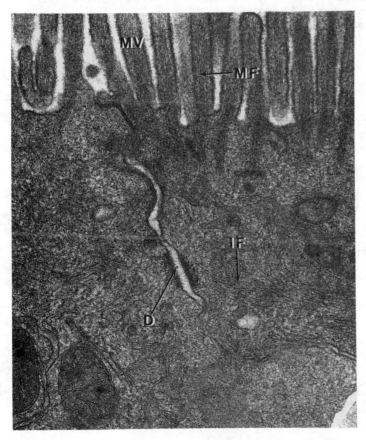

Figure 5-5. Electron micrograph of intermediate filaments (*IF*) associated with a desmosome (*D*). Notice that the intermediate filaments are slightly larger than the microfilaments (*MF*) in the microvilli (*MV*) core.

B. Location

1. The cortex of most cells is semi-rigid because it contains a thick bundle of microfilaments.

2. Microvilli and stereocilia project from the apical surface of many absorptive epithelial cells and contain many actin-rich microfilaments (see Figure 5-5).

3. The cleavage furrow formed during cytokinesis is rich in microfilaments.

4. Microfilaments are abundant in muscle cells as the **thin filaments** (see Chapter 13 I B 1 b).

C. Function. Microfilaments contain **actin,** which interacts strongly with **myosin** to generate the contractile force (see Chapter 13 I C) used in cell locomotion, cytokinesis, and movement of microvilli.

VI. INTERACTIONS BETWEEN CYTOSKELETAL ELEMENTS

A. The cytoskeleton. In many cells, cytoplasm is an amalgam of microtubules, intermediate filaments, and microfilaments linked together just as muscles, bones, and ligaments are linked together in the entire body.

B. Cytoskeletal function

1. The cytoskeleton provides rigid support for most cells and helps determine the cell's shape.

2. Cytoskeletal elements also interact strongly during movement.
   a. Cytoskeletal elements are crucial for organelle translocation. When cell metabolic needs require that mitochondria move from one part of the cell to another, cytoskeletal elements coordinate the process.
   b. Cytoskeletal elements are crucial for cell movement. For example, when a leukocyte moves, cytoskeletal elements mediate this locomotion.
   c. Coordinated cytoskeletal activity also produces muscle contraction and, thus, gross deformation of organs and movement of the entire organism. Table 5-1 contains a summary of cytoskeletal elements and their functions.

Table 5-1. Summary of the Structure and Function of Cytoskeletal Elements

Structure	Diameter	Subunit	Composition	Function
Microtubule	24 nm	α- and β-tubulin	Hollow tube consisting of 13 protofilaments	Cell motility, chromosome movement, ciliary beating, cell shape maintenance
Intermediate filament	10 nm	Desmin, keratin, vimetin	Hollow tube with 4–5 protofilaments	Integration of contractile units in muscle, cytoskeletal integration in non-muscle tissue
Microfilament	7–9 nm	G-actin	Solid thread of polymerized G-actin to form F-actin	Muscle contraction, changes in cell shape, cytokineses, cytoplasmic streaming

STUDY QUESTIONS

Directions: The groups of questions below consist of lettered choices followed by several numbered items. For each numbered item select the **one** lettered choice with which it is **most** closely associated. Each lettered choice may be used once, more than once, or not at all.

Questions 1–4

Match each description below with the appropriate structures.

(A) Cilia
(B) Flagella
(C) Basal bodies
(D) Centrioles
(E) Centrosomes

1. Have a 9 + 2 microtubule arrangement and can be 50 μm long

2. Have nine peripheral triplets but no central microtubule doublet; associated with striated rootlets

3. Move mucus in the respiratory system; some have sensory functions

4. Help organize the mitotic spindle

Questions 5–9

Match each cell type or types below with the filament most likely to appear there.

(A) Desmin filament
(B) Tonofilament
(C) Vimetin filament
(D) Neurofilament
(E) Glial filament

5. Epidermal cell

6. Motor neuron

7. Fibroblast

8. Skeletal and smooth muscle cell

9. Oligodendrocyte and astrocyte

Questions 10–14

Match each statement below with the protein it describes.

(A) Tubulin
(B) Dynein
(C) Actin
(D) Desmin
(E) Keratin

10. Present in epithelial tonofilaments

11. Associated with subunit-A, a protein with ATPase activity in the ciliary axoneme

12. Present in skeletal muscle cell intermediate filaments

13. The most abundant chemical constituent of muscle cell microfilaments and thin filaments

14. Exists in two classes, each weighing 55 kD, which are assembled into 13 protofilaments in microtubules

Questions 15–18

Match each description with the appropriate lettered structure in the micrograph below.

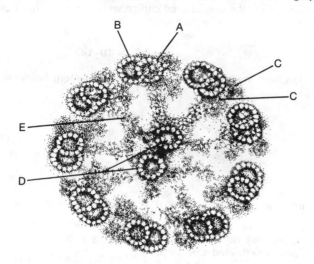

15. A complete microtubule with outer and inner arms attached

16. A structure that contains dynein

17. A structure that does not exist in basal bodies or centrioles

18. An incomplete microtubule attached to subunit-A

ANSWERS AND EXPLANATIONS

1–4. The answers are: 1-B, 2-C, 3-A, 4-D. [*III*] Cilia, flagella, basal bodies, centrioles, and centrosomes all are morphologically related to one another but have different functions.

Cilia and flagella have closely related structures and functions. Cilia are about 10 μm long and are present in the respiratory system and the female reproductive tract. In the respiratory system, the action of cilia moves mucus. In the female reproductive tract, cilia move ova and mucus. The outer segments of cones and rods are modified cilia. Cilia also have a sensory function in the organ of Corti and the olfactory epithelium. Flagella are about 50 μm long and are present in spermatozoa. They propel spermatozoa in the female reproductive tract.

Basal bodies have peripheral triplets and lack a central microtubular doublet. They are anchored in the apical surfaces of ciliated cells by striated rootlets.

Centrioles are the paired structures in the centrosome. During mitosis, centrioles migrate to the poles of the dividing cell and organize the mitotic spindle, which contains microtubules.

Centrosomes are pairs of centrioles near the nucleus of interphase cells.

5–9. The answers are: 5-B, 6-D, 7-C, 8-A, 9-E. [*IV*] There are at least five classes of intermediate filaments. All classes are morphologically similar, but biochemically distinct. All intermediate filaments are hollow tubes 10 nm in diameter and consist of four or five protofilaments.

Desmin filaments are abundant in smooth, skeletal, and cardiac muscle cells. They also are present in other motile systems.

Tonofilaments are abundant in many epithelial cells (e.g., epidermal cells). They contain keratin and often are associated with desmosomes.

Vimetin filaments are abundant in fibroblasts. Vimetin is a protein with a molecular weight of 50–54 kD. It is immunologically distinct from desmin and keratin.

Neurofilaments are the intermediate filaments found in neurons. Glial filaments are the intermediate filaments found in glial cells, such as oligodendrocytes (which are responsible for myelination in the central nervous system) and astrocytes (a class of central nervous system glial cell).

10–14. The answers are: 10-E, 11-B, 12-D, 13-C, 14-A. [*II A 3; III A 1 b (1) (d); IV B 1, 2; V A, B 4*] Tubulin exists in two classes (α-tubulin and β-tubulin) and forms the 13 protofilaments of microtubules.

Dynein is a protein with ATPase activity. It is found in subunit-A in the ciliary axoneme and also is present in flagella.

Actin is a globular protein that is assembled into long microfilaments and the thin filaments of muscle cells.

Desmin and keratin are components of intermediate filaments. Desmin exists in muscle tissue, and keratin exists in skin.

15–18. The answers are: 15-A, 16-C, 17-D, 18-B. [*III A 1*] This is an electron micrograph of a cilium axoneme, which consists of a central pair of microtubules (*D*) surrounded by nine peripheral microtubule doublets. The central pair terminates at the point where the cilium inserts into the basal body and, therefore, is absent from basal bodies. It is absent from centrioles as well.

The peripheral microtubule arrays consist of a subunit-A (*A*), which is a complete microtubule comprised of 13 protofilaments and attached arms, and a subunit-B (*B*), which is an incomplete microtubule. The outer and inner arms (*C*) contain dynein, the protein associated with ATPase activity necessary for ciliary bending.

Histologists believe that radial spokes (*E*) link peripheral doublets to the central pair.

6

Nuclear Cell Biology

I. THE NUCLEUS. Most human cells contain a nucleus. A notable exception is the adult erythrocyte, which lacks a nucleus. Erythrocyte precursors in the bone marrow are nucleated cells; however, the nucleus is extruded late in erythrocyte development.

A. Structure

1. The nucleus is separated from the cytoplasm by the **nuclear envelope,** a double layer of unit membranes. The nuclear envelope is perforated by **nuclear pores,** which allow large macromolecules synthesized in the nucleus to pass into the cytoplasm.

2. **In the light microscope,** the nucleus often appears as a round or ellipsoid structure with a diameter of roughly 1–3 μm.
    a. The location of the nucleus within the cell depends on the cell type. It may appear in the center of the cell (e.g., in leukocytes), near the basal surface (e.g., in tall columnar cells), or near the peripheral surface (e.g., in skeletal muscle cells).
    b. The nucleus **stains intensely** with many **basophilic dyes** (e.g., hematoxylin). Many cells exhibit peripheral, darkly stained clumps of **heterochromatin** and lightly stained masses of **euchromatin** (Figure 6-1).
        (1) **Heterochromatin** is abundant in inactive nuclei (e.g., the nuclei of small lymphocytes and spermatozoa).
        (2) **Euchromatin** is abundant in active nuclei (e.g., plasma cell nuclei). However, active nuclei contain heterochromatin as well as euchromatin.

B. Functions

1. The nucleus **contains DNA** in the form of linear arrays of **genes.**
    a. **Genes** determine most of an organism's characteristics.
    b. **Chromosomes** are discrete collections of genes and nuclear proteins.

2. The nucleus contains **the entire human karyotype,** which consists of 46 chromosomes (see V).

3. Not all genes are located on nuclear chromosomes. For example, mitochondria have a separate genome that encodes some functions (see Chapter 4 II E 6).

C. Ultrastructure

1. **The nuclear envelope**
    a. The nuclear envelope is 7–8 nm thick (Figure 6-2) and contains several hundred **nuclear pores** (3–40 nuclear pores/μm² of surface area; Figure 6-3).
        (1) Nuclear pores are 60 nm in diameter and consist of eight subunits.
        (2) Large macromolecules (e.g., mRNA, rRNA) pass from the nucleus to the cytoplasm through nuclear pores.
    b. The nuclear envelope **outer membrane** often is continuous with the membranes of the rough endoplasmic reticulum.
    c. The nuclear envelope **inner membrane** has numerous filaments that attach chromatin and other structures used for pore diameter control to the nucleoplasmic matrix side of the inner membrane.

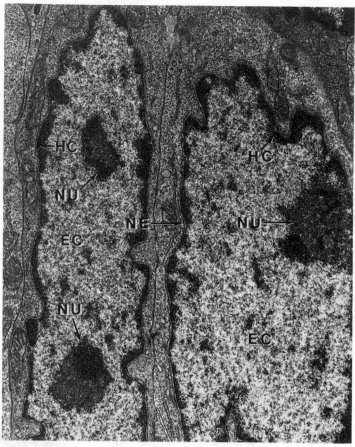

Figure 6-1. Electron micrograph of an adult cell nucleus. The nuclear envelope (*NE*) is a pair of unit membranes that separate the nucleus from the cytoplasm. The nucleolus (*NU*), heterochromatin (*HC*), and euchromatin (*EC*) are also shown.

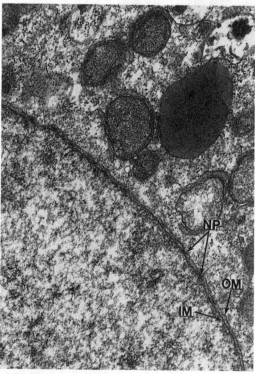

Figure 6-2. Electron micrograph of the nuclear envelope. The nuclear envelope consists of two unit membranes, the inner membrane (*IM*) and the outer membrane (*OM*). The nuclear envelope has nuclear pores (*NP*), through which it allows macromolecules synthesized in the nucleus to move into the cytoplasm.

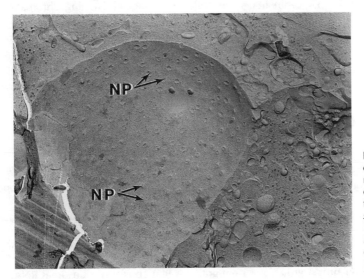

Figure 6-3. Freeze-fracture-etch of a cell nucleus. The outer surface of the nuclear envelope has large numbers of nuclear pores (*NP*). (Reprinted from Johnson KE: *Histology: Microscopic Anatomy and Embryology,* New York, John Wiley, 1982, p 14. Micrograph courtesy of Dr. J.D. Robertson, Department of Anatomy, Duke University.)

2. **Chromatin.** The nucleoplasmic matrix consists of a large amount of fibrous material called chromatin.
   a. Chromatin fibers are approximately 20 nm in diameter and consist of straight smooth areas of DNA interspersed with **nucleosomes**.
   b. DNA is tightly coiled in chromatin fibers. One length of chromatin contains 30 lengths of DNA.
3. **The nuclear matrix.** The form and function of the nucleus is controlled by the nuclear matrix located between the nuclear envelope and the chromatin. If intact nuclei are isolated from cells and have their DNA, RNA, and nuclear envelope extracted, a **nuclear protein matrix** remains.
   a. The matrix is a protein mixture that contains three major protein components with molecular weights in the 60–70 kD range.
   b. Nuclear matrix proteins contract and expand with different concentrations of divalent cations, although this activity is not related to the actin-myosin function (see Chapter 13 I B, C).
   c. Nuclear matrix proteins may control nuclear functions such as DNA replication, transcription, post-translational RNA metabolism, and RNA transport.

II. CHROMOSOME STRUCTURE

A. **Nuclear structure.** At higher resolution, chromatin consists of many nucleosomes, arranged like beads on a string.

1. Each nucleosome is 10 nm in diameter and consists of a tight DNA spiral of 146 base pairs wrapped around an octameric core of **histones**.

2. Histones are a family of five or more basic proteins arranged as an octamer.

B. **Condensed nucleosomes.** Nucleosomes form thin fibers that are arranged into a tight helical spiral 30 nm in diameter. Each turn of this helix contains six nucleosomes.

C. **Higher order coiling**

1. The 30-nm fibers of the nucleosome are arranged into another level of coiling, forming extensively coiled regions called **chromomeres** interspersed with relatively straight regions called **interbands**.

2. The alternating extensively coiled fibers and less coiled fibers may cause the banding seen on stained metaphase chromosomes.

3. The densely coiled masses of 30-nm fibers are coiled once again, creating chromatids of individual, spiralized chromosomes.

4. This model describes how large amounts of DNA are packed into short chromosomes (Figure 6-4).

III. THE NUCLEOLUS

A. Structure

1. The nucleolus is a dense mass in the center of the nucleus. Cells typically have one or two nucleoli; however, the number of nucleoli varies as the cell's physiologic condition changes.

2. Nucleoli are most prominent during interphase, disappear during metaphase, and reappear during telophase.

3. The **fibrillar and granular components** of the nucleolus are visible in the electron microscope (Figure 6-5).

    a. The fibrillar component is a mass of 5-nm diameter fibrils that are the DNA strands of the **nucleolar organizer region**.

    b. The granular component is a group of 15-nm granules that consist of **ribosomal RNA**.

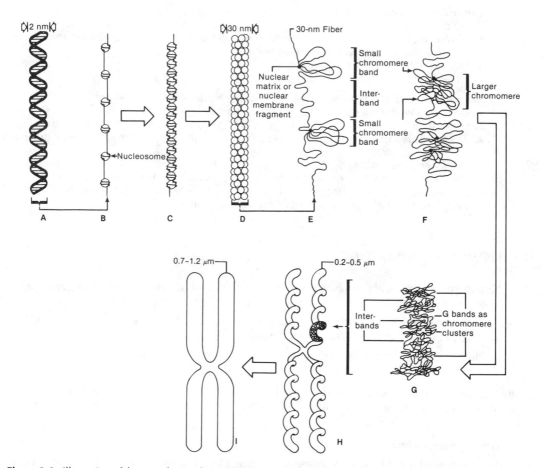

Figure 6-4. Illustration of the complex packing of DNA into chromosomes: (*A*) DNA, (*B*) extended nucleosomes, (*C*) 10-nm fiber, (*D*) condensed nucleosomes forming a 30-nm fiber, (*E*) chromomeres and interband chromatin, (*F*) clustering of chromomeres, (*G*) chromosome bands, (*H*) spiralized chromosomes, and (*I*) compact chromosomes. (Reprinted from Thorpe NO: *Cell Biology*. New York, John Wiley, 1984, p 513.)

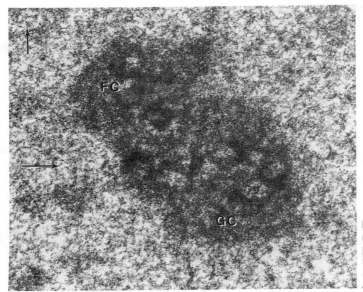

Figure 6-5. Electron micrograph of a nucleolus, showing the fibrillar component (*FC*) and the granular component (*GC*). The fibrillar nature of the nuclear chromatin is evident at the *arrows*.

B. Function

1. **Ribosomal RNA (rRNA) synthesis** occurs in the nucleolus. The nucleolar organizer region contains genes that encode rRNA. Evidence of this includes:
   a. The enzymes necessary for rRNA synthesis are present in the nucleolar organizer region.
   b. Nucleoli irradiated with ultraviolet microbeams cease rRNA synthesis and the nucleoli disappear.
   c. Mutant cells that lack the nucleolar organizer region also lack nucleoli and do not synthesize rRNA.

2. After rRNA is synthesized and assembled into ribonuclear protein particles (**ribosome precursors),** they are transported through nuclear pores into the cytoplasm. In the cytoplasm, ribosomes bind to rough endoplasmic reticular membranes or assemble into **polysomes**.

IV. MITOSIS

A. Functions. Mitosis divides chromosomes equally between daughter cells, maintains the number of diploid chromosomes, and increases the number of cells.

B. The cell cycle (Figure 6-6)

1. **Interphase** is the period between mitotic divisions when cells rest.
   a. Interphase begins with the **first gap phase (G_1)**. DNA is not synthesized during G_1.
   b. During the ensuing **S-phase,** DNA synthesis occurs.
   c. After the S-phase, DNA synthesis ceases and the **second gap phase (G_2)** occurs.

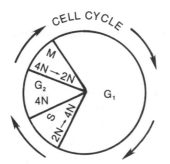

Figure 6-6. Diagram of the cell cycle. The cell cycle consists of interphase (*G_1, S, and G_2*) and mitosis (*M*). During G_1, there is a gap period of no DNA synthesis. DNA synthesis occurs during S, and the DNA content of the nucleus is doubled from 2N to 4N. There is a brief gap period (G_2) between S and M when the cell nucleus is 4N. During M, cell division occurs and the nuclear DNA content is again reduced to 2N. Two daughter nuclei are produced by the end of M. Each daughter nucleus has 2N DNA. (Reprinted from Johnson KE: *Histology: Microscopic Anatomy and Embryology.* New York, John Wiley, 1982, p 21.)

2. Mitosis, or the **M-phase,** is the period of cell division.
   a. The M-phase occurs after G_2. The nuclear envelope disappears, and chromosomes condense **(prophase)** and then assemble on the **metaphase plate (metaphase).**
   b. Next, chromosomes move to opposite poles **(anaphase)** and then decondense as the nuclear envelope reassembles **(telophase).**

C. The mitotic spindle

1. **Structure**
   a. There are two **centrioles** in the mitotic spindle, one located at each opposite pole of the spindle.
   b. **Microtubules** radiate away from each spindle toward the chromosomes assembled on the metaphase plate.
      (1) **Polar microtubules** project away from the centriole toward the metaphase plate. They do not attach directly to chromosomes.
      (2) **Kinetochore microtubules** have the same polarity as polar microtubules, but they attach to the chromosomes at the **kinetochore** (centromere).
      (3) The two kinds of microtubules overlap extensively in an interdigitated fashion. When they slide past one another during anaphase, chromosomes move away from the metaphase plate toward the opposite spindle poles.

2. **Function**
   a. The mitotic spindle is the engine that moves chromosomes during cell division.
   b. It begins to assemble after chromosome condensation and becomes completely assembled during metaphase.

V. KARYOTYPE ANALYSIS

A. **The normal human karyotype.** The karyotype is the ensemble of chromosomes included in the diploid chromosomal complement.

1. **Most cells have a single copy of the diploid karyotype** in each nucleus, although **polyploidy** (integral multiples of the normal diploid complement) can also be observed under certain circumstances.
   a. Liver cells can be tetraploid or octaploid.
   b. Some individuals are born with an abnormal triploid karyotype.

2. **The normal female karyotype** consists of 22 pairs of **autosomes** and a pair of identical X **sex chromosomes.** Females are **homomorphic** with respect to sex chromosomes.

3. **The normal male karyotype** consists of 22 pairs of distinct autosomes and a pair of sex chromosomes designated X and Y because they are morphologically different. Males are **heteromorphic** with respect to sex chromosomes.

4. Three characteristics facilitate precise identification of each pair of autosomes:
   a. Differences in position of the centromere
   b. Chromatid length
   c. Differences in banding patterns exhibited by different regions of the chromatids when stained

B. Methods of karyotype analysis

1. First, samples of living tissue are gathered. Methods of gathering tissue samples include:
   a. Punch biopsy of skin
   b. Amniocentesis or chorionic villus biopsy for fetal tissues
   c. Collecting the foreskin after circumcision

2. Next, tissue samples are dispersed into single cell suspensions and then grown in vitro using tissue culture methods. Accurate analysis requires a substantial population of dividing cells.

3. Cultured cells are treated with the antimitotic agent **colchicine,** which inhibits microtubule function, thus preventing mitotic spindle formation.

4. Cells with chromosomes arrayed on the metaphase plate accumulate in the culture because they are blocked at metaphase.

5. Groups of colchicine-treated cells are removed from the culture medium, treated with hypotonic media, squashed to spread their chromosomes, and then stained to reveal chromosome banding patterns.

6. The spread chromosomes are photographed and then standard techniques are used to arrange the chromosomes based on morphologic differences.

STUDY QUESTIONS

Directions: Each question below contains five suggested answers. Choose the **one best** response to each question.

1. Each of the following statements concerning the nucleolus is true EXCEPT

(A) electron microscopy reveals its homogeneous granular structure
(B) DNA strands comprise its nucleolar organizer region
(C) microbeam irradiation suppresses rRNA synthesis in the nucleolus
(D) rRNA synthesis and assembly occur in the nucleolus
(E) a cell nucleus may contain more than one nucleolus

2. Each of the following statements concerning the structure of the nuclear envelope is true EXCEPT

(A) it consists of two asymmetrical unit membranes
(B) it contains nuclear pores, which have a hexameric substructure
(C) macromolecules pass from the nuclear matrix into the cytoplasm through nuclear pores
(D) it has 3–40 nuclear pores/μm^2
(E) unit membranes end abruptly at nuclear pores

3. Each of the following statements concerning the nucleolar organizer region is true EXCEPT

(A) it is located in the nucleolus
(B) it is located in the fibrillar component of the nucleolus
(C) it contains DNA
(D) it contains the genes necessary for rRNA synthesis
(E) ultraviolet microbeam irradiation does not affect rRNA synthesis

4. Each of the following statements concerning the mitotic spindle is true EXCEPT

(A) it contains many microtubules
(B) its components attach to the chromosome at the centromere
(C) it is responsible for chromosome movement during anaphase
(D) it is assembled before chromosome condensation
(E) the mitotic spindle forms from the centriole

5. Each of the following statements concerning the human karyotype is true EXCEPT

(A) karyotype analysis can be performed with histologic sections
(B) male sex chromosomes are heteromorphic
(C) female sex chromosomes are homomorphic
(D) there are 22 pairs of autosomes
(E) a normal human karyotype contains 46 chromosomes

Directions: The groups of questions below consist of lettered choices followed by several numbered items. For each numbered item select the **one** lettered choice with which it is **most** closely associated. Each lettered choice may be used once, more than once, or not at all.

Questions 6–11

Match each characteristic described below with the most appropriate form or forms of chromatin.

(A) Heterochromatin
(B) Euchromatin
(C) Both
(D) Neither

6. Contains nucleosomes and DNA

7. Abundant in spermatozoa

8. Abundant in plasma cells

9. Stains darkly with acid dyes

10. Stains darkly with toluidine blue or hematoxylin

11. Abundant in cells with nuclei actively synthesizing rRNA or mRNA

Questions 12–17

Match each mitotic action described below with the appropriate phase of mitosis.

(A) Interphase
(B) Prophase
(C) Metaphase
(D) Anaphase
(E) Telophase

12. Extensive chromosome coiling

13. DNA synthesis

14. Chromosomes move to the spindle poles

15. The nuclear envelope disintegrates

16. Chromosomes assemble on the equatorial plane

17. The nuclear envelope reappears and chromosomes uncoil

ANSWERS AND EXPLANATIONS

1. The answer is A. [*III A*] The nucleolus is a structure within the nucleus of many cells. An electron microscopic view of the nucleolus reveals that it has a fibrillar component that contains strands of DNA and a granular component that contains ribosome precursors. The DNA strands comprise the nucleolar organizer region, which contains the genes necessary for rRNA synthesis. Within the nucleolus, rRNA is synthesized and assembled into ribonuclear protein particles (ribosome precursors), which, in turn, are assembled into ribosomes. Then, ribosomes are transported through nuclear pores into the cytoplasm. When the nucleolar organizer region is subjected to ultraviolet microbeam irradiation, rRNA synthesis stops and the nucleolus disappears.

2. The answer is B. [*I A 1, C 1*] The nuclear envelope completely surrounds the nucleus, forming a boundary between the nucleoplasmic matrix and the cytoplasm. It consists of two closely apposed asymmetrical unit membranes, which contain several hundred octameric nuclear pores. The nuclear envelope ends abruptly at each nuclear pore. The outer membrane is studded with ribosomes and is continuous with the rough endoplasmic reticulum. The inner membrane is thicker than the outer membrane. The inner face of the inner membrane has a fibrous substructure composed of strands of chromatin, which is involved in regulating the diameter of nuclear pores. Each square micrometer of nuclear envelope surface contains 3–40 nuclear pores.

3. The answer is E. [*III*] The nucleolar organizer is the fibrillar portion of the nucleolus that contains the genes that encode for the synthesis of rRNA. It is located in the DNA strands that make up the fibrillar component of the nucleolus. The nucleolar organizer region contains genes encoding for rRNA synthesis and, therefore, contains DNA. Ultraviolet microbeam irradiation of the nucleolar organizer region causes a cessation of rRNA synthesis and disappearance of the nucleolus.

4. The answer is D. [*IV C 2 b*] The mitotic spindle consists of a large collection of microtubules. The microtubules are closely associated at one pole with centrioles and also attach to individual chromosomes at the centromere. As microtubules slide past one another, chromosomes move during anaphase of mitosis. The mitotic spindle is assembled from centrioles during metaphase of mitosis, after chromosomal condensation, which occurs during prophase (i.e., before metaphase).

5. The answer is A. [*V B*] The karyotype is the ensemble of chromosomes, arranged according to size and banding pattern, that are characteristic of a species. Karyotype analysis is performed using cultured cells treated with colchicine to arrest cells at metaphase. During analysis, chromosomes are spread, photographed, and assembled by size and banding characteristics into a karyotype. The human karyotype contains 23 pairs of chromosomes for a total of 46 chromosomes. There are 22 pairs of autosomes in both sexes. Female sex chromosomes are homomorphic because they contain two morphologically identical X chromosomes. Male sex chromosomes are heteromorphic because they contain an X chromosome and a Y chromosome, two morphologically distinct chromosomes.

6–11. The answers are: 6-C, 7-A, 8-C, 9-D, 10-A, 11-B. [*I A 2, C 2*] Chromatin is a constituent of the nucleoplasmic matrix; it exists as either heterochromatin or euchromatin. Neither form stains with acid dyes. However, when a basic dye such as hematoxylin or toluidine blue is applied, heterochromatin stains darkly and euchromatin stains lightly. Heterochromatin is not active in nucleic acid synthesis and, therefore, is abundant in the nuclei of cells not actively synthesizing rRNA and mRNA, such as spermatozoa. Euchromatin is active in nucleic acid synthesis and, therefore, is abundant in the nuclei of cells actively synthesizing rRNA or mRNA, such as plasma cells. Plasma cells, however, contain both heterochromatin and euchromatin.

12–17. The answers are: 12-B, 13-A, 14-D, 15-B, 16-C, 17-E. [*IV B*] Interphase is a three-stage period when the cell rests between mitoses; DNA synthesis occurs during the S-phase of interphase. During prophase, the nuclear envelope disintegrates and chromosomes become extensively coiled. During metaphase, chromosomes become aligned on the metaphase plate in the equatorial plane of the cell. During anaphase, chromosomes move from the metaphase plate to the spindle poles and cytokinesis begins. During telophase, chromosomes uncoil, the nuclear envelope reappears, and cytokinesis concludes.

7

The Extracellular Matrix

I. INTRODUCTION

A. Extracellular matrix general features

1. The extracellular matrix consists of fibrous proteins embedded in amorphous ground substance.

2. Connective tissues consist of cells, fibers, and extracellular matrix. Connective tissues encase vital organs (e.g., brain, lungs) and provide attachment sites for muscles.

B. Extracellular matrix macromolecules

1. Macromolecules are crucial to the formation of extracellular matrix. Fibrous proteins such as collagen provide the extracellular matrix with tensile strength.

2. Other macromolecules (collectively called **amorphous ground substance**) are glycoproteins and proteoglycans firmly bound to extracellular matrix fibrillar components. The glycoproteins and proteoglycans cement the fibrillar components together, increasing the tensile strength of the extracellular matrix.

3. Cells become firmly attached to the extracellular matrix, allowing the formation of multicellular arrays.

II. CONNECTIVE TISSUE

A. The four basic types of human tissue

1. **Epithelial tissues** coat surfaces and line cavities with continuous sheets of cells. A **basement membrane** underlies epithelial tissues.

2. **Connective tissues** consist of cells, fibers, and amorphous ground substance. They bind epithelial tissues to the subjacent structures of the body and connect organs in the body.

3. **Muscular tissues** move organs and the entire organism.

4. **Nervous tissues** coordinate the activities of many tissues and organs.

B. Cardinal features of connective tissue

1. All connective tissues contain isolated cells surrounded by extracellular matrix.

2. Connective tissue cells synthesize their own extracellular matrix and hold organs together.
   a. For example, the pancreas consists of many epithelial cells connected by delicate slips of connective tissue to form an entire organ. A connective tissue capsule surrounds the organ as well. The liver, spleen, and kidneys have similar connective tissue capsules.
   b. Skeletal muscles are surrounded by connective tissue capsules and are attached to bones by tough connective tissue **tendons**.
   c. Individual bones are surrounded by a connective tissue capsule called the **periosteum**. Bones are joined at joints by connective tissues.

3. Connective tissue viscosity and tensile strength vary according to the arrangement of their extracellular matrix fibrous and amorphous components.
   a. For example, blood is a fluid connective tissue. Its fibers are potential fibers of the protein **fibrinogen,** which become actual fibers of **fibrin** only when blood clots.
   b. In contrast, bone is a semi-rigid connective tissue with the tensile strength of reinforced concrete. Its fibers are tough strands of **collagen** embedded in an amorphous extracellular matrix of proteoglycans and glycoproteins infiltrated with calcium salts.

III. COLLAGEN

A. Distribution. Collagen is the most abundant nonaqueous component of the extracellular matrix, constituting more than 70% (dry weight) of tendons and the dermis of skin.

B. Composition

1. Collagen contains three polypeptide chains bound into a superhelix.
   a. Several kinds of polypeptide chains are synthesized. Synthesis is directed by distinct genes.
   b. Some collagens have two kinds of polypeptide chains interwoven in the superhelix, while others contain only one kind of polypeptide chain.
   c. Collagen polypeptide chains are predominantly glycine, glutamic acid, aspartic acid, proline, hydroxyproline, and an assortment of other amino acids including arginine, lysine, hydroxylysine, and leucine.

2. Polypeptide chain triple helices are assembled into collagen microfibrils that can be viewed microscopically.

C. Types of collagen. At present, 11 different types of collagen have been identified (Types I through XI.) Types I–V are the most abundant varieties of collagen. Types VI–XI are sometimes called minor components, as they occur in relatively small quantities; however, they may have important functions. Table 7-1 summarizes the major features of collagen Types I–V.

1. **Type I collagen** is found in bone, tendon, skin, and the cornea.
   a. Type I collagen consists of two $\alpha1(I)$ chains and one $\alpha2(I)$ chain and is designated by the notation $[\alpha1(I)]_2\alpha2(I)$.
   b. Type I collagen contains very little carbohydrate, and the lysines present are not highly hydroxylated.
   c. Type I collagen is synthesized by connective tissue fibroblasts, osteoblasts, smooth muscle cells, and some epithelial cells.
   d. Type I collagen forms microscopically visible fibrils that have a characteristic 67-nm periodicity in the electron microscope (Figure 7-1) due to the asymmetrical arrangement of collagen molecules (Figure 7-2). Similar arrangements occur in Types II and III collagen.

2. **Type II collagen** is found in cartilage, the cornea, and the vitreous body of the eye.
   a. Type II collagen consists of three $\alpha1(II)$ chains and is designated by the notation $[\alpha1(II)]_3$.
   b. Type II collagen contains a higher degree of lysine hydroxylation than Type I collagen.
   c. Type II collagen is synthesized by chondroblasts and chondrocytes, neural retinal cells, and in the notochord.

Table 7-1. Characteristics of Major Types of Collagen

Collagen Type	Extension Peptides	Glycosylation	Distribution
Type I	Cleaved	Slight	Bone, tendon, skin, cornea
Type II	Cleaved	Slight	Cartilage, cornea, vitreous body
Type III	Cleaved	Slight	Fetal skin, blood vessels, within organs
Type IV	Uncleaved	Abundant	Basement membranes
Type V	Uncleaved	Abundant	Around blood vessels and smooth muscle cells

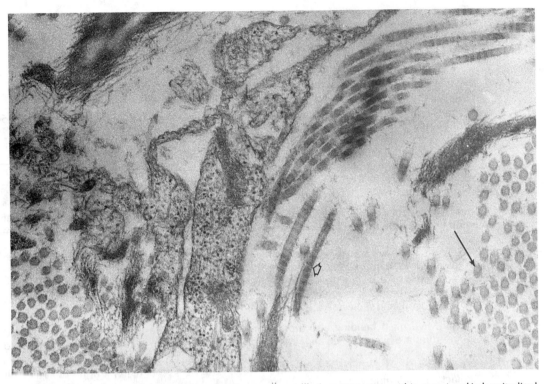

Figure 7-1. Transmission electron micrograph of Type I collagen fibrils in cross section (*thin arrow*) and in longitudinal section (*fat, open arrow*), showing the characteristic 67-nm periodicity. (Micrograph courtesy of Dr. E.N. Albert, Department of Anatomy, George Washington University.)

3. **Type III collagen** is found in fetal dermis, around blood vessels, and in many organs.
    a. Type III collagen consists of three $\alpha1(III)$ chains and is designated by the notation $[\alpha1(III)]_3$.
    b. Type III collagen contains little lysine hydroxylation and some cysteine. It is synthesized by fibroblasts and myoblasts.

4. **Type IV collagen** is an important structural component of the basement membrane—the fibrous meshwork of extracellular matrix on the basal surface of almost all epithelial layers.
    a. Two varieties of Type IV collagen exist. They are designated by the notations $[\alpha1(IV)]_3$ and $[\alpha2(IV)]_3$.
    b. Type IV collagen has highly hydroxylated lysine residues and is highly glycosylated.
    c. A variety of epithelial cells synthesize Type IV collagen.

5. **Type V collagen** is abundant around blood vessels and smooth muscle cells.
    a. Type V collagen consists of two $\alpha1(V)$ chains and one $\alpha2(V)$ chain and is designated by the notation $[\alpha1(V)]_2\alpha2(V)$.

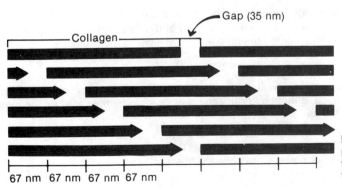

Figure 7-2. Diagram of the end-to-end packing of polarized collagen molecules that creates the 67-nm periodicity characteristic of collagen fibrils.

b. Type V collagen is synthesized by smooth muscle cells and chondrocytes.

6. Types IV and V collagen are amorphous and probably exist as diffuse networks of polypeptide chains rather than as assembled, microscopically recognizable fibrils.

D. Collagen synthesis: early stages

1. All collagen molecules have **structural similarities**.
   a. Each is composed of three α chains intertwined in a triple helix about 300 nm long and 1.5 nm thick.
   b. Multiple triple helices are bundled into collagen fibrils many micrometers long, with diameters of 10–300 nm.
   c. Collagen fibrils assemble into much larger collagen fibers that are several micrometers in diameter. Collagen fibers are a prominent component of many connective tissues and are visible in the light microscope.

2. Collagen polypeptide chains are **synthesized on ribosomes** bound to endoplasmic reticular membranes. The chains elongate and pass into the lumen of the rough endoplasmic reticulum as large polypeptide chains called **pro-α chains**.
   a. Pro-α chains have signal peptides, which help the polypeptide chain pass through the ribosome into the rough endoplasmic reticulum.
   b. Pro-α chains also have **extension peptides** at either end, which help form the triple helix during the assembly of **procollagen** (Figure 7-3). Extension peptides on the carboxyl terminus of the procollagen molecule are cross-linked by interchain disulfide bonds.

3. **Interchain hydrogen bonding** also aids in procollagen molecule assembly. The proline and lysine residues are hydroxylated after translation but before the pro-α chains are assembled into procollagen.
   a. This procollagen modification is important for collagen cross-linking and requires several cofactors, including vitamin C.
   b. **Scurvy,** the deficiency of dietary vitamin C, results in abnormal pro-α chain formation, indirectly causing skin and blood vessel fragility.

4. **Lysine hydroxylation** is important for glycosylation, a second post-translational modification of procollagen. Procollagens have unusual disaccharide residues that usually contain glucose and galactose but lack the sialic acid characteristic of many other glycoprotein secretion products. Glycosylation in collagens varies. Type I has very little carbohydrate and Type V has abundant carbohydrate.

E. Collagen synthesis: late stages

1. After procollagen is synthesized and before it is released from the cell, a **procollagen peptidase** cleaves the extension peptides, leaving a collagen molecule (tropocollagen).

2. Collagen molecules have a tendency to assemble spontaneously into collagen fibrils, but they will do so only after the extension peptides are cleaved by procollagen peptidase. This prevents intracellular fibrillogenesis.

3. Types I, II, and III procollagen are converted to collagen; Types IV and V are not. This may explain why the former assemble into fibrils while the latter do not.

4. Collagen molecules are 300 nm long and have a head and a tail end.
   a. Numerous collagen molecules are packed end-to-end and side-to-side, with regular gaps between heads and tails and an approximate one-third stagger to the molecules.
   b. This configuration results in a regular 67-nm periodicity that is clearly visible in the electron microscope.

5. The three-dimensional arrangement of collagen molecules and the factors that regulate fibrillogenesis are poorly understood, although fibrillogenesis appears to be regulated by the interaction of molecules such as glycosaminoglycans and fibronectin with collagen.

6. In tissue, collagen fibers may be arranged randomly (in skin), in parallel arrays (in tendons), or in nearly orthogonally overlapping arrays (in the cornea). The cells that secrete collagen may be involved in the order or disorder observed in collagen fibers in situ.

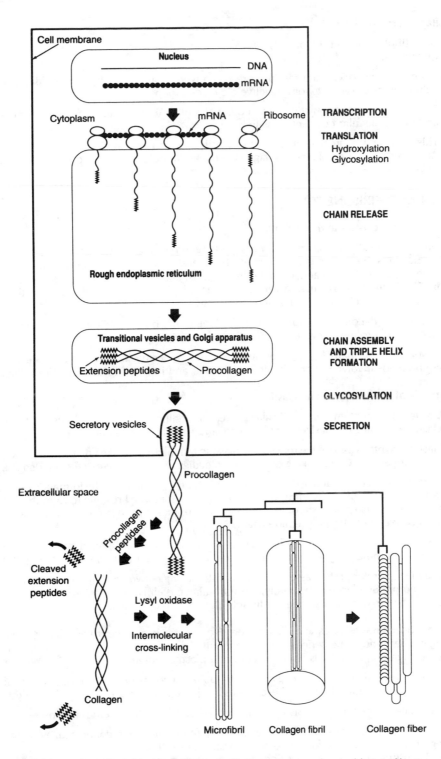

Figure 7-3. Collagen synthesis begins inside cells, but the final processing and assembly into fibers occurs outside cells after secretion. The intracellular events include synthesis of pro-α chains in the rough endoplasmic reticulum, hydroxylation and glycosylation in the rough endoplasmic reticulum and Golgi apparatus, and assembly of pro-α chains into triple helices in the Golgi apparatus and secretory vesicles. The extracellular events include cleavage of extension peptides, fibrillogenesis and cross-linking, and assembly of fibrils into mature fibers.

F. Collagen cross-linking

1. After fibrils form outside of cells, extensive cross-linking occurs within and between collagen molecules.

2. Lysine and hydroxylysine residues are oxidatively deaminated by **lysyl oxidase,** an unusual extracellular enzyme, to form reactive aldehyde groups.

3. The aldehyde groups react with each other and other amino acids to form both intramolecular and intermolecular covalent cross-links.

4. This mechanism greatly increases the tensile strength of individual collagen fibrils so that structures composed of many collagen fibers and fibroblasts (e.g., tendons connecting muscles to bones) are especially durable and strong.

IV. ELASTIN AND FIBRONECTIN

A. Elastin characteristics and distribution

1. Elastin is an extracellular matrix polypeptide with peculiar elastic properties.
   a. Elastin is the main component of **elastic fibers,** which are found in skin, blood vessels, the nose, the external ear, gastrointestinal organs, and the lungs.
   b. Elastic fibers allow these organs to resume their original shape after a distorting force has temporarily changed their shape.

2. Elastin is a glycoprotein with a molecular weight of about 70 kD. Like collagen, it is rich in glycine and proline; however, it lacks the hydroxyproline and hydroxylysine found in collagen.

3. Elastin contains many hydrophobic amino acids and **desmosine** and **isodesmosine,** two unique amino acids involved in creating intramolecular and intermolecular cross-links (Figure 7-4).

B. Synthesis of elastic fibers and elastin

1. Elastic fibers contain a **microfibrillar protein** in addition to the amorphous elastin molecules. Elastic fibers are synthesized by fibroblasts and smooth muscle cells.

2. Elastin is synthesized on the rough endoplasmic reticulum and, like collagen, is packaged in the Golgi apparatus. Cross-linking of elastin molecules occurs in the extracellular space.

3. Lysyl oxidase creates aldehyde groups on three lysines, resulting in the formation of three allysyl residues. Then, three allysyl residues condense with one lysyl residue to form the heterocyclic ring characteristic of desmosine. The resulting desmosyl residues form intermolecular cross-links between elastin polypeptide chains.

4. Elastin molecules exist as cross-linked random coils. When cross-linked, they form a rubber-like network of molecules that returns to its original shape after distortion (see Figure 7-4).

5. The microfibrillar protein assembles into anastomosing networks of fibrous proteins surrounded by amorphous elastin. Apparently, microfibrillar protein is synthesized and secreted before elastin and may help organize elastin into deformable networks.

C. Fibronectin characteristics and distribution. Fibronectin is a glycoprotein that is found around collagen fibers and in the pericellular environment of many connective tissues. Also, blood plasma contains high concentrations of fibronectin, as a plasma protein.

1. **Fibronectin** is a dimer consisting of two 60- to 70-nm polypeptide chains that have a molecular weight of 230–250 kD each. These chains are joined at one end by several disulfide bonds.

2. Extracellular fibronectin exists as fibrils that are visible in the electron microscope.

3. Fibronectin consists of a series of elaborately folded domains connected by short unfolded regions. Three types of folded domains exist: Types I, II, and III.
   a. Each functional domain of fibronectin contains one or more repeating unit domains. For example, the collagen-binding domain of fibronectin consists of several Type II and Type I coiled regions, while the cell-binding domain may consist of as few as one Type III unit (Figure 7-5).
   b. Recent studies have shown that the arginine-glycine-aspartic acid-serine tetrapeptide of the Type III region is sufficient to recognize cells. This tetrapeptide is specifically recognized by

integrin, an integral membrane glycoprotein. Integrin recognizes extracellular fibronectin and binds to it. Also, integrin is anchored to cytoplasmic microfilaments by several intermediate proteins including **vinculin** and **talin**. Thus, it serves as a bridge between the extracellular matrix and the cytoskeleton.

D. Fibronectin functions (Figure 7-6)

1. Fibronectin in the extracellular matrix serves as an intermediate protein, linking cells to other extracellular matrix components, most notably collagen and certain glycosaminoglycans (e.g., heparin).

2. Fibronectin promotes the adhesion of isolated cells to plastic petri dishes that are coated with layers of this glycoprotein.

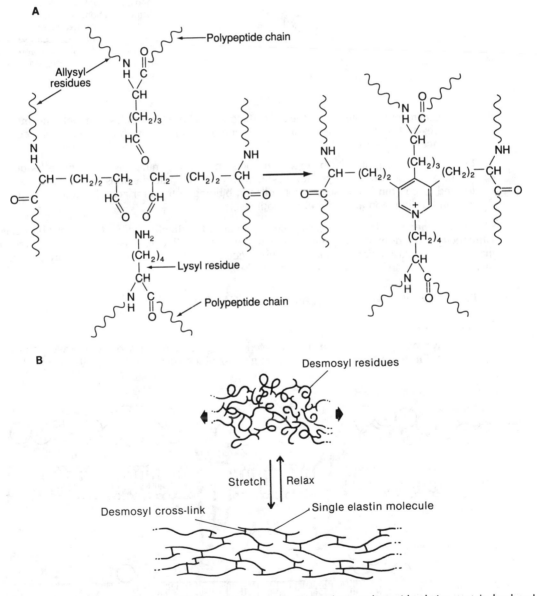

Figure 7-4. (*A*) Elastin synthesis and the assembly of elastic fibers. Elastin polypeptide chains contain lysyl and allysyl residues that become covalently linked to form intermolecular bridging desmosyl residues. (*B*) During stretching and relaxing, these desmosyl residues hold the elastin chains together.

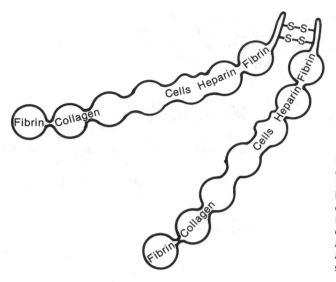

Figure 7-5. The fibronectin molecule consists of two polypeptides linked together by covalent disulfide bonds. Each polypeptide has several distinct functional domains for binding to cells and extracellular molecules. Thus, fibronectin can serve as an adhesive bridge between cells and extracellular matrix molecules such as collagen.

3. Fibronectin receptors (integrin) on the cell surface are anchored in the plasma membrane. Integrin recognizes and binds to extracellular fibronectin-containing fibrils, thus anchoring the cells to the extracellular matrix.

4. Research into the molecular biology of fibronectin has determined its polypeptide sequence. Careful study of fibronectin peptide fragments reveals separate domains in the molecule: one devoted to cell binding, a second for collagen binding, a third for actin binding, a fourth for heparin binding, and a fifth for fibrin binding.

E. **Plasma fibronectin.** Plasma contains a large amount of a slightly different form of fibronectin. Fibronectin has a domain for specific interaction with fibrin. Platelets also have a type of integrin that recognizes plasma fibronectin. Plasma fibronectin binds to fibrin that is formed during clotting, thereby serving as an intermediate cell adhesion protein between platelets and fibrin fibers in a clot.

V. PROTEOGLYCANS AND GLYCOSAMINOGLYCANS

A. Characteristics

1. Proteoglycans are synthesized and secreted by the resident cells of connective tissues. A proteoglycan has a protein backbone with glycosaminoglycans covalently attached to it.

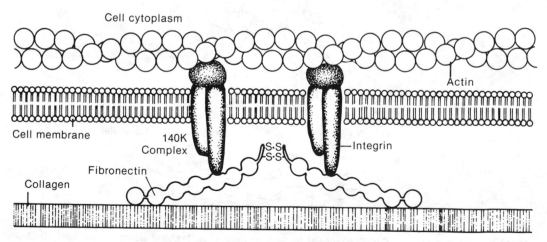

Figure 7-6 The integrin molecule is an integral membrane protein capable of binding to actin inside cells and fibronectin outside cells. Fibronectin binds to collagen outside cells and integrin in cell membranes.

2. Five major varieties of glycosaminoglycans exist; all consist of polysaccharides composed of repeating disaccharide units of sugar acids (uronic acids) and neutral sugars, uronic acids and acetylated amino sugars, or neutral sugars and acetylated amino sugars.

3. The glycosaminoglycans are sulfated to a variable extent. Due to their polyanionic nature, glycosaminoglycan molecules bind large amounts of water.

4. Glycosaminoglycans are large space-filling macromolecules that create an aqueous diffusion pathway through connective tissues. In solution, they form gelatinous, semi-rigid domains.

5. Many glycosaminoglycans are linked to proteins by a linker trisaccharide. This consists of one xylose and two galactose residues attached to the serine residue of the protein at one end, and the repeating disaccharide that is characteristic of that glycosaminoglycan at the other end.

B. Hyaluronic acid contains multiple repeating disaccharides consisting of **N-acetylglucosamine** and **D-glucuronic acid**. The macromolecule has a molecular weight on the order of 1000 kD. It is not sulfated and is not covalently bound to proteins, like other glycosaminoglycans. Hyaluronic acid is abundant in most loose connective tissues, the vitreous body of the eye, cartilage and skin, and **Wharton's jelly** in the umbilical cord.

C. Chondroitin sulfate consists of the repeating disaccharide **N-acetylgalactosamine** and **D-glucuronic acid**. It is highly sulfated, usually in carbons 4 and 6 of the sugars. Approximately 60 repeating units are covalently bonded into a molecule with a molecular weight of 30 kD. Chondroitin sulfate is a glycosaminoglycan that is abundant in cartilage, bone, skin, and the cornea.

D. Dermatan sulfate consists of repeating units of the disaccharide **N-acetylgalactosamine-4-sulfate** and **L-iduronic acid**. It has a molecular weight of 30 kD and is abundant in skin and present in blood vessels, tendons, and lung connective tissues.

E. Keratan sulfate contains the disaccharide **N-acetylglucosamine-6-sulfate** and **galactose** and may contain galactosamine. The degree of sulfation is highly variable. Unlike other glycosaminoglycans, keratan sulfate lacks uronic acids.

F. Heparan sulfate and the related compound **heparin** have a repeating unit of **N-acetylglucosamine** and either **D-glucuronic acid** (common) or **iduronic acid** (rare). Heparan sulfate is present on the surface of many cell types and is abundant in the basement membrane. Heparin is a potent anticoagulant.

G. Cartilage proteoglycans. The proteoglycan of **hyaline cartilage** (found in embryonic bone precursors and in several places in adults, including at the ends of long bones) has been particularly well characterized by cell biologists.

1. Large cartilage **proteoglycan aggregates** contain a single long hyaluronic acid molecule at their core and are shaped like a bottle brush. The hyaluronic acid backbone represents the central wire core of the brush and **proteoglycan subunits** represent the bristles. Link proteins attach the proteoglycan subunits to the hyaluronic acid.

2. Each proteoglycan subunit contains alternating keratan sulfate and chondroitin sulfate glycosaminoglycans attached to core proteins (Figure 7-7).

H. Amorphous ground substance glycoproteins. In addition to collagen and fibronectin, the extracellular matrix contains many glycoproteins that have not been studied in great detail. Their structure and function are poorly understood; however, they probably fill gaps between fibrous proteins and proteoglycans and may help bind different components.

VI. BASEMENT MEMBRANES

A. Structure

1. The basal surface of most epithelia has a thin layer of material lying between it and the subjacent connective tissue domains.

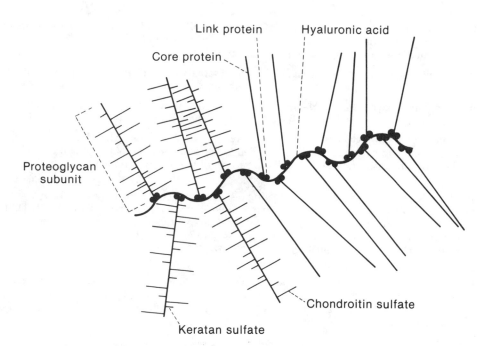

Figure 7-7. Cartilage proteoglycan aggregates consist of a long hyaluronic acid molecule and numerous proteoglycan subunits. Each proteoglycan subunit consists of a core protein with chondroitin sulfate and keratan sulfate attached. The proteoglycan subunits are attached to hyaluronic acid by link proteins.

2. This layer is visible in the light microscope. In the electron microscope a substructure may be apparent.

3. In some locations, the basement membrane is an exceedingly thin, gossamer network around an epithelium (e.g., around capillaries). In locations where epithelial basal surfaces oppose one another (e.g., in the renal glomerulus), the basement membrane can be several micrometers thick.

4. Basement membranes also surround muscle and Schwann cells.

B. Four important functions of basement membranes

1. The most important and ubiquitous function of the basement membrane is to provide a surface that epithelial cells (above) and connective tissue cells (below) can attach to, thus holding these two distinct tissues together.
   a. For example, skin consists of many layers of epithelial cells **(epidermis)** and an underlying connective tissue **(dermis)**. The outer epidermal layers are dead, dehydrated cells containing hydrophobic proteins that prevent desiccation of the body.
   b. The lower epidermal layers consist of highly proliferative cells that continuously replace cells in the upper epidermal layer as they are lost to abrasion.
   c. This layered or stratified epithelium provides an important protective barrier for the body's outer surface.

2. The basement membrane is a barrier that prevents microorganisms from entering the organism's inner domain.

3. The basement membrane prevents the loss of cells and fluids from the body.

4. The basement membrane performs selective filtration. For example, the glomerular basement membrane filters wastes from blood passing through the kidneys.

C. Four important constituents of the basement membrane. Peripheral parts of the basement membrane appear to be rich in laminin and proteoglycans, while the central portion is relatively rich in Type IV collagen.

1. **Laminin**
   a. Laminin is a glycoprotein with a molecular weight of 1000 kD. It is shaped like a cross and is composed of two different subunits that have molecular weights of 220 kD (A chain) and 440 kD (B chain) and are joined together by disulfide bonds.
   b. Laminin molecules contain distinct domains including a 50-kD heparin-binding domain at the end of the B chain and cell-binding and collagen-binding domains in the A chain.
   c. Like fibronectin, laminin is an extracellular matrix glycoprotein that binds to cells and components of the extracellular matrix.
   d. Laminin molecules have no obvious fibrous substructure in the electron microscope.

2. **Type IV collagen.** The pro-α chains of Type IV collagen have unusually long extension peptides that are not cleaved from the molecule. Therefore, Type IV collagen does not assemble into microscopic microfibrils in vivo. Rather, it exists in an amorphous form.

3. **Heparan sulfate** proteoglycans are abundant in basement membranes.

4. **Fibronectin** is also present in some basement membranes.

STUDY QUESTIONS

Directions: Each question below contains five suggested answers. Choose the **one best** response to each question.

1. Each of the following statements concerning collagen synthesis is true EXCEPT

(A) cross-linking occurs outside the cell
(B) glycosylation occurs outside the cell
(C) polypeptide chains are assembled inside the cell
(D) lysyl oxidase mediates cross-linking
(E) collagen is assembled into microfibrils inside the cell

2. Each of the following statements concerning elastin is true EXCEPT

(A) it contains glycine and desmosine
(B) it is amorphous
(C) lysyl oxidase is involved in its cross-linking
(D) it is abundant in dermal elastic fibers
(E) it is abundant in basement membranes

3. All of the following statements concerning cartilage proteoglycan are true EXCEPT

(A) it contains hyaluronic acid
(B) its molecular weight is less than 100 kD
(C) it is a space-filling macromolecule
(D) it contains chondroitin sulfate
(E) it binds water

4. Each of the following statements concerning basement membranes is true EXCEPT

(A) they do not contain elastin
(B) they contain abundant Type IV collagen
(C) they underlie epithelial cells
(D) they are not present in nerve and muscle cells
(E) they contain little Type I collagen

Directions: The groups of questions below consist of lettered choices followed by several numbered items. For each numbered item select the **one** lettered choice with which it is **most** closely associated. Each lettered choice may be used once, more than once, or not at all.

Questions 5–8

Match each description below with the most appropriate glycosaminoglycan or glycosaminoglycans.

(A) Hyaluronic acid
(B) Chondroitin sulfate
(C) Both
(D) Neither

5. A component of cartilage proteoglycan

6. Present in extracellular matrix amorphous ground substance

7. A polymer of the disaccharide repeating unit *N*-acetylglucosamine and galactose

8. A polymer of the disaccharide repeating unit *N*-acetylgalactosamine and glucuronic acid

Questions 9–13

Match each description below with the most appropriate type of collagen.

(A) Type I collagen
(B) Type II collagen
(C) Type III collagen
(D) Type IV collagen
(E) Type V collagen

9. Abundant around smooth muscle cells

10. Abundant in skin and bone

11. Abundant in basement membranes

12. Abundant in cartilage and the vitreous body

13. Contains three identical α1(III) chains wound into a superhelix

Questions 14–18

Match each description below with the most appropriate macromolecule.

(A) Collagen
(B) Fibronectin
(C) Integrin
(D) Elastin
(E) Chondroitin sulfate

14. An extracellular glycoprotein involved in cell adhesion; contains domains for binding glycosaminoglycans and collagen

15. An amorphous protein containing the unusual amino acid desmosine

16. A fibrous extracellular matrix protein that is extensively cross-linked by lysine residues

17. A glycosaminoglycan that is abundant in cartilage

18. An integral membrane protein that binds to extracellular fibronectin and intracellular microfilaments

ANSWERS AND EXPLANATIONS

1. The answer is B. [*III D*] Collagen is synthesized in cells such as fibroblasts. Collagen α chains are synthesized inside cells on the rough endoplasmic reticulum, and, like all proteins destined to be secreted from the cell, they are glycosylated in the Golgi apparatus. Lysyl oxidases are extracellular enzymes that mediate the intermolecular cross-linking that occurs after collagen molecules are secreted into the extracellular compartment.

2. The answer is E. [*IV A, B*] Elastin is an extracellular polymer with the ability to return to its original shape after being distorted. Elastin is a polypeptide that contains glycine, an amino acid present in many polypeptides. It also contains the unusual amino acids desmosine and isodesmosine, which are involved (via lysyl oxidase) in creating intramolecular and intermolecular cross-links. Elastic fibers contain an amorphous component rich in elastin. Elastin is abundant in dermal elastic fibers but is not present in basement membranes.

3. The answer is B. [*V B, G*] Cartilage proteoglycan is an abundant component of hyaline cartilage extracellular matrix. It has a molecular weight that exceeds 1 million (1000 kD) and consists of a core protein and a long thread of hyaluronic acid. Linker proteins bind hyaluronic acid molecules to the core protein and attach glycosaminoglycans (e.g., chondroitin sulfate) to the hyaluronic acid. Numerous chondroitin sulfate molecules are attached to the hyaluronic acid, forming an aggregate that has a molecular weight in the millions. This aggregate binds water and cations and accounts for a large volume of the extracellular matrix amorphous ground substance.

4. The answer is D. [*VI A 4*] Basement membranes lie underneath the basal surface of most epithelial cells, and they surround muscle and nerve cells. Often, cells are firmly attached to the basement membrane. Only connective tissue cells lack basement membranes. Basement membranes contain Type IV collagen, some proteoglycans, and, in some cases, fibronectin. They do not contain elastin.

5–8. The answers are: 5-C, 6-C, 7-D, 8-B. [*V B, C*] Hyaluronic acid and chondroitin sulfate both are glycosaminoglycans consisting of disaccharide repeating units. The repeating unit of hyaluronic acid is *N*-acetylglucosamine and glucuronic acid. The repeating unit of chondroitin sulfate is *N*-acetyl-galactosamine and glucuronic acid. Both glycosaminoglycans exist in the amorphous ground substance of the extracellular matrix. In cartilage proteoglycan, they are associated with keratan sulfate—a glycosaminoglycan with the repeating unit *N*-acetylglucosamine and galactose—and core protein.

9–13. The answers are: 9-E, 10-A, 11-D, 12-B, 13-C. [*III C 1–5*] Type I collagen is synthesized by connective tissue fibroblasts, osteoblasts, smooth muscle cells, and some epithelial cells. Type I collagen is most abundant in skin and bone.

Type II collagen is synthesized by chondroblasts and chondrocytes, neural retinal cells, and in the notochord. This type of collagen is most abundant in cartilage and the vitreous body.

All types of collagen consist of three α helices wound together in a tight spiral. Type III collagen contains three identical α chains, designated $[\alpha 1(III)]_3$, and is synthesized by fibroblasts and myoblasts.

Type IV collagen is synthesized by a variety of epithelial cells and is most abundant in basement membranes.

Type V collagen is synthesized by smooth muscle cells and chondrocytes and is abundant in basement membranes that surround smooth muscle cells and blood vessels.

14–18. The answers are: 14-B, 15-D, 16-A, 17-E, 18-C. [*III A, F; IV A, C, D; V C*] Collagen is a fibrous protein and a prominent constituent of the fibrous component of the connective tissue extracellular matrix. It has high tensile strength due, in part, to the extensive cross-linking of lysine residues.

Fibronectin is a fibrous glycoprotein and a component of the extracellular matrix. It is bound to the cell surface and extracellular matrix by domains that specifically bind integrin, collagen, fibrin, or glycosaminoglycans of other extracellular matrix macromolecules.

Integrin is an integral membrane protein that binds cells to the extracellular matrix components, such as fibronectin.

Elastin is a protein in elastic fibers. Like collagen, elastin molecules are extensively cross-linked. Elastin contains unusual amino acids such as desmosine and isodesmosine, which help form intermolecular crosslinks.

Chondroitin sulfate is a glycosaminoglycan. Chondroitin sulfate, hyaluronic acid, core protein, and linker proteins are all part of the large multimolecular cartilage proteoglycan aggregate.

8
Epithelial Tissue

I. INTRODUCTION

A. Cells, tissues, and organs. Humans are composed of billions of cells. A tissue is a collection of cells with similar structural characteristics. Organs are groups of tissues. Most organs are complex groupings of different tissue types. An organism is composed of organs that are grouped together and functionally integrated.

B. The four types of human tissue

1. **Epithelium.** Epithelial cells form sheets of cells that are tightly joined to one another. These cells cover surfaces, line cavities, and form the secretory portions of many glandular structures, which are complex invaginations of the body surface.

2. **Connective tissues** join epithelial structures to other parts of the body. They exist under all epithelial layers and have a cellular component and an extracellular matrix. Bone, cartilage, and blood are specialized connective tissues.

3. **Muscle tissue**
   a. **Skeletal muscle** tissue is found in the gross muscles that cause skeletal movement.
   b. **Cardiac muscle** tissue is found in the heart wall and proximal portions of the aorta.
   c. **Smooth muscle** tissue is a prominent component of blood vessel walls and visceral organs such as those of the gastrointestinal, urinary, and reproductive systems.

4. **Nerve tissue.** The brain, spinal cord, autonomic ganglia, peripheral nerves, and portions of sensory organs are composed of nerve tissue.

C. Organ composition. The liver exemplifies that most organs have at least some of each basic type of tissue.

1. The functional **(parenchymal)** cells of the liver—the cells that secrete products into the gastrointestinal tract—are epithelial cells. Epithelium also lines ducts and blood vessels in the liver.

2. Connective tissue forms a complex network of cells and extracellular matrix that supports liver parenchymal cells and joins them to blood vessels. Also, the liver is surrounded by a connective tissue capsule, which is coated with an epithelial layer.

3. The liver contains little muscle tissue; however, it is present in the walls of blood vessels.

4. Similarly, the liver contains little nerve tissue; however, blood vessels in the liver contain small branches of nerve fibers.

D. Tissue derivation. The **three primary germ layers** in embryology are **ectoderm** (the outer germ layer), **mesoderm** (the middle germ layer), and **endoderm** (the inner germ layer).

1. Epithelium is derived from each of these layers.
   a. Ectoderm forms epidermis.
   b. Mesoderm forms mesothelium.
   c. Endoderm forms the lining of the gastrointestinal tract.

2. Most connective tissue and muscle tissue is derived from mesoderm.

3. Most nerve tissue is derived from ectoderm.

E. Cardinal features of epithelium

1. Epithelial cells line surfaces. Epithelia line and protect virtually all free surfaces in the human body except joint cavities and the anterior surface of the iris, which is a naked connective tissue domain.

   a. The outer surface of the body is covered by the epithelial epidermis of skin. The hair follicles and glands in skin also have epithelial components.

   b. Epithelium lines the digestive system and its diverticula, such as the respiratory system, liver, pancreas, and gallbladder.

   c. The cardiovascular system is lined by an epithelium called **endothelium.**

   d. Body cavities derived from the intraembryonic coelom (the pericardial cavity, thoracic cavity, and peritoneal cavity) are lined by an epithelium called **mesothelium.**

   e. The urogenital system is lined by a layer of epithelial cells as well.

2. Epithelial cells have tight lateral adhesions. An epithelium is one or more layers of cells that are tightly joined together. The adhesions hold the epithelial cells together into a coherent barrier tissue. The apical junctions between cells in many epithelia have a sealing and adhesive structure called the **junctional complex,** which isolates the internal milieu of the organism and tightly joins epithelial cells together.

3. Epithelial cells are polarized. The epithelial apical surface typically faces a free surface of the body or the lumen of an organ or blood vessel and may be covered by microvilli or cilia. The basal surface rests on an extracellular layer of fibrils and glycoproteins called the **basement membrane,** or **basal lamina,** which is the boundary between the epithelium and the underlying connective tissue.

4. Epithelia are avascular. In most organs, the connective tissue beneath or around the epithelium contains blood vessels and lymphatics, which nourish the epithelium by diffusion.

II. EPITHELIAL CLASSIFICATIONS

A. General classifications (Table 8-1)

1. Epithelia are classified by the **number of cell layers** they contain.

   a. Simple epithelia have one cell layer. All cells rest on the basement membrane and reach the apical surface.

   b. Stratified epithelia have more than one cell layer, consequently not all cells rest on the basement membrane or reach the apical surface.

Table 8-1. Types of Epithelia and Their Locations

Type of Epithelium	Sample Locations
Simple squamous	Lining of peritoneal and pleural cavities; endothelial lining of all normal blood vessels
Simple cuboidal	Proximal convoluted tubules; sweat glands; epithelium of small and large intestines
Pseudostratified ciliated columnar epithelium with goblet cells	Nasopharynx; trachea; bronchi
Pseudostratified columnar with stereocilia	Epididymis; ductus deferens
Stratified squamous	Vagina; parts of the oral cavity; pharynx; esophagus; anal canal
Stratified squamous keratinized	Epidermis; hard palate; gingiva
Stratified cuboidal	Sweat gland ducts
Stratified columnar	Male urethra
Transitional	Renal pelvis; ureters; urinary bladder

    c. Pseudostratified epithelia are simple epithelia that appear to be stratified. In these epithelia, all cells rest on the basement membrane; however, not all cells reach the apical surface. The stratified appearance occurs because nuclei lie at different levels in the epithelia.

  2. Epithelia are further classified by **cell thickness.**
    a. Squamous cells are flat.
    b. Cuboidal cells have approximately equal height and width.
    c. Columnar cells are taller than they are wide.

  3. In stratified epithelia, the apical layer is diagnostic.
    a. The epithelium of the skin (epidermis) is stratified, squamous (outer layer cells are flat), keratinized (apical cells are converted to keratin and nuclei are not visible in the outermost cell layer) epithelium.
    b. Tracheal epithelium is pseudostratified, ciliated, columnar epithelium.
    c. The epithelium lining the proximal convoluted tubules of the kidneys is simple cuboidal epithelium.

B. Simple squamous epithelium is a single layer of flat cells.

  1. Simple squamous epithelium coats, or partially coats, the stomach, liver, gallbladder, and other visceral organs. It also is found in respiratory system alveoli and in the thin limbs of the loop of Henle in the kidney.

  2. Mesothelium is the simple squamous epithelium that lines serous cavities (peritoneal, pleural, and pericardial cavities) and coats many of the organs in these cavities.

  3. Endothelium is the simple squamous epithelium that lines the lumen of the cardiovascular system.

C. Simple cuboidal epithelium is a single layer of cells that are equal in height and width. It typically exists in areas where ion transport occurs (e.g., in kidney tubules, sweat glands, and some glandular ducts). It also covers the choroid plexus—the four clusters of capillaries in the walls of the ventricles of the brain that help produce cerebrospinal fluid (CSF).

  1. Simple cuboidal epithelium is abundant in the kidney, in the proximal and distal tubules and parts of the collecting-duct system.

  2. Many glands are composed of simple cuboidal epithelial cells assembled into round **acini.** In the configuration, the cuboidal cells are distorted into a rough pyramidal shape.

  3. Follicular epithelial cells in the thyroid gland can assume a cuboidal shape, depending on their function.

  4. Often, cuboidal epithelial cells have many apical microvilli and mitochondria, which facilitate ion pumping and fluid transport.

D. Simple columnar epithelium is a single layer of tall cells and is present in areas where absorption occurs.

  1. Simple columnar epithelium composed of absorptive cells and goblet cells covers the lining of the small and large intestines (Figure 8-1).

  2. It is found in some glandular ducts and the gallbladder, and it lines papillary collecting ducts in the urinary system.

  3. Uterine epithelium is a simple columnar epithelium.

E. Pseudostratified columnar epithelium is a layer of cells, in which all of the cells rest on the basement membrane but only some extend to the apical surface of the epithelium.

  1. Cell height varies in pseudostratified epithelium, as does the position of the nuclei with respect to the apical and basal limits of the epithelium. The apparent layering of nuclei makes the epithelium appear stratified.

  2. Pseudostratified columnar epithelium exists in many sections of the respiratory system, such as the nasal cavities, nasopharynx, trachea, and bronchi (Figure 8-2).

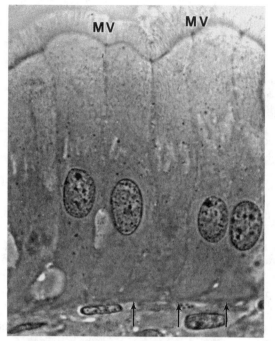

Figure 8-1. Simple columnar epithelium from the duodenal portion of the small intestine. These tall, absorptive cells rest on a basement membrane (*arrows*) and have an apical brush border of microvilli (*MV*).

3. Pseudostratified columnar epithelium is abundant in the male reproductive system. It is the luminal epithelium of the epididymis, vas deferens, prostate gland, seminal vesicles, and the prostatic urethra.

F. Stratified squamous epithelium contains many layers of cells, including an apical layer of flat cells. These epithelia are classified according to the characteristics of the apical layer.

1. Stratified squamous epithelium exists in locations subjected to chronic abrasion. The apical layers are continuously sloughed and then replaced by cell division in the basal layer.

2. The body is covered by a stratified squamous epithelium called epidermis. This epithelium is keratinized because cells in the outer layer contain the protein keratin (Figure 8-3).

3. The digestive system contains some stratified squamous epithelium. Stratified squamous epithelium on the tongue and lining the esophagus resists the abrasion of mastication, swallowing, and the passage of food from the oral cavity to the stomach. Stratified squamous epithelium in the anal canal resists the abrasion of passing semi-solid feces.

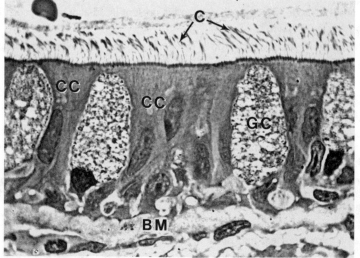

Figure 8-2. Pseudostratified, ciliated, columnar epithelium with goblet cells from the trachea. Notice the thick basement membrane (*BM*). Goblet cells (*GC*) and cilia (*C*) on columnar ciliated cells (*CC*) also are evident.

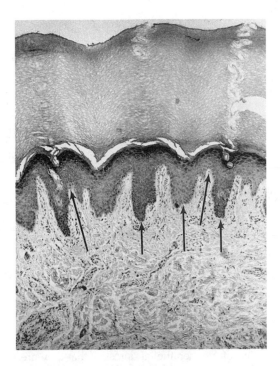

Figure 8-3. Stratified, squamous, keratinized epithelium from the epidermis of skin. The basement membrane (*arrows*) is the basal limit of this thick multilayered epithelium. The cells of the apical layers are dead and lack nuclei, and are completely keratinized.

4. In the female reproductive system, stratified squamous epithelium is abundant in the vagina and covering the cervix, where it resists abrasion during sexual intercourse.

5. Stratified squamous epithelium in the female reproductive system and digestive system is not extensively keratinized, except for patches on the hard palate and gingiva (gums).

G. **Transitional epithelium** is a stratified epithelium found exclusively in the urinary passages of the urinary system. It contains many layers of polyhedral cells and an outer apical layer of round pillow-shaped cells. It is found in the minor calyces, major calyces, renal pelvis, ureters, urinary bladder, and proximal urethra. Transitional epithelium in the bladder undergoes a reversible morphologic change during bladder distension and evacuation (Figure 8-4).

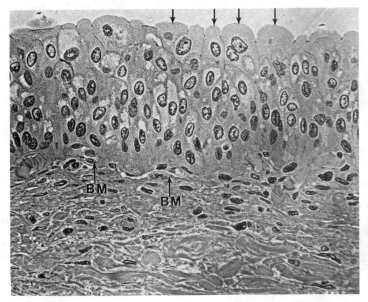

Figure 8-4. Transitional epithelium from the urinary bladder. This is the condition of the epithelium in an empty bladder. In a distended bladder, the epithelium would be much thinner. Note the pillowy apical cells (*arrows*) and the basement membrane (*BM*).

H. Minor types of epithelium

1. Stratified cuboidal epithelium lines the ducts of epidermal sweat glands.

2. Stratified columnar epithelium is found in the male prostatic urethra.

3. Some epithelia are highly specialized and, therefore, defy simple classification or description.
    a. For example, chorionic villi in the placenta consist of a syncytial epithelial layer (syncytiotrophoblast) lying on a proliferative cuboidal epithelial layer (cytotrophoblast), which, in turn rests on a basement membrane.
    b. Seminiferous epithelium in the male is stratified, highly proliferative, and, after puberty, sheds live haploid gametes from its apical surface.
    c. Follicular epithelial cells in the ovary are surrounded by a basement membrane and make intimate contact with the developing oocyte.

4. Some epithelia, such as those in the thyroid follicular epithelium, are in a constant state of flux. They may be squamous, cuboidal, or columnar, depending on the amount of thyroglobulin contained in the follicle lumen.

5. Many sensory epithelia (e.g., organ of Corti in the cochlear duct; retina) are highly complex epithelia.

III. EPITHELIAL POLARITY AND CELL REPLACEMENT

A. Apical modifications

1. **Apical surfaces are not adhesive.**
    a. An unusual, but functionally crucial, feature of most epithelial apices is that they are not adhesive. For example, when you clap your hands they do not adhere, no matter how hard you clap or how long you hold your hands together.
    b. Similarly, the apices of the mesothelium lining the thoracic cavity, pericardial cavity, and peritoneal cavity do not adhere to one another, and organs in these cavities move past one another without impediment or pain. However, if infection or surgery destroys the mesothelium, the organs will stick together, resulting in painful adhesions.
    c. The apex of luminal epithelial cells also exhibits this nonadhesive property, even in the smallest tubules in the body. Consequently, a lumen is maintained in all tubules, including the ductus deferens and capillaries, because walls cannot collapse and adhere to one another.
    d. In many instances, especially in the small intestines, the apical portion of the cell is coated with a thick, glycoconjugate-rich layer external to the outer leaflet of the plasma membrane called the **glycocalyx,** which makes apices nonadhesive (Figure 8-5).

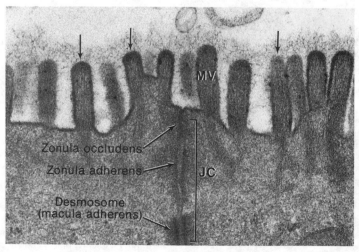

Figure 8-5. The apical surface of some epithelial cells have microvilli (*MV*) and a thick layer of glycocalyx (*arrows*), which renders the epithelial surface nonadhesive. This property maintains the lumen of an organ. The junctional complex (*JC*) that joins two adjacent cells consists of an apical zonula occludens, a zonula adherens, and a desmosome (macula adherens).

2. Apical protrusions. The epithelial apex often contains protrusions from the cell surface. The protrusions may be scattered, as in mesothelium and endothelium, and may take the form of microvilli.

    a. Microvilli are cylindrical, cell-surface projections, 80 nm wide and 1–2 μm long, which increase the cell surface area for absorbing materials from the lumen.

        (1) Microvilli can be sparse (as in mesothelium), numerous (as in the syncytiotrophoblast), or very dense (as in the **brush border** of kidney tubules and intestinal epithelium).

        (2) Microvilli are motile. They have many actin-containing microfilaments and large amounts of myosin.

    b. Stereocilia are long microvilli present in the male reproductive tract and in the membranous labyrinth of the inner ear. They are similar to microvilli except that they are longer and constricted at the point where they join the cell apex. Little is known about their movements in vivo. Sensory stereocilia contain many actin-containing microfilaments and large amounts of myosin.

    c. Cilia. Many epithelia apices are ciliated in some locations (see Chapter 5 III A for information about ciliary ultrastructure).

        (1) In the trachea, certain columnar epithelial cells have many apical cilia that move mucus along the apical surface of the tracheal lumen (airway).

        (2) Epithelial cells in the uterine tubes have apical cilia that help move the fertilized ova into the reproductive tract.

        (3) The apical portions of many sensory epithelial cells have modified cilia or cilia-like structures involved in the energy transduction function of the cell. However, these cilia are not motile like the cilia in the respiratory and female reproductive tracts.

3. Apical-basal polarity. Many epithelial cells have a polarized ultrastructure that reflects their apical specializations.

    a. Cilia and microvilli move rapidly in their respective tasks of movement and absorption. These activities require large amounts of ATP.

    b. Many ciliated cells have numerous mitochondria in the cell apex, close to the basal bodies that anchor the cilia. These mitochondria produce ATP from oxidative phosphorylation.

B. Basal modifications

1. Basement membranes

    a. Most epithelia have a basement membrane (lymphatic capillaries are a notable exception), which ranges from scant to thick, depending on the epithelium.

        (1) Small capillaries have a thin or fenestrated basement membrane.

        (2) The basement membrane in the trachea is quite thick and serves as a defense against bacterial invasion.

        (3) The epidermal basement membrane is thick and rugged to endure the constant stress sustained by the epidermis.

    b. Some basement membranes have an amorphous **basement lamina** adjacent to the epithelium and a fibrous **reticular lamina** below the basement lamina.

    c. The fine structure of the basement membrane varies considerably with the location and function of the epithelia.

        (1) For example, capillary basement membranes are very thin and have no obvious substructure.

        (2) In contrast, the glomerular basement membrane for the double epithelium in the Bowman's capsule is thick but lacks a reticular lamina.

2. Basement membrane composition

    a. Basement membranes contain highly glycosylated Type IV collagen, which is rich in the amino acid hydroxylysine and is composed of three α1(IV) subunits. Type IV collagen appears to be highly cross-linked and rich in carbohydrate units.

    b. Many basement membranes contain other glycoproteins such as **laminin** and proteoglycans such as **heparan sulfate proteoglycan.**

    c. These biochemical characteristics have been established from the thicker, unusual basement membranes and may not apply to all basement membranes present in humans.

3. Adhesion of basement membranes and epithelial cells
   a. Basal cells in the **stratum germinativum** of the epidermis are firmly anchored to the basement membrane by rivet-like intracellular specializations called **hemidesmosomes**.
   b. Hemidesmosomes are similar to the desmosomes present in many junctional complexes.
   c. Desmosomes and hemidesmosomes seem to be sites where a cell strongly adheres to another cell or to an extracellular basement membrane.

4. Basement membrane functions
   a. The basement membrane is the basal limit of an epithelium. In many luminal epithelia, the basement membrane is the boundary between the epithelium and subjacent connective tissue.
   b. The basement membrane anchors the epithelium and may be a substratum for epithelial cell and connective tissue cell attachment.
   c. The basement membrane maintains the shape of acini and branched ducts and tubules.
   d. In some instances, the basement membrane is a selective barrier (see Chapter 26 II C 2 a (2) concerning the function of the glomerular basement membrane).

5. Other basal modifications. The basal membrane of cells where ion transport is occurring often is elaborately folded into numerous intercalated, finger-like projections, which increase the surface area of the basal plasma membrane. The many mitochondria that produce the ATP necessary for ion transport are located near these projections.

C. The junctional complex

1. General characteristics. Epithelial tissue defines boundaries and separates functional compartments in the body. For example, the lumen of the small intestines contains a complex mixture of digestive enzymes capable of turning the gastrointestinal tract wall into a soup of amino acids. However, the junctional complex, an apical and lateral membrane specialization, isolates the gastrointestinal tract lumen from the sensitive wall of the gut.

2. Ultrastructure (see Figure 8-5)
   a. Short sections of the membrane outer leaflets appear to fuse, forming an occluding junction called the **zonula occludens** (tight junction), which extends around the apex of the columnar epithelial cells in a belt, creating a seal between the lumen and the lateral extracellular fluid environment.
   b. In freeze-fracture-etch, the zonula occludens in some epithelial tissues appears as an anastomosing network of ridges (points of membrane fusion) that impede the movement of molecules from the lumen to the lateral extracellular compartment.
   c. Below the zonula occludens, plasma membranes diverge to form a separation of 10–15 nm called the **zonula adherens** (intermediate junction), a structure thought to be an adhesive junction.
      (1) The zonula adherens is a simple membrane apposition with varying amounts of electron-dense material in the intervening gap.
      (2) Intermediate (10-nm) filaments radiate from the zonula adherens into the cytoplasmic matrix of apposed cells.
   d. The **macula adherens** (desmosome) is located below the zonula adherens. As its name implies, the macula adherens is thought to be a structure that holds cells together.
      (1) Each of the apposed cells contributes half of the desmosome. An epithelial cell abutting on a basement membrane sometimes forms a hemidesmosome.
      (2) At the macula adherens, the plasma membranes separate to 25–30 nm. An intermediate dense line runs between the cells in this separation.
      (3) The inner aspect of apposed plasma membranes is covered with a punctate electron-dense material. Long bundles of **tonofilaments** (a kind of 10-nm intermediate filament) radiate from the plaque of electron-dense material.
   e. Many epithelia contain a structure called the **gap junction** (nexus), which is a specialized region where the outer leaflets approach each other, leaving a small (2-nm) but definite gap.
      (1) Gap junctions appear to be composed of hexagonal arrays of barrel-shaped structures, composed of six subunits arranged around an electron lucid central core. The central core is an aqueous channel between closely apposed cells.
      (2) Ions and other small molecules pass freely between closely apposed cells, presumably through the aqueous channels in the gap junction.

D. Epithelial cell replacement. Some epithelial cells are replaced as they are lost by the division of other cells within the epithelium. This is especially true in stratified squamous epithelia such as the epidermis, where apical squamous cells are lost to friction. Without replenishment, the entire epithelium would soon wear away. Epidermal basal cells constantly divide to produce the daughter cells that replace sloughed apical cells.

1. **Intestinal epithelium.** A slightly different process occurs in the gastrointestinal tract. Here, cells are constantly subjected to the destructive influence of digestive enzymes, and intestinal villi are highly motile.
   a. As intestinal villi move up and down, cells on the villi tips are sloughed into the lumen.
   b. Cell division occurs at the base of intestinal villi in deep pits called the **crypts of Lieberkühn**.
   c. Daughter cells migrate up the crypt walls to the top of the villi where they are lost from the epithelium.

2. **Stem cells**
   a. In instances of epithelial turnover, the basal portion of the epithelium invariably contains a population of **stem cells**.
   b. Like all mitotic cells, stem cells divide into two cells. One daughter cell remains as a stem cell while the other differentiates and moves toward the apex.
   c. Proliferative stem cells are undifferentiated and usually rest on the basement membrane.
   d. The cell cycle varies tremendously among stem cell populations in different epithelia, ranging from days to years. For example, the epidermis turns over every 27 days, while seminiferous epithelium turns over every 64 days.

IV. GLANDULAR TISSUES.
Two types of glands exist: endocrine glands and exocrine glands. Both types of glands are composed of and properly classified as epithelial tissues.

A. Endocrine glands

1. **Endocrine glands secrete products into the bloodstream.** They are derived from surface epithelium during development; however, they soon lose their connection to the body surface.

2. **Endocrine tissue retains many epithelial characteristics.** For example, thyroid follicular epithelial cells form sheets, are surrounded by a basement membrane, secrete hormone precursors apically, exhibit apical-basal polarity, and contain junctional complexes.

B. Exocrine glands

1. **Exocrine glands secrete products onto the body surface through ducts.** The skin and digestive tract, which are continuous with each other, receive exocrine secretions from glands such as sweat glands (skin) and the liver and pancreas (digestive tract).

2. **Exocrine gland classification.** Exocrine glands are classified in several ways according to their mode of cellular secretion and arrangement of cells and ducts. They can be simple glands or compound glands.
   a. **Straight tubular glands** consist of a simple straight tubule and exist in the small intestine.
   b. **Coiled tubular glands** consist of a coiled tubule. Sweat glands are one example of coiled tubular glands.
   c. **Branched tubular glands** have branches deep in the gland. This type of gland exists in the stomach and endometrium.
   d. **Simple tubuloalveolar glands** have a single duct leading to a cluster of alveoli, or **acini**. Examples of this type of gland include the small salivary glands in the oral cavity and Brunner's glands in the duodenum.
   e. **Simple alveolar glands** have several acini attached to a single duct. The sebaceous glands of the skin are this type of gland.
   f. **Compound tubular glands** have numerous tubules connected to multiple ducts. These glands exist in the testes.
   g. **Compound tubuloalveolar glands** have numerous secretory acini that drain into numerous efferent ducts. These ducts typically merge into a smaller number of main ducts. The parotid salivary glands and the pancreas are examples of this type of gland.

   h. Compound alveolar glands are compound glands because they have numerous draining ducts. However, they terminate in acini that have flat squamous cells rather than acini with cuboidal or pyramidal cells. The lungs are compound alveolar glands.

C. Modes of secretion

   1. Merocrine secretion involves the release of membrane-bound packets of secretion product. The packets are formed as membranes derived from the endoplasmic reticulum and Golgi apparatus surround the secretion product. The thyroid gland and pancreas exhibit merocrine secretion.

   2. Apocrine secretion products include a portion of the apical cytoplasm from the secretory cell. Mammary glands exhibit apocrine secretion.

   3. Holocrine secretion is characterized by whole cells bursting open to become the secretory product. Epidermal sebaceous glands exhibit holocrine secretion.

STUDY QUESTIONS

Directions: Each question below contains five suggested answers. Choose the **one best** response to each question.

1. Each of the following statements concerning simple columnar epithelium is true EXCEPT

(A) it is found in intestinal absorptive epithelium
(B) it is found in epididymal epithelium
(C) all of its cells rest on the basement membrane
(D) all of its cells reach the apical level
(E) it contains well-developed junctional complexes

2. Each of the following statements concerning stratified epithelium is true EXCEPT

(A) all of its cells rest on the basement membrane
(B) some of its cells reach the apical level
(C) it is found in the esophagus
(D) it can be keratinized
(E) it is found in the epidermis

Questions 3–5

The following three questions refer to the epithelium shown in the micrograph below. *L* = lumen.

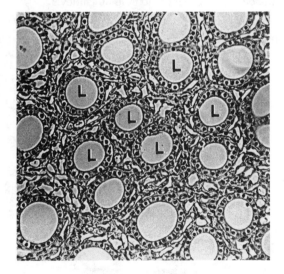

3. The epithelium surrounding the limina in the above micrograph is best described as

(A) simple squamous
(B) simple cuboidal
(C) simple columnar
(D) transitional
(E) pseudostratified columnar

4. Which one of the organs listed below would contain the type of epithelium pictured in this micrograph?

(A) Lungs
(B) Trachea
(C) Testis
(D) Kidney
(E) Epidermis

5. Each of the following statements concerning the epithelium shown in the micrograph is true EXCEPT

(A) all of its cells rest on the basement membrane
(B) all of its cells reach the lumen
(C) all of its cells have a nonadhesive apical surface
(D) all of its cells are taller than they are wide
(E) none of its cells are ciliated

6. Each of the following statements concerning stratified, squamous, keratinized epithelium is true EXCEPT

(A) some cells rest on the basement membrane
(B) apical cells are squamous and dead
(C) all cells reach the apical surface
(D) it is found in the epidermis of skin
(E) hemidesmosomes anchor basal cells to the basement membrane

7. Each of the following statements concerning mesothelium is true EXCEPT

(A) it is derived from mesoderm
(B) it is a simple squamous epithelium
(C) it lines the pleural cavity
(D) it secretes serous fluids
(E) it lines blood vessels derived from mesoderm

8. Each of the following statements concerning epithelial tissues is true EXCEPT

(A) they maintain a constant thickness
(B) their apical surface is nonadhesive
(C) they have strong apical-basal polarity
(D) they have strong lateral adhesions
(E) they do not contain blood vessels

9. Each of the following statements concerning the basement membrane is true EXCEPT

(A) it contains laminin
(B) it is thin in tracheal epithelium
(C) in some cases, it performs selective filtration
(D) in some cases, hemidesmosomes attach it to the basal surface of epithelial cells
(E) it surrounds capillaries

10. Each of the following statements concerning the apical surface of epithelial layers is true EXCEPT

(A) often, it is covered by a glycocalyx
(B) it is nonadhesive
(C) often, it has microvilli
(D) it has junctional complexes
(E) often, it has cilia

Directions: The groups of questions below consist of lettered choices followed by several numbered items. For each numbered item select the **one** lettered choice with which it is **most** closely associated. Each lettered choice may be used once, more than once, or not at all.

Questions 11–15

Match each statement below with the most appropriate junctional complex structure or structures.

(A) Zonula occludens
(B) Zonula adherens
(C) Both
(D) Neither

11. Present in cuboidal epithelium

12. Present in pseudostratified epithelium

13. Present in basement membranes

14. Present in desmosomes

15. Forms an anastomosing network of ridges in freeze-fracture-etch

Questions 16–20

Match each description below with the most appropriate epithelium or epithelia.

(A) Transitional epithelium
(B) Pseudostratified epithelium
(C) Both
(D) Neither

16. Abundant in the trachea and bronchi

17. Abundant in the renal calyces

18. All cells rest on the basement membrane

19. Some cells do not reach the apical level

20. Often contains cilia or microvilli

ANSWERS AND EXPLANATIONS

1. The answer is B. [*II A, D*] Simple columnar epithelium consists of a single layer of cells that are taller than they are wide. All of its cells rest on a basement membrane and reach the apical level. Simple columnar epithelia often are specialized for absorption and are found in the stomach, small intestines, and colon. The epididymis, like most excurrent ducts in the male reproductive system, contains pseudostratified columnar epithelium.

2. The answer is A. [*II A, F*] A stratified epithelium contains more than one layer of cells. The basal cells rest on the basement membrane but do not reach the apical level. Conversely, apical cells reach the apical level but do not rest on the basement membrane. Stratified squamous epithelium is found in the esophagus, oral cavity, anal canal, vagina, cervix, and in the epidermis of keratinized skin. In these locations, it is specialized to resist abrasion.

3. The answer is B. [*II C*] This is a micrograph of a simple cuboidal epithelium in the kidney. The micrograph is taken from the renal medulla and displays a field of collecting tubules. It can be identified as simple cuboidal epithelium by the shape of the cells and the number of cell layers. Simple cuboidal epithelium has a single layer of cells that are equal in height and width, and all cells rest on the basement membrane and reach the apical surface at the lumen. Simple squamous epithelium contains flat cells. Simple columnar epithelium contains cells that are taller than they are wide. Transitional epithelium contains layers of cells. In pseudostratified columnar epithelium, all cells rest on the basement membrane, but only some cells reach the apical surface.

4. The answer is D. [*II C*] This is a micrograph of a simple cuboidal epithelium. This type of epithelium is found in areas where ion transport occurs, such as the kidneys, sweat glands, and some glandular ducts. It also covers the choroid plexus in the brain. The alveoli of the lungs are lined by a mixture of squamous and cuboidal cells. The trachea is lined by a pseudostratified ciliated epithelium with goblet cells. The epididymis contains pseudostratified epithelium with stereocilia. The seminiferous tubules of the testis have a peculiar stratified epithelium of gametogenic cells.

5. The answer is D. [*II C*] This is a micrograph of simple cuboidal epithelium. Its cells are equal in height and width. Like all simple epithelia, all of its cells rest on the basement membrane and reach the nonadhesive apical surface at the lumen. These cells do not have cilia. The nonadhesive quality of the apical layer of these epithelia is a feature that results in the formation of a lumen in many tubular organs.

6. The answer is C. [*II F*] In stratified squamous epithelium, some but not all cells rest on the basement membrane where they are sometimes anchored by hemidesmosomes. In keratinized epithelium, the apical cells are flat, dead, and nonadhesive. Since it is stratified, not all cells reach the apical surface. Stratified, squamous, keratinized epithelium is found in the epidermis, gingiva, and hard palate and in the lower anal canal. It is specialized to resist abrasion.

7. The answer is E. [*I D, E 1*] Mesothelium is the simple squamous epithelium that lines the four serous cavities in the body. These cavities include two pleural cavities, the pericardial cavity, and the peritoneal cavity. Mesothelium cells are derived from mesoderm and have apical surfaces that are nonadhesive. They secrete serous lubricating fluids into the four serous cavities. Blood vessels are lined by a simple squamous epithelium called endothelium, which is derived from mesoderm.

8. The answer is A. [*I E 1–4; II G*] Epithelial tissues often change thickness in response to changing physiologic conditions. For example, the urinary bladder epithelium is thin when the bladder is distended but becomes thick when the bladder empties. Also, thyroid follicular epithelial cells are squamous when follicles are full, cuboidal when they are filling, and columnar when they are empty. Apical surface nonadhesiveness, apical-basal polarity, and strong lateral adhesions are important features of epithelial tissue. Also, epithelial tissues are avascular and are supplied by blood vessels and lymphatics in adjacent or surrounding connective tissues.

9. The answer is B. [*III B 1 a (2)*] The basement membrane is an extracellular matrix that underlies or surrounds almost all epithelial layers (one exception is the epithelium of lymphatic capillaries). The basement membrane contains laminin and heparan sulfate proteoglycan. It is especially thick in the trachea, where it filters bacteria from inspired air, and in the renal glomerulus, where it filters some blood proteins during urine production. Receptors for basement membrane components and hemidesmosomes anchor epithelial cells to the basement membrane.

10. The answer is D. [*III A 1, 2*] The apical surface of many epithelial layers is covered by a thick, extracellular glycoprotein coat called the glycocalyx. The glycocalyx renders the apical surface non-adhesive, thus assuring that the lumen of the organ remains patent during functional activity. Apical surfaces of many epithelial cells have microvilli (e.g., epithelial cells in the bronchi and uterine tubes). Epithelial apical surfaces do not have junctional complexes. These structures form on the lateral surfaces of apposed epithelial cells. They separate the luminal compartment of an organ from the surrounding subepithelial connective tissues and hold epithelial cells together.

11–15. The answers are: 11-C, 12-C, 13-D, 14-D, 15-A. [*III C 2*] The zonula occludens and the zonula adherens are parts of the junctional complex and are found in cuboidal and pseudostratified epithelia. The zonula occludens forms an anastomosing network of ridges in freeze-fracture-etch. The basement membrane underlies epithelia but is not part of the junctional complex. Desmosome is another name for the macula adherens, a third junctional complex structure.

16–20. The answers are: 16-B, 17-A, 18-B, 19-C, 20-B. [*II A 1 c, E, G*] All cells in pseudostratified epithelium rest on the basement membrane, but not all cells reach the apical level. Pseudostratified epithelium is abundant in the respiratory system and the male reproductive tract. Tracheal and bronchial pseudostratified epithelium is ciliated. Epididymal pseudostratified epithelium contains modified microvilli called stereocilia.

Transitional epithelium is found exclusively in the minor calyces, major calyces, renal pelvis, ureters, urinary bladder, and proximal urethra of the urinary system. Transitional epithelium is stratified, and thus some cells do not rest on the basement membrane and some cells do not reach the apical level.

9

Connective Tissue

I. INTRODUCTION

A. Connective tissue locations and functions. Connective tissues are ubiquitous in the body. Every organ is composed of, or ensheathed by, at least some connective tissue. Cartilage, bone, and peripheral blood are specialized connective tissues (see Chapters 10 and 11).

1. Visceral organs (e.g., kidneys, lungs) contain an abundance of connective tissue that holds the parenchymal epithelial cells together to form the organ.

2. The cardiovascular system is rich in connective tissues. Here, they tie muscle cells and endothelial cells together into a functionally integrated system.

3. Skeletal muscles are bound together by connective tissues and attached to bones by ligaments and tendons, which are types of specialized connective tissue.

4. The central nervous system (CNS) contains less connective tissue than other systems. Little connective tissue is associated with the nervous system parenchyma of neurons and glial cells; however, collagenous fibers and connective tissue cells form a sheath around the brain and spinal cord. Also, some connective tissue exists in and around peripheral nerves and ganglia.

B. Connective tissue components. All connective tissues are composed of **cells, extracellular fibers,** and an **amorphous ground substance**.

1. **Resident and immigrant cells**
   a. Connective tissue resident cells vary considerably and may include fibroblasts, chondroblasts and chondrocytes, osteoblasts and osteocytes, adipocytes, and macrophages.
   b. Connective tissue immigrant cells can include all of the formed cellular elements of the blood with the exception of erythrocytes. When injury or inflammation damages tissue, leukocytes (e.g., monocytes), lymphocytes, and phagocytic granulocytes (e.g., neutrophils, eosinophils, basophils) leave the circulation and join fibroblasts and other connective tissue resident cells to repair damage and combat microorganisms that cause inflammation.

2. **Extracellular fibers**
   a. Connective tissue extracellular fibers include collagen fibers, elastic fibers, and reticular fibers.
   b. Blood contains an extracellular fibrous component called fibrinogen, which is a potential group of fibers. When blood clots, the potential fibers of fibrinogen are converted into actual fibers of fibrin.

3. **Amorphous ground substance**
   a. Connective tissue amorphous ground substance is composed of proteoglycans, glycosaminoglycans, and glycoproteins (see Chapter 7 V for more information about amorphous ground substance).
   b. Often, proteoglycans and glycoproteins are bound by weak, noncovalent interactions in the extracellular matrix.
   c. Frequently, amorphous ground substance macromolecules are bound to fibrous components.
   d. Mineralization often occurs in connective tissue amorphous ground substance. The minerals contribute hardness to mineralized connective tissues such as bone, dentin, and enamel.

Table 9-1. Types of Connective Tissues and Their Locations

Types of Connective Tissues	Sample Locations
Connective tissues proper	
General connective tissues proper	
Loose (areolar) connective tissue	Lamina propria, mesenteries
Dense regular connective tissue	Tendons
Dense irregular connective tissue	Dermis
Special connective tissues proper	
Adipose tissue	Subcutaneous fat, anatomic fat around kidneys
Reticular tissue	Network around sinusoids in spleen, lymph nodes, and liver
Specialized connective tissues	
Bone	Femur
Cartilage	Tracheal cartilage
Blood	Peripheral blood

C. **Connective tissue classification.** Table 9-1 lists the general classifications of connective tissues and examples of their locations in the body.

1. **General connective tissues proper** are subdivided into loose (areolar) and dense connective tissues. Dense connective tissues are further classified as regular or irregular.

2. **Special connective tissues proper** are adipose tissue and reticular tissue.

3. **Specialized connective tissues** include bone, cartilage, and blood.

II. GENERAL CONNECTIVE TISSUES PROPER

A. **Classifications**

1. **Loose (areolar) connective tissues** are relatively rich in cells and often have a watery extracellular matrix containing relatively few extracellular fibers. Examples include the mesenteries, lamina propria (i.e., the connective tissue domain beneath moist epithelia; Figure 9-1), and the adventitial layer of blood vessels.

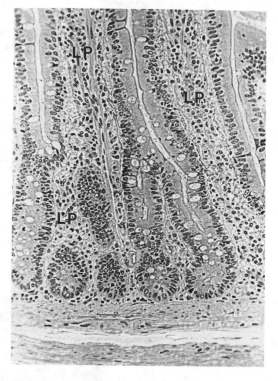

Figure 9-1. Loose areolar connective tissue in the lamina propria (*LP*) of the gastrointestinal tract mucosal surface.

    a. Loose connective tissue contains a variety of cells, which can include resident fibroblasts, immigrant macrophages, mast cells, and any of the formed elements of blood except red blood cells and platelets. It invariably contains some resident macrophages and mast cells.

    b. Loose connective tissue contains sparse collagen and elastic fibers randomly scattered in overlapping patterns.

    c. The typically sparse amorphous ground substance contains some proteoglycans and glycoproteins.

2. Dense connective tissues have more fibers per unit volume than loose connective tissues and are classified as regular or irregular according to the orientation of the extracellular fibers in the connective tissue. They have a relatively sparse cellular component and a dominant fibrous component.

    a. Dense regular connective tissues have numerous extracellular fibers arranged in regular arrays. Dense masses of fibers run in parallel or orthogonally arranged layers. The regular texture of this connective tissue is visible in the light microscope. Dense regular connective tissue exists in ligaments, tendons, and in the stroma of the cornea.

    b. Dense irregular connective tissues have numerous extracellular fibers in dense random arrays (Figure 9-2). This form of connective tissue exists in the dermis of the skin, capsules of organs such as the liver and spleen, and in the periosteum surrounding bones.

B. Cell biology of connective tissue cells

1. Fibroblasts are the most important and widely distributed connective tissue cells.

    a. Functions. Fibroblasts secrete extracellular matrix, bind extracellular matrix constituents to form tissue, and facilitate wound healing. For example, when the skin is cut, fibroblasts proliferate and migrate toward the wound to fill the gaps in the tissue. As proliferation continues, they secrete large amounts of extracellular matrix. Fibroblasts form the scar that closes the wound.

    b. Morphology and distribution. Fibroblast morphology varies from fusiform to stellate, depending on its packing arrangement relative to extracellular fibrils.

        (1) In dense regular connective tissues, fibroblasts have a fusiform or spindle shape because they are sandwiched between regular arrays of collagen fibers.

        (2) In areolar connective tissues, fibroblasts may adopt a stellate shape due to the irregular arrangement of collagen fibers.

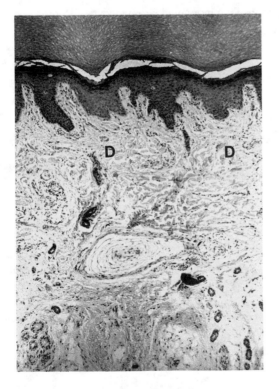

Figure 9-2. Dense irregular connective tissue from the dermis (*D*) of skin.

(3) Individual fibroblasts are pleomorphic, perhaps because they can move about in connective tissues.

c. **Functional cell biology.** Fibroblasts have a deceptively simple nuclear morphology. They have a variable cell shape and a single nucleus with several nucleoli. The morphology varies from spherical in stellate fibroblasts to cigar-shaped in fusiform fibroblasts.

(1) Fibroblasts are differentiated for extracellular matrix synthesis and secretion. The cytoplasm contains an abundance of rough endoplasmic reticulum, many mitochondria, and a well-developed Golgi apparatus (Figure 9-3). Also, lysosomes and vacuoles of secretion product are prominent features of fibroblasts.

(2) Fibroblast secretions include collagen, fibronectin, glycoproteins, and proteoglycans.

(3) Fibroblasts contain intracellular glycogen, because they help synthesize large amounts of extracellular glycoconjugates.

(4) The Golgi apparatus is unusually rich in glycosyl transferase enzymes, which add carbohydrate moieties to glycoconjugates.

(5) Fibroblasts have many actin-containing microfilaments because they are highly motile cells.

(6) Fibroblasts also contain numerous microtubules, which probably help maintain the fibroblastic cell morphology.

2. Macrophages

a. **Functions.** Macrophages are dedicated phagocytes and are distributed throughout the body. They are derived from bone marrow stem cells and can assume several different forms.

(1) Phagocytosis is part of the normal economy of many different cell types. For example, follicular epithelial cells in the thyroid gland use phagocytosis to engulf thyroglobulin when thyroid-stimulating hormone stimulates the release of thyroxine, the active hormone of the thyroid gland. Phagocytes such as neutrophils and eosinophils engulf microorganisms that enter the body.

(2) Phagocytosis is a principal function of macrophages. They are highly mobile populations of cells that move rapidly from one compartment to another in the body as the need for phagocytosis arises. For example, the alveolar macrophage population increases when the amount of dust and debris entering the lungs increases.

(3) Macrophages have an extended life span in the peripheral tissues of the body. Effete macrophages are continuously replaced by differentiation of bone marrow stem cells into monocytes, which can leave the closed circulation and enter connective tissues, lymphoid glands, and serous cavities.

(4) Macrophages and related cells comprise the mononuclear phagocyte system.

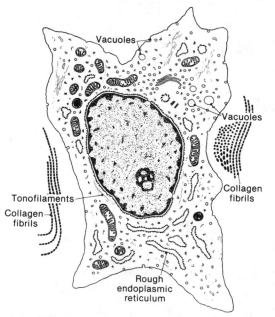

Figure 9-3. Illustration of a fibroblast as seen in the electron microscope. Collagen fibrils are synthesized in the rough endoplasmic reticulum of fibroblasts and transported through the cell in vacuoles. Fibroblasts also contain tonofilaments near the nucleus. (Reprinted from Lentz TL: *Cell Fine Structure*. Philadelphia, WB Saunders, 1971, p 65.)

b. **Mononuclear phagocyte system.** Three criteria distinguish cells of the mononuclear phagocyte system from other phagocytic cells: They have a characteristic morphologic appearance similar to macrophages; they are derived from bone marrow; and their phagocytic activity is modulated by immunoglobulins and complement system components. Mononuclear phagocyte system constituents include:

(1) **Bone marrow stem cells**

(2) **Monocyte precursors** in bone marrow

(3) **Monocytes** in peripheral blood

(4) **Fixed macrophages** in connective tissues (histiocytes)

(5) Phagocytic **Kupffer cells** lining the liver sinusoids

(6) **Alveolar macrophages** and **free macrophages** in serous cavities

(7) **Free and fixed macrophages** in the spleen, lymph nodes, and thymus

(8) **Microglia** in the CNS. (Other CNS glia such as astrocytes and oligodendroglial cells are derived from embryonic neuroectoderm; however, convincing evidence exists that microglial cells are derived from mesenchyme and are part of the mononuclear phagocyte system.)

(9) **Osteoclasts** in bone. (Osteoclasts are derived from monocytes and share many of the morphologic and functional characteristics of other mononuclear phagocyte system components.)

c. **Morphology**

(1) Macrophages have a highly variable shape because they move throughout connective tissues. They have a diameter of approximately 20 μm and a single irregularly shaped nucleus with one or two prominent indentations and a conspicuous mass of euchromatin.

(2) The cytoplasm contains many phagocytic vacuoles and may contain engulfed microorganisms and other cellular debris.

(3) Macrophages are intracellular organelles that synthesize hydrolytic acids and lysosomes. Other prominent features of macrophage cytoplasm include primary lysosomes, phagosomes, secondary lysosomes, multivesicular bodies, and residual bodies.

(4) Macrophages have abundant rough endoplasmic reticulum and a prominent Golgi apparatus near the nucleus.

(5) Macrophages also have a prominent cortical array of actin-containing microfilaments and many microtubules and intermediate filaments for locomotion and phagocytosis (Figure 9-4).

d. **Functional cell biology.** Macrophages have a complex functional capacity. They are actively involved in phagocytosis and pinocytosis, and they have an essential role in marshaling the immunologic response (see Chapter 12 VII). Their plasticity enables them to ingest other cells whole by attaching to a cell and then spreading over its surface.

(1) The **nonspecific phagocytic functions** of macrophages are especially evident in alveolar macrophages, which can engulf soot, asbestos, airborne industrial pollutants, cigarette smoke, and cotton fibers.

(2) Macrophages also have several **specific phagocytic functions,** such as the phagocytosis of bacteria. Macrophage cell-surface receptors recognize certain carbohydrate residues in bacteria cell walls, then the specific phagocytosis of the bacteria occurs by **opsonization.**

(a) During opsonization, microorganisms are coated with immunoglobulins or complement components. The cell-surface receptors recognize a part of the immunoglobulin molecule called the **Fc fragment.** (Macrophages recognize the immunoglobulin coating on the opsonized bacterium rather than a component of the bacterial cell wall per se.)

(b) This process engenders a highly selective phagocytic system with a memory. Bacteria entering the body elicit an immunologic response, because their cell surface properties are recognized as foreign by the defending organism.

(c) During an immunologic response, immunoglobulins are synthesized as lymphocytes. Then, the bacterium is opsonized and destroyed. Memory lymphocytes remain to defend the organism against subsequent bacterial assaults.

(3) **Consequences of the opsonization-lysosomal function**

(a) Macrophages adhere to opsonized bacteria and engulf them in membrane-delimited

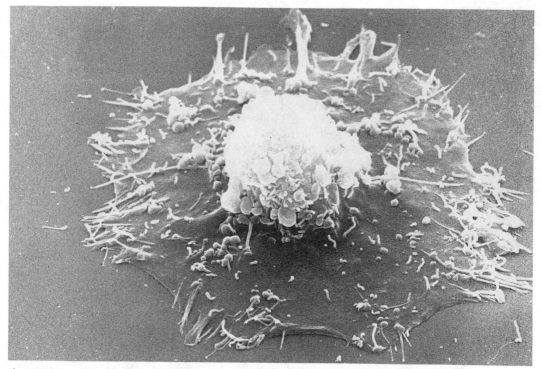

Figure 9-4. Scanning electron micrograph of a macrophage. The irregular outline indicates that it is a highly motile cell. (Reprinted from Johnson KE: *Histology: Microscopic Anatomy and Embryology*. New York, John Wiley, 1982, p 129. Micrograph courtesy of Dr. E.N. Albert, Department of Anatomy, George Washington University.)

      sacs called **phagosomes**. Primary lysosomes fuse with phagosomes to form secondary lysosomes called **phagolysosomes,** a variety of heterolysosomes (see Chapter 4 III for lysosome terminology). During fusion, lysosomal hydrolases are released into the phagolysosome and begin to degrade the bacterium.

    (b) Macrophage lysosomes contain specific bacteriocidal agents including **lysozyme,** a hydrolytic enzyme for degrading bacteria cell walls containing oligosaccharide, and **superoxide dismutase,** an enzyme that helps generate bacteriocidal oxygen-free radicals and hydrogen peroxide.

    (c) Macrophages also exhibit **chemotaxis.** They can move up a concentration gradient of certain complement system components (materials released during an inflammatory response) and bacterial components. Chemotaxis is important for attracting phagocytic cells and increasing their number in areas of bacterial invasion and inflammation.

3. Mast cells are a distinct cell type but are morphologically similar to basophils in peripheral blood.

    a. Distribution. Mast cells are widely distributed in loose areolar connective tissues and are especially plentiful in the **lamina propria,** which is the areolar connective tissue beneath the moist mucosal epithelium of many visceral organs.

    b. Mast cell granules. Mast cells are large residents of connective tissue and contain hundreds of granules. The granules are membrane-delimited organelles (modified lysosomes). Mast cell granules contain large amounts of the sulfated proteoglycan **heparin** (an anticoagulant), which causes them to stain metachromatically when treated with certain dyes.

    c. Functional cell biology. Mast cells are involved in inflammatory reactions and immediate hypersensitivity allergic reactions and are relatively common in the lamina propria of the respiratory and digestive systems.

        (1) Mast cells bind a portion of the immunoglobulin E (IgE) molecules released into the serum during exposure to antigens such as ragweed pollen.

        (2) Upon re-exposure to the antigen, the IgE bound to the surface of the mast cells facilitates release of **histamine** (which promotes capillary leakage and edema), **slow reacting**

substance (which promotes smooth muscle contraction and blood vessel leakage), **eosinophil chemotactic factor** (which attracts eosinophils), and **heparin** (which inhibits coagulation).

C. Extracellular fibers

1. **Collagen fibers**
    a. **Distribution.** Collagen fibers are composed of collagen macromolecules, the most abundant protein in the human body. Collagen is a major constituent of connective tissues proper and cartilage and bone. The collagen fibers in tendons are composed of numerous smaller fibers of **tropocollagen**. In the electron microscope, collagen fibers exhibit an alternating pattern of dark and light bands with a periodicity of 67 nm (see Chapter 7 III for more information about collagen).
    b. **Collagen synthesis** is a complex process performed by mesodermally derived fibroblasts, osteoblasts, chondroblasts, smooth muscle cells, and epithelial cells.
        (1) Type IV collagen is contained in basement membranes and is secreted by the epithelium above.
        (2) Collagen is a stable protein under the physiologic conditions that exist in connective tissues; however, it is constantly degraded and replaced by collagen-secreting cells.
        (3) Fibroblasts synthesize collagen de novo and secrete it into the extracellular matrix. They also have the ability to degrade collagen with several specific degradative enzymes called **collagenases**. Several varieties of collagenases exist and each has an affinity for a specific type of collagen.
        (4) Collagenases also exist in macrophages and neutrophils, which enter collagen-rich connective tissue by crossing the endothelium and basement membrane of capillaries. Collagenases may help these cells cross the capillary basement membranes.
        (5) Collagen turnover accelerates during tissue inflammation and repair, and the collagenases of fibroblasts, macrophages, and neutrophils may help this process.
    c. **Diseases caused by abnormal collagen structure**
        (1) **Marfan syndrome** involves defective collagen cross-linking and results in abnormally long and distensible extremities, abnormally distensible skin, and a high incidence of vascular aneurysms.
        (2) **Ehlers-Danlos syndrome** refers to a group of inherited disorders that share some of the clinical features Marfan syndrome and may be caused by defective collagen cross-linking.
        (3) **Scurvy** is caused by a vitamin C deficiency. Vitamin C is required for collagen cross-linking. Patients with scurvy exhibit defects in bone and tooth formation, fracture repair, and wound healing.

2. **Elastic fibers** exist in many areas of the body in many kinds of connective tissue. These fibers distend when stressed and then return to their original shape when the stress is released (see Chapter 7 IV for information about elastin biochemistry).
    a. Elastic fibers are prominent in parts of the body that expand and contract regularly. For example, they are a major component of many blood vessel walls. In the wall of the aorta, elastic fibers form numerous concentric fenestrated elastic laminae (Figure 9-5).
    b. Scattered elastic fibers exist in the loose areolar connective tissue of mesenteries and in the dermis.
    c. Elastic fibers also exist in elastic cartilage (e.g., pinna of the ear, epiglottis), the lungs and pleural membranes, the vocal cords, and elastic ligaments (e.g., the ligamentum flavum, which runs down the dorsal surface of the vertebral column).
    d. Special stains (e.g., Weigert's stain) must be used to differentiate elastic fibers from collagenous fibers in the light microscope.

D. Amorphous ground substance

1. **Characteristics.** The extracellular matrix of all connective tissue contains amorphous ground substance. The characteristics of the amorphous ground substance are as varied as the connective tissues themselves.
    a. In loose areolar connective tissue, it is sparse and watery.
    b. The amorphous ground substance in cartilage contains proteoglycan (see Chapter 7 V G), which provides structural rigidity and flexibility yet permits nutrients to diffuse from surrounding blood vessels.

Figure 9-5. Electron micrograph of the elastic laminae in the wall of the aorta. (Micrograph courtesy of Dr. E.N. Albert, Department of Anatomy, George Washington University.)

    c. In bones and teeth, the amorphous ground substance is calcified, providing these specialized connective tissues with increased tensile strength.

    d. In blood, the amorphous ground substance is a complex mixture of glycoproteins dissolved in liquid.

  2. Functions. Amorphous ground substance is a watery open mesh that allows water and metabolites to diffuse freely through connective tissues. It contains proteins that bind cells to fibers, and it provides connective tissues with tensile strength and flexibility appropriate for their functions.

III. SPECIAL CONNECTIVE TISSUES PROPER

A. Adipose tissue

  1. Composition and distribution

    a. Two types of adipose tissue exist: **white fat** and **brown fat**. White fat contains primarily **unilocular adipocytes** (one large fat vacuole). Brown fat contains **multilocular adipocytes** (many small fat vacuoles).

    b. Adipose tissue also contains scattered collagen fibers, a rich blood supply, fibroblasts, leukocytes, and macrophages.

    c. Adipocytes comprise the bulk of adipose tissue; however, they represent only about 20% of the total number of cells in the tissue.

    d. White fat is common throughout the subcutaneous compartment and is particularly prevalent over the abdomen and around the hips and buttocks.

    e. There is a sexual dimorphism in fat deposition. Adipose tissue accounts for about 22% of the weight of a normal female and up to 15% of the weight of a normal male.

2. **Functions.** Adipose tissue is a specialized connective tissue that fulfills several functions: It stores excess food, provides protective padding for the body, and assists thermoregulation.

   a. Adipose tissue is an important anatomic packing material. Substantial pads of fat protect the kidney and eyes, and adipose tissue infiltrates the mesenteries and the omentum in varying degrees, depending on an individual's nutritional status.

   b. In infants, brown fat generates heat.

3. **Adipocyte functional cell biology**

   a. White fat adipocytes are large cells that contain a single central fat vacuole. The cell cytoplasm is displaced peripherally but otherwise contains the usual array of organelles. The nucleus is flat and unremarkable.

   b. The fat vacuole is not bounded by a phospholipoprotein bilayer. Instead, it appears to be bounded by a regularly spaced array of 9-nm diameter filaments of unknown composition or length.

   c. The fat vacuole is amorphous and is composed of triglycerides and traces of cholesterol, cholesterol esters, monoglycerides, and phospholipids.

   d. During periods of nutritional excess, adipocytes contain enzymes that catalyze fatty acid synthesis from glucose. In the endoplasmic reticulum, fatty acids are synthesized into triglycerides, which are transported directly or via lysosomes into the central fat vacuole.

   e. During periods of fasting, large fat vacuoles decrease in size and break up into smaller vacuoles or disappear. Also, fasting seems to cause a proliferation of smooth endoplasmic reticulum and a tremendous increase in cell surface area—changes that relate to the cell's attempts to mobilize and export stored triglycerides.

   f. Adipocytes that lack a central fat vacuole are similar to fibroblasts; however, recent evidence suggests that they are distinct types of cells.

B. **Reticular tissue**

1. **Distribution.** Reticular tissue is a special connective tissue that exists only in the small vascular channels of the liver, spleen, lymph nodes, and bone matrix (Figure 9-6).

2. **Reticular fibers and cells**

   a. Reticular connective tissue is composed of **reticular fibers**—a lacework reticulum of collagen fibers with a rich coat of glycoproteins.

   b. Reticular fibers are **argyrophilic fibers**. When treated with silver salts, they reduce the silver salts, creating silver metal deposits that stain the fibers black.

   c. Reticular cells resemble fibroblasts. The reticular cells in lymph nodes are associated with lymph channels; however, their function is poorly understood.

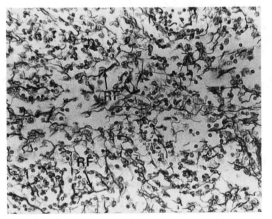

Figure 9-6. Reticular fibers (*RF*) surrounding the liver sinusoids. These fibers were stained with silver salts.

STUDY QUESTIONS

Directions: Each question below contains five suggested answers. Choose the **one best** response to each question.

1. Dense regular connective tissue contains all of the following components EXCEPT

(A) numerous extracellular fibers
(B) numerous cells
(C) numerous collagen fibers
(D) fibroblasts
(E) amorphous ground substance

2. Which of the following symptoms is most often associated with defective collagen cross-linking?

(A) Fever
(B) Edema
(C) Poor wound healing
(D) High blood pressure
(E) Blindness

3. Each of the following statements concerning mast cells is true EXCEPT

(A) they are found in the lamina propria
(B) they are distinct from basophils
(C) their granules stain orthochromatically with basic dyes
(D) their granules contain heparin
(E) they are involved in inflammatory responses

4. Which of the following statements best characterizes adipose tissue?

(A) It is one type of general connective tissue proper
(B) It has a poor blood supply
(C) Its cells contain fat vacuoles surrounded by a unit membrane
(D) It may contain unilocular or multilocular adipocytes
(E) It is equally distributed in males and females

5. Each cell type listed below is a component of the mononuclear phagocyte system EXCEPT

(A) histiocytes
(B) macrophages
(C) microglia
(D) neutrophils
(E) osteoclasts

6. Each organ listed below contains connective tissue embedded among parenchymal cells EXCEPT

(A) skin
(B) brain
(C) pancreas
(D) liver
(E) spleen

Questions 7–10

The following four questions refer to the cell pictured below.

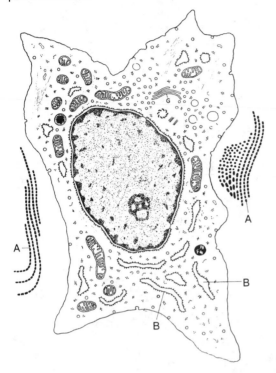

7. Which of the following cells is depicted in the illustration?

(A) Macrophage
(B) Mast cell
(C) Fibroblast
(D) Monocyte
(E) Adipocyte

8. The cell depicted in the illustration would be found in each of the locations listed below EXCEPT

(A) brain parenchyma
(B) mesenteries
(C) lamina propria
(D) dermis
(E) liver capsule

9. The structures labeled *A* in the illustration are

(A) elastic fibers
(B) reticular fibers
(C) argyrophilic fibers
(D) collagen fibers
(E) muscle fibers

10. Of the functions listed below, which one best describes the function of the structures labeled *B* in the illustration?

(A) Contains collagenase for degradation of collagen fibers
(B) Is involved in the assembly of collagen pro-α-chain polypeptides
(C) Contains many glycosyl transferases for collagen glycosylation
(D) Is a phagolysosome for digesting bacteria
(F) Is a lipid droplet boundary that isolates triglycerides in adipocytes

Directions: The group of questions below consists of lettered choices followed by several numbered items. For each numbered item select the **one** lettered choice with which it is **most** closely associated. Each lettered choice may be used once, more than once, or not at all.

Questions 11–15

Match each tissue location listed below with the most appropriate type of connective tissue.

(A) Loose areolar tissue
(B) Adipose tissue
(C) Dense irregular connective tissue
(D) Dense regular connective tissue
(E) Reticular connective tissue

11. Dermis

12. Orbital space in the skull

13. Lymphatic system

14. Tendons

15. Lamina propria

ANSWERS AND EXPLANATIONS

1. The answer is B. [*II A 2*] All connective tissue contains cells, fibers, and amorphous ground substance. Dense regular connective tissue is a type of general connective tissue proper that contains few cells and many fibers. The cells in dense regular connective tissue are fibroblasts. In this type of connective tissue, fibers and cells are arranged in simple, regular geometric patterns. In tendons, the cells and fibers of dense regular connective tissue are arranged in parallel arrays.

2. The answer is C. [*II B 1, C 1 c*] As a wound heals, fibroblasts in the wound area proliferate. Then, they secrete large amounts of collagen, which becomes extensively cross-linked. Diseases that disrupt collagen cross-linking usually result in poor wound healing. For example, scurvy results from a deficiency of vitamin C, which is an important cofactor for lysyl oxidase activity. When lysyl oxidase is low, collagen cross-linking is reduced. Patients with scurvy exhibit poor wound healing.

3. The answer is C. [*II B 3 b*] Mast cells are common components of many of the body's connective tissue domains. They are especially abundant in the loose connective tissue of the mesenteries and lamina propria. Mast cells are granulocytes that participate in the inflammatory response. Their granules contain heparin—a glycosaminoglycan with a highly negative charge because it contains numerous sulfate groups. Consequently, mast cell granules stain metachromatically with basic dyes such as toluidine blue. Although basophils and mast cells are morphologically similar, they are distinct cell types.

4. The answer is D. [*III A 1*] Adipose tissue is one type of specialized connective tissue proper, which has a rich blood supply. Adipocytes can have one large fat vacuole (unilocular) or many small fat vacuoles (multilocular). Fat vacuoles are not surrounded by a unit membrane but are bounded by a meshwork of fibrils that are 9 nm in diameter. Females have more adipose tissue than males.

5. The answer is D. [*II B 2 b*] Cells of the mononuclear phagocyte system are dedicated phagocytes. They include monocytes, histiocytes, macrophages, and microglia. These cells are derived from bone marrow, and their function is mediated by immunoglobulins. Osteoclasts, another type of cell in this system, are derived from the fusion of monocytes. Osteoclasts actively phagocytose bone. Neutrophils are phagocytic cells but are not part of the mononuclear phagocyte system.

6. The answer is B. [*I A 4*] Connective tissue is an important component of almost every organ in the body. Generally speaking, connective tissue holds the epithelial parenchymal cells and blood vessels together. For example, the connective tissue of skin (dermis) holds the epidermis and hypodermis together. Similarly, connective tissue in the pancreas and liver connects the parenchymal and vascular elements. Connective tissue also forms a capsule around the pancreas, liver, and spleen. In the brain, connective tissue is not embedded among the parenchymal cells; however, the meninges surrounding the CNS have connective tissue elements.

7. The answer is C. [*II B 1 c*] This cell is a fibroblast. Fibroblasts synthesize and secrete extracellular matrix components, such as collagen and proteoglycan. Consequently, they have abundant rough endoplasmic reticulum and a well-developed Golgi apparatus. A macrophage would contain many lysosomes and an irregular nucleus, and a monocyte would have a similar appearance. The cytoplasm of this cell would contain specific granules if it were a mast cell; fat vacuoles would exist in the cytoplasm of an adipocyte.

8. The answer is A. [*I A 4; II B 1 b*] Fibroblasts are the most widely distributed connective tissue cells. They are found in the mesenteries, lamina propria, dermis, and liver capsule, but not in brain parenchyma. Brain parenchyma does not contain connective tissue and, thus, contains no fibroblasts. The meninges, which surround the brain, do contain fibroblasts.

9. The answer is D. [*II B 1 c (2), C 1 b; Figure 9-3*] The structures labeled *A* are collagen fibers. Fibroblasts secrete collagen fibers, which are recognized in the electron microscope by their 67-nm periodicity. Collagen fibers are assembled in the extracellular compartment after collagen polypeptides are synthesized and secreted by fibroblasts.

10. The answer is B. [*II B 1 c (1), C 1 b; Figure 9-3*] The structures labeled *B* are areas of rough endoplasmic reticulum in the fibroblast. Rough endoplasmic reticulum is involved in the assembly of the nascent polypeptide chains of proteins (e.g., the pro-α-chains of collagen) for secretion from the fibroblast. After the collagen polypeptides are secreted, they become cross-linked and are assembled into the larger collagen fibrils.

11–15. The answers are: 11-C, 12-B, 13-E, 14-D, 15-A. [*II A 1, 2; III A 2, B 1; Table 9-1*] The dermis is a dense irregular connective tissue that provides a firm anchor for the epidermis. It also attaches the epidermis to deep hypodermal adipose tissue.

Adipose tissue is a food store or an anatomic packing material. It fills in the space between the eyeball and the orbit, cushioning the eyeball in the orbit.

Reticular connective tissue is an open meshwork of reticular cells and reticular fibers. It is abundant in lymph nodes, where it forms numerous channels for lymph circulation and provides a structural framework for lymphocytes.

Tendons are a classic example of dense regular connective tissue. They consist of numerous collagen fibers arranged parallel to one another with a paucity of interspersed fibroblasts. This arrangement creates a connective tissue with exceptional tensile strength.

The lamina propria is the loose areolar connective tissue beneath moist epithelial layers in the body. It is the connective tissue component of the mucosa in organs of the gastrointestinal, respiratory, and urogenital tracts. The lamina propria contains many of the formed elements of blood, such as neutrophils, lymphocytes, and plasma cells. These cells move freely between mucosal capillaries and the lamina propria.

10
Cartilage and Bone

I. INTRODUCTION

A. Functions

1. Cartilage and bone are specialized connective tissues that provide the body with mechanical support and protection.

2. Bones are the principal supports for the body. They provide a rigid structure for muscle attachment and constitute a system of levers that turn muscle contraction into purposeful movements. Cartilage augments this structure by providing flexible support.

3. Bones protect vital organs in the skull and thoracic cavity.

4. Bones contain a large store of calcium (Ca^{2+}), which can be mobilized, and bone marrow for hematopoiesis.

B. Common features of cartilage and bone. The histologic organization of cartilage and bone provides a unique combination of rigid, yet plastic, living tissue.

1. Cartilage and bone have many characteristics in common. In **endochondral ossification,** one type of bone formation, cartilage models are converted to bone.

2. Mature gross bones in adults have a cartilage component on their articular surfaces.

3. Like all connective tissues, cartilage and bone are composed of **cells, fibers,** and **amorphous ground substance**.

II. CARTILAGE. Adults have cartilage on the articular surfaces of long bones and in the trachea, bronchi, nose, ears, larynx, and intervertebral disks.

A. General characteristics

1. **Structure.** Cartilage contains cells (fibroblasts, chondroblasts, and chondrocytes), fibers (collagen and elastin), and amorphous ground substance (chondroitin sulfate and hyaluronate). It is dominated by the acellular elements and is devoid of blood vessels and nerves.

2. **Functions**
   a. Cartilage characteristics make it an excellent skeletal tissue for the fetus. Most adult bones existed as cartilaginous models during fetal life.
   b. Cartilage can grow rapidly and still provide a degree of support.

3. **Types.** The three varieties of cartilage are **hyaline cartilage, elastic cartilage,** and **fibrocartilage**.

B. Hyaline cartilage

1. **Distribution**
   a. Hyaline cartilage is a bluish, opalescent tissue that is widely distributed in the body. It is the most abundant type of cartilage in the adult.
   b. Hyaline cartilage exists on the ventral ends of ribs; in the larynx, trachea, and bronchi; and on the articular surface of bones.

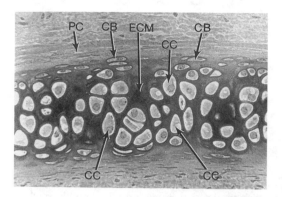

Figure 10-1. Light micrograph of hyaline cartilage in the trachea. Numerous chondrocytes (*CC*) are present in lacunae, which are surrounded by extracellular matrix (*ECM*). The perichondrium (*PC*) consists of flat peripheral fibroblasts and deep chondroblasts (*CB*).

   c. Hyaline cartilage also exists at the epiphyseal plates in the bones of fetuses and growing children.

2. **Appearance in the light microscope** (Figure 10-1)
   a. In the light microscope, hyaline cartilage appears as discrete masses of tissue surrounded by a dense irregular connective tissue layer called the **perichondrium**. The perichondrium is a layer of fibroblasts that merges with the fibroblasts and extracellular matrix of the connective tissue immediately surrounding the cartilage.
   b. The perichondrium has an inner layer of flattened **chondroblasts,** which are especially evident in growing cartilages. Chondroblasts are differentiated cells that secrete cartilaginous extracellular matrix. Chondroblasts surrounded by their secretion are called **chondrocytes**.
   c. Hyaline cartilage is **avascular** and lacks nerves.

3. **Hyaline cartilage extracellular matrix**
   a. The extracellular matrix of hyaline cartilage is amorphous; however, it is intensely **metachromatic** when stained with toluidine blue. Extracellular matrix metachromasia is caused by large concentrations of **cartilage proteoglycan,** a molecular complex consisting of core and linker proteins attached to the glycosaminoglycans—hyaluronic acid, chondroitin sulfates (several varieties), and keratan sulfate (see Chapter 7 V E).
   b. Cartilage extracellular matrix is rich in collagen fibers composed predominantly of Type II collagen.
   c. Chondrocytes are surrounded by a heterogeneous extracellular matrix consisting of an intensely metachromatic **territorial matrix** and a less metachromatic **intercellular matrix**.

4. **Chondrocyte functional cell biology.** In the electron microscope, chondroblasts and chondrocytes have the fine structure typical of cells that synthesize large amounts of protein and polysaccharide for secretion from the cell (Figure 10-2).
   a. The **cytoplasm** of chondroblasts and chondrocytes is **intensely basophilic** because it contains large amounts of rough endoplasmic reticulum. In cartilage, the polypeptide chains of collagen and proteoglycans are synthesized on ribosomes bound to the endoplasmic reticulum membranes.
   b. Chondrocytes contain a **large euchromatic nucleus** with a prominent nucleolus, numerous mitochondria, and a well-developed Golgi apparatus.
      (1) Mitochondria supply adenosine triphosphate (ATP) for energy-consuming functions such as protein synthesis and the transport of amino acids and sugar precursors for extracellular matrix constituents.
      (2) The Golgi apparatus synthesizes complex extracellular glycoconjugates.
   c. Chondrocytes have many cytoplasmic **vacuoles** that bud from the distal face (maturing face) of the Golgi apparatus. The **vacuoles** contain fibrous materials (collagen precursors) and amorphous materials (cartilage proteoglycan aggregates and glycoproteins).
   d. Chondrocytes store large amounts of nutrients such as glycogen granules and large lipid vacuoles because they exist so far from blood vessels.

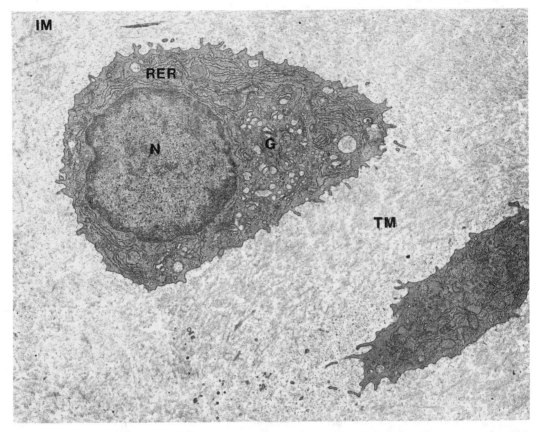

Figure 10-2. Electron micrograph of chondrocytes in hyaline cartilage. The chondrocyte nucleus (*N*) is euchromatic. The cell contains an abundance of rough endoplasmic reticulum (*RER*) and a prominent Golgi apparatus (*G*). The territorial matrix (*TM*) surrounding the cell has a high concentration of proteoglycans; the intercellular matrix (*IM*) has a low concentration of proteoglycans. Both locations have an abundance of collagen fibers. (Reprinted from Johnson KE: *Histology: Microscopic Anatomy and Embryology.* New York, John Wiley, 1982, p 61. Micrograph courtesy of Dr. D.P. DeSimone.)

C. Elastic cartilage

1. Distribution. Elastic cartilage exists in the pinna (auricle) of the external ear, the walls of the external auditory meatus, the auditory (eustachian) tube, the epiglottis, and parts of the larynx.

2. Histology
  a. In the light microscope, untreated elastic cartilage appears similar to hyaline cartilage. However, special stains (e.g., **Verhoeff's stain** or **Weigert's stain**) reveal that elastic cartilage contains **elastic fibers** composed of **elastin** and collagen fibers (Figure 10-3).
  b. As in hyaline cartilage, the extracellular matrix in elastic cartilage is metachromatic due to high concentrations of glycosaminoglycans.
  c. Elastic cartilage elasticity ensures the patency of the lumina of tubes that are surrounded by this variety of cartilage.

D. Fibrocartilage

1. Distribution. Fibrocartilage exists in the annulus fibrosus of intervertebral disks (where vertebral bodies attach to one another), the symphysis pubis, and the junctions between large tendons and articular cartilage in large joints. It is especially prominent where large tendons attach to bones.

2. Histology. Fibrocartilage is quite distinct from hyaline and elastic cartilage.
  a. Fibrocartilage consists of many regularly arranged collagen fibers, much like a tendon and, therefore, looks like a type of tissue intermediate to tendon and cartilage.

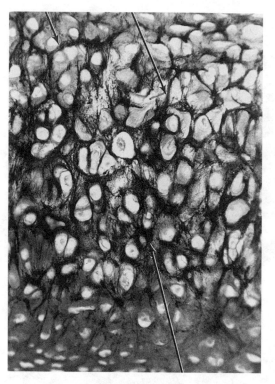

Figure 10-3. Light micrograph of elastic cartilage from the epiglottis. *Arrows* indicate darkly stained elastic fibers.

b. Fibrocartilage is less cellular than hyaline or elastic cartilage. Usually, fibrocartilage chondrocytes are sparsely scattered among large arrays of collagen fibers (Figure 10-4).

c. Fibrocartilage extracellular matrix exhibits less metachromasia than hyaline cartilage because it contains fewer glycosaminoglycans and many more collagen fibers.

E. Intervertebral disks

1. Distribution. Intervertebral disks exist between vertebral bodies in the spinal column.

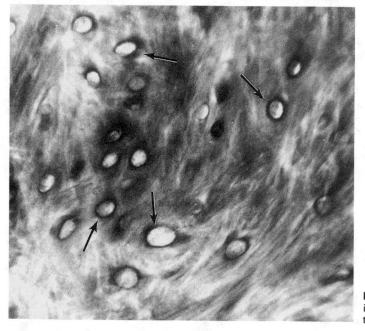

Figure 10-4. Fibrocartilage from the intervertebral disk. Note the relatively few chondrocytes (*arrows*).

2. **Structure.** The intervertebral disks consist of an outer layer of fibrocartilage called **annulus fibrosus** and a core of a special liquid connective tissue called the **nucleus pulposus**.
   a. Both sides of the annulus fibrosus are firmly attached to adjacent vertebral bodies.
   b. The nucleus pulposus is a viscous liquid containing a few cells and a large concentration of hyaluronic acid, which serves as a shock absorber for intervertebral disks.
   c. In children and young adults, the annulus fibrosus is relatively thin but quite strong. With age, the tensile strength of the annulus fibrosus decreases. When the annulus fibrosus ruptures, the semiliquid nucleus pulposus oozes onto nearby nerve roots, causing pain and increasing the potential for dysfunction (herniated disks).

III. BONE STRUCTURE

A. **Macrostructure.** The bulk of this discussion focuses on long bones, the predominant bones in the human body.

1. **Long bones** (Figure 10-5)
   a. The middle tubular section of a typical long bone (e.g., the tibia) is called the **diaphysis** and is composed of dense **compact bone**.
   b. Each end has an expanded portion called the **metaphysis,** which includes a region where compact bone forms a shell around a mass of spongy or **cancellous bone**.
   c. The ends of long bones are capped by an **epiphysis**. Articular surfaces of bones (areas where bones abut) have a thin layer of hyaline cartilage covering the epiphyseal compact bone. This articular cartilage is extremely slippery in its normal state.
   d. The junction between the epiphysis and the metaphysis is called the **epiphyseal plate**. In growing bones, the epiphyseal plate is a proliferative plate of hyaline cartilage. This is the region of bone that grows longer prior to puberty and during the pubertal growth spurt.

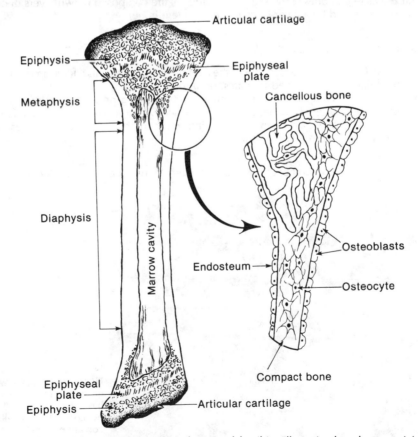

Figure 10-5. Illustration of the major macroscopic features of the tibia. (Illustration based on an original drawing in Johnson KE: *Histology: Microscopic Anatomy and Embryology.* New York, John Wiley, 1982, p 62.)

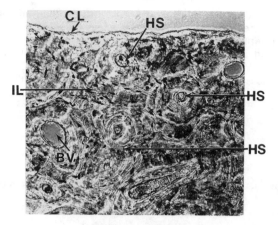

Figure 10-6. Low power light micrograph of compact bone showing circumferential lamellae (*CL*), haversian systems (*HS*), and interstitial lamellae (*IL*). Blood vessels (*BV*) run through the haversian canals.

    e. The **periosteum,** a dense irregular layer of connective tissue with the ability to form new bone, covers all bone surfaces except articular surfaces. The inner layer of the periosteum is composed of **osteoblasts,** which are cells that have the ability to secrete bone constituents.
    f. The marrow cavity and the surface of spongy bone spicules are lined with a thin layer of osteoblasts called **endosteum,** which also has osteogenic potential. The periosteum and endosteum account for growth in the diameter of bones.

  2. Short bones (e.g., bones of the carpus and tarsus) are composed of a core of cancellous bone completely surrounded by compact bone.

  3. Flat bones (e.g., bones of the skull, the scapulae) are composed of two layers of compact bone (**plates**) separated by a layer of cancellous bone (**diploë**).

B. Bone histology (Figure 10-6)

  1. Lamellae. Examining bone sections in the light microscope reveals a large amount of mineralized extracellular matrix arranged in plates, or **lamellae**.
    a. Lamellae contain small **lacunae** and an anastomosing network of minute **canaliculi**. Live **osteocytes** occupy lacunae, and delicate osteocyte processes fill the canaliculi.
    b. The mineralized matrix prevents free diffusion; consequently, the osteocytes are connected to vascular spaces by canaliculi.

  2. Compact bone
    a. Bone lamellae in compact bone are arranged in three common **patterns**.

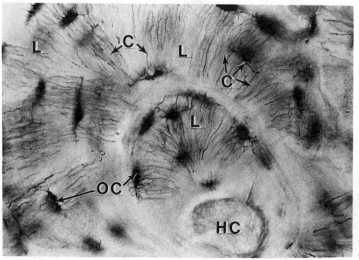

Figure 10-7. High power light micrograph of a portion of an osteon. The haversian canal (*HC*) is surrounded by concentric lamellae (*L*), which contain osteocytes (*OC*) that are interconnected by processes in minute channels called canaliculi (*C*).

     (1) Circumferential lamellae are parallel to the free surfaces of the periosteum and endosteum.

     (2) Haversian systems (osteons) are parallel to the long axis of compact bone. In this configuration, 4–20 lamellar layers are concentrically arranged around a vascular space (Figure 10-7).

     (3) Interstitial systems are irregular arrays of lamellae, often roughly triangular or quadrangular.

  b. Compact bones have two kinds of **vascular channels** (Figure 10-8).

     (1) Haversian canals are surrounded by concentrically arranged lamellae. The long axis of the lamellar cylinder is parallel to the long axis of the blood vessel.

     (2) Volkmann's canals are less common than haversian canals. Blood vessels from the periosteum penetrate the compact bone and cross haversian systems as they extend into the bone. The long axes of Volkmann's canals are perpendicular or nearly perpendicular to the haversian system lamellae.

3. Cancellous bone is composed of small quantities of lamellae, which lie close to blood vessels and the endosteum and receive nutrition by direct diffusion. The lamellae are not arranged into haversian systems.

4. The periosteum

  a. While bones grow in diameter, the osteoblast layer of the periosteum is a cuboidal layer of cells loosely arranged in a sheet. The osteoblasts are forming new bone by secreting uncalcified bone matrix.

  b. When the bone reaches its full diameter, osteoblasts become quiescent and indistinguishable from the other densely packed fibroblasts in the periosteum. When needed (e.g., when a bone fractures) these cells again become osteoblasts and express their osteogenic potential.

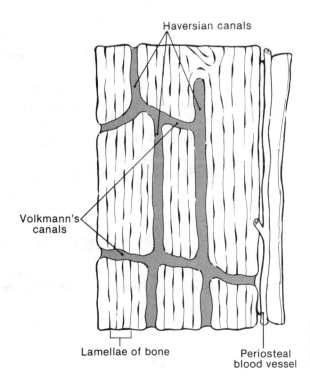

Haversian canals

Volkmann's canals

Lamellae of bone

Periosteal blood vessel

Figure 10-8. Illustration of the blood supply of bone. Notice that haversian canals run parallel to lamellae and that Volkmann's canals cross lamellae.

   c. Sharpey's fibers (i.e., dense bundles of collagen fibers trapped in the bony matrix of growing bone) firmly anchor the periosteum to the underlying bone.

   5. The endosteum is a layer of osteoblasts that covers all free bony surfaces inside the bone. Endosteal osteoblasts can secrete osteoid tissue. They also form a boundary between the bone and the marrow cavity.

IV. BONE EXTRACELLULAR MATRIX. The extracellular matrix of bone is dominated by inorganic salts but also contains substantial quantities of organic matrix.

A. Organic components

   1. Indirect histochemical evidence suggests that bone matrix contains some poorly characterized glycoproteins and some glycosaminoglycans (e.g., keratan sulfate, chondroitin sulfate, hyaluronic acid).

   2. Bone extracellular matrix contains a large amount of Type I collagen in the form of cross-banded fibers. These fibers have a periodicity of 67 nm and diameters of 50–70 nm.

B. Inorganic components. Bone extracellular matrix contains complex calcium phosphate that is similar to **hydroxyapatite** $[Ca_{10}(PO_4)_6(OH)_2]$. It also contains calcium carbonate, citrate, fluoride, magnesium ions, and sodium ions.

C. Relative importance of extracellular matrix components

   1. If the organic matrix is removed from a bone, the remaining mineralized bone is extremely brittle.

   2. If the mineral component of the bone is removed by prolonged exposure to acid and chelating agents, the bone becomes rubbery.

   3. Both bone components work synergistically to produce the extraordinary tensile strength and flexibility of bone.

V. BONE CELLS

A. Types and derivations of bone cells. Growing bones contain four kinds of cells: **osteoprogenitor cells, osteoblasts, osteocytes,** and **osteoclasts.**

   1. These four seemingly distinct cell types all stem from undifferentiated embryonic mesenchyme.

   2. Bone cells are probably capable of interconversion (e.g., osteoprogenitor cells can become osteoblasts, which, in turn, can become osteocytes).

   3. Osteoclasts are derived from the fusion of monocyte-like precursors and, therefore, may be modified versions of cells of the mononuclear phagocyte system.

   4. Two stem cell populations yield bone cells: **osteoprogenitor cells** and **osteoclast precursors.**

B. Osteoprogenitor cells are undifferentiated cells that are similar to fibroblasts. They can be highly proliferative or have latent mitotic potential.

   1. Osteoprogenitor cells are common in embryos during bone formation and comprise part of the periosteum in mature bones.

   2. Osteoprogenitor cells exist near all free surfaces of bone (e.g., periosteum, endosteum, and the lining of haversian canals) and on the trabeculae of cartilage that degenerates as the epiphyseal plate grows.

   3. They have a poorly stained oval nucleus and acidophilic or faintly basophilic cytoplasm in the light microscope.

C. Osteoblasts secrete bone extracellular matrix and are similar to fibroblasts and chondroblasts.

1. The cytoplasm of osteoblasts is intensely basophilic because they contain a large amount of rough endoplasmic reticulum.

2. Osteoblasts have a prominent nucleus with a large basophilic nucleolus and a well-developed Golgi apparatus, which prepares collagen for export, glycosylates collagen, and produces glycosaminoglycans and glycoproteins for the extracellular matrix.

D. Osteocytes are similar to osteoblasts in some respects but are less active in matrix secretion. Therefore, the rough endoplasmic reticulum and Golgi apparatus are less prominent (Figure 10-9).

1. Osteocytes respond to **parathormone** (see VI B) to help regulate blood calcium and can secrete new bone extracellular matrix.

2. They occupy lacunae in the solid matrix and are attached to each other by the slender cellular processes that occupy canaliculi.

3. Osteocyte processes are joined by **gap junctions,** which facilitate the exchange of small molecules and probably expedite the conduction of hormones from the vascular haversian canals to osteocytes in distant lamellae.

E. Osteoclasts are syncytial cells (i.e., large multinuclear cells formed by the fusion of many mononuclear cells). They arise from stem cells other than the undifferentiated mesenchymal cells that produce the osteoprogenitor-osteoblast-osteocyte cell line.

1. Osteoclasts are considered to be part of the mononuclear phagocyte system because they are derived from monocytes and are involved in bone matrix phagocytosis.
   a. Osteoclasts have a conspicuous ruffled, folded border that is closely applied to bone fragments as they are broken down.
   b. Osteoclasts secrete acid hydrolases (from lysosomes) and ions that disintegrate bone; however, this process is poorly understood.

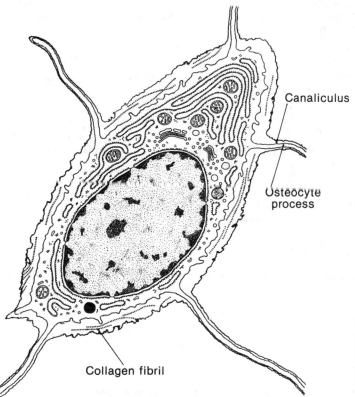

Canaliculus

Osteocyte process

Collagen fibril

Figure 10-9. Illustration of the fine structure of an osteocyte. Notice the abundance of rough endoplasmic reticulum and the osteocyte processes in the canaliculi. (Reprinted from Lentz TL: *Cell Fine Structure.* Philadelphia, WB Saunders, 1971, p 77.)

2. Osteoclast cytoplasm is acidophilic and stains intensely with eosin. It contains numerous mitochondria, many prominent Golgi bodies, and numerous lysosomes and residual bodies.

3. Osteoclasts respond to **parathormone** and **calcitonin** to help regulate serum calcium levels.

4. As osteoclasts erode bone lamellae, they create small pockets in the bone called **Howship's lacunae.**

VI. HORMONAL CONTROL OF CALCIUM METABOLISM

A. Calcium storage in the mineralized matrix

1. Calcium is essential for many functions including cell adhesion, membrane permeability, muscle contraction, and blood clotting.

2. More than 99% of the body's calcium is stored in bones. Every minute, 25% of all blood-borne calcium atoms are replaced by calcium atoms from bone.

3. A dual, antagonistic system of endocrine secretions regulates serum calcium.

B. Parathormone (PTH) is a polypeptide secreted by **chief cells** in the parathyroid glands.

1. Low serum calcium stimulates the secretion of parathormone, which binds to specific membrane receptors in bone and renal cells.

2. In renal tissue, parathormone stimulates the synthesis of the active vitamin D metabolite 1,25-dihydroxyvitamin D_3 [1,25-$(OH_2)D_3$], which stimulates resorption of calcium in the gastrointestinal tract and promotes bone resorption.

3. Parathormone also stimulates osteocytic and osteoclastic **osteolysis** and renal calcium resorption.

4. When serum calcium levels rise, a feedback mechanism prevents further parathormone secretion.

C. Calcitonin (thyrocalcitonin) is a polypeptide hormone synthesized by thyroid gland **parafollicular cells (C cells),** which helps regulate serum calcium levels. Calcitonin is antagonistic to parathormone and depresses bone resorption.

D. Other hormones

1. Gonadal steroids regulate the rate of skeletal maturation and turnover and the form of the skeleton, causing distinct sexual dimorphism of the skeleton.
   a. For example, male and female pelvises are grossly different.
   b. Postmenopausal women can suffer from **osteoporosis** as estrogen production drops, causing the bone resorption rate to exceed the bone deposition rate.

2. The anterior pituitary gland secretes **growth hormone,** which directly affects proliferation of chondrocytes in the epiphyseal plate in children and adolescents.

VII. NUTRITIONAL INFLUENCES ON BONE FORMATION. Normal bone growth requires adequate nutrition, especially dietary calcium, as well as vitamins D and C.

A. Vitamin D promotes normal calcification and collagen synthesis.

B. Vitamin C is an essential cofactor for enzymes involved in collagen synthesis. Vitamin C deficiency causes **scurvy,** which disturbs bone and tooth formation and causes symptoms of defective collagen synthesis.

VIII. JOINTS

A. Synarthroses are joints with little mobility.

1. **Synostoses** are joints in which one bone directly joins another (e.g., joints in the skull).

   2. Syndesmoses are joints (e.g., in which bone and ligaments join intervertebral joints).

B. Diarthroses are joints with great mobility.

   1. These joints have a fluid-filled cavity enclosed by a fibrous connective tissue capsule. The joint cavity is not lined by an epithelium; instead, it has a peculiar layer of cells called **synovial cells**.

   2. In some diarthroses, the capsule has a synovial membrane containing a rich vascular supply that leaks lubricating synovial fluid into the joint cavity.

   3. Articular surfaces are covered by hyaline cartilage and are devoid of perichondrium. Chondrocytes on the articular surface are very flat.

STUDY QUESTIONS

Directions: Each question below contains five suggested answers. Choose the **one best** response to each question.

1. Which one of the following characteristics is attributable to osteocyte processes?

(A) They secrete elastin
(B) They are joined by gap junctions
(C) They contain a Golgi apparatus
(D) They contain many tonofilaments
(E) They disappear when osteocytes divide

2. Which one of the following statements best describes chondroblasts?

(A) They are endosteal cells capable of secreting proteoglycan
(B) They are perichondrial cells capable of secreting Type II collagen
(C) They are periosteal cells capable of secreting Type I collagen
(D) They show little mitotic activity
(E) They are filled with rough endoplasmic reticulum but lack a Golgi apparatus

3. Which one of the following statements best describes osteoblasts?

(A) They secrete Type I collagen
(B) They secrete Type II collagen
(C) They have heterochromatinized nuclei
(D) They do not contain nucleoli
(E) They contain little rough endoplasmic reticulum

4. Each of the following statements concerning osteocytes is true EXCEPT

(A) they cause osteolysis
(B) they have many processes
(C) they are nourished by capillaries in canaliculi
(D) they are joined to each other by gap junctions
(E) they are sensitive to parathormone stimulation

5. Each of the following statements concerning osteoclasts is true EXCEPT

(A) they are large acidophilic cells
(B) they contain many cytoplasmic mitochondria
(C) they are part of the mononuclear phagocyte system
(D) they are found in Howship's lacunae
(E) they secrete bone extracellular matrix

6. Each of the following statements concerning parathormone is true EXCEPT

(A) it has the same effect on serum Ca^{2+} as calcitonin
(B) it stimulates osteocytes to resorb bone
(C) it stimulates osteoclasts to resorb bone
(D) it stimulates renal tubule retention of Ca^{2+}
(E) its secretion is regulated by a negative feedback mechanism

7. Each of the following statements concerning calcitonin is true EXCEPT

(A) it is a steroid hormone
(B) it inhibits osteoclastic activity
(C) it stimulates bone deposition
(D) it is antagonistic to parathormone
(E) it is secreted by thyroid gland parafollicular cells

Directions: The groups of questions below consist of lettered choices followed by several numbered items. For each numbered item select the **one** lettered choice with which it is **most** closely associated. Each lettered choice may be used once, more than once, or not at all.

Questions 8–11

For each tissue attribute described below, select the tissue type with which it is most closely associated.

(A) Hyaline cartilage
(B) Bone
(C) Both
(D) Neither

8. Contains collagen fibers and glycoproteins

9. Contains an abundance of hydroxyapatite

10. Has a rich blood supply

11. Contains cells that are nourished by diffusion through canaliculi

Questions 12–15

For each attribute described below, select the type of bone with which it is most closely associated.

(A) Compact bone
(B) Spongy bone
(C) Both
(D) Neither

12. Contains haversian systems (osteons)

13. Lacks a rich blood supply

14. Contains numerous osteocytes and osteoclasts

15. Found in the femoral metaphysis

ANSWERS AND EXPLANATIONS

1. The answer is B. [*V D 3*] Osteocytes are the principal cell type in bone. They help regulate blood calcium and secrete bone extracellular matrix; however, they do not secrete elastin. Osteocyte processes occupy canaliculi and are joined to each other by gap junctions. Osteocytes contain a Golgi apparatus, but it is located in the cell body, not in the slender processes. Tonofilaments are present where desmosomes join cells together, not where cells are joined by gap junctions. Osteocytes do not divide.

2. The answer is B. [*II B 2 b, 3*] Chondroblasts are cartilage cells located deep in the perichondrium, the dense connective tissue capsule that surrounds cartilage. As chondroblasts secrete extracellular matrix rich in Type II collagen and cartilage proteoglycan, they become surrounded by their own extracellular matrix and differentiate into chondrocytes. Both chondroblasts and chondrocytes are capable of cell division by mitosis. The endosteum is a layer of osteoblasts in bone. The periosteum is the outer connective tissue covering of bone.

3. The answer is A. [*IV A; V C*] Osteoprogenitor cells are derived from embryonic mesenchyme. They can differentiate into osteoblasts and osteocytes. Osteoblasts are secretory cells capable of secreting uncalcified bone extracellular matrix. Bone matrix contains amorphous ground substance, which calcifies after secretion, and an abundance of Type I collagen. Osteoblasts contain abundant rough endoplasmic reticulum, a Golgi apparatus, and prominent nucleoli with abundant euchromatin but little heterochromatin. These ultrastructural features are commonly found in cells specialized for secretion of proteins and polysaccharides into the extracellular matrix.

4. The answer is C. [*III B 1; V D*] Osteocytes are mature bone cells trapped in the lacunae of bone lamellae. They are interconnected by numerous processes that radiate from the central portion of the cell through canaliculi. Canaliculi are minute diffusion channels that carry nutrients from capillaries in the haversian canals to osteocytes in the lamellae. Osteocyte processes in canaliculi are joined to each other by gap junctions. When stimulated by parathormone, osteocytes cause osteolysis, which releases Ca^{2+}.

5. The answer is E. [*V C, E*] Bone extracellular matrix is secreted by osteoblasts, not by osteoclasts. Osteoclasts are large syncytial cells formed by the fusion of many monocytes. They are part of the mononuclear phagocyte system; however, scientists have not determined whether or not their function is mediated by immunoglobulins. Osteoclast cytoplasm contains an abundance of mitochondria, which makes the cytoplasm acidophilic, as well as many lysosomes, which are involved in bone degradation. Lysosomes contribute to cytoplasmic acidophilia. Mitochondria form the ATP necessary for transporting ions out of the cell, which, in turn, results in an acidic osteolytic milieu in Howship's lacunae.

6. The answer is A. [*VI B, C*] Parathormone is a polypeptide hormone secreted by chief cells in the parathyroid glands. When serum calcium (Ca^{2+}) is low, parathormone production increases, stimulating osteocytes and osteoclasts to degrade bone and promoting renal retention of Ca^{2+}, thus increasing serum Ca^{2+}. When serum calcium rises, a negative feedback mechanism reduces parathyroid secretion of parathormone. Calcitonin is another polypeptide hormone that helps regulate serum calcium; however, its effects are antagonistic to parathormone (i.e., it depresses bone resorption). Together, parathormone and calcitonin precisely regulate serum calcium.

7. The answer is A. [*VI C*] Calcitonin is a polypeptide hormone secreted by parafollicular cells (neural crest derivatives) in the thyroid gland. Parathyroid gland oxyphilic cells are not active in hormone secretion. Calcitonin stimulates bone deposition and inhibits bone degradation.

8–11. The answers are: 8-C, 9-B, 10-B, 11-B. [*I A, B; II A 1, B 3; III B; IV A 1, B; V D*] Bone and cartilage are specialized connective tissues that support and protect the body. Bone contains cells that are nourished by diffusion through canaliculi surrounding osteocyte processes.

Bones are the primary supports for the body and translate muscle contraction into purposeful movement. They contain many blood vessels and Type I collagen. Bone extracellular matrix is rich in glycoproteins and glycosaminoglycans and is infiltrated with hydroxyapatite.

Cartilage is a semirigid tissue that provides flexible support for the body. It is avascular and contains Type II collagen. Cartilage extracellular matrix contains an abundance of proteoglycans and some glycoproteins.

12–15. The answers are: 12-A, 13-D, 14-C, 15-C. [*III A 1, B*] Compact bone and spongy bone are similar tissues arranged in different ways. Both contain many blood vessels and numerous osteocytes and osteoclasts. Cortical compact bone and marrow-filled, medullary spongy bone exist in the metaphysis of long bones such as the femur.

Compact bone in the shaft of long bones has many haversian systems. These haversian systems consist of many concentric lamellae of bone arranged around a central blood vessel. The long axes of the haversian systems are parallel to the bone long axis.

Spongy bone in the metaphysis of long bones consists of an anastomosing network of blood vessels surrounded by bone lamellae. The bony lamellae in spongy bone form an anastomosing network of interwoven lamellae (spongy bone is sometimes called woven bone).

11
Peripheral Blood, Bone Marrow, and Hematopoiesis

I. INTRODUCTION

A. Blood composition. Blood is a liquid connective tissue that is ubiquitous in the body. It contains cells, (potential) extracellular fibers, and extracellular amorphous ground substance.

1. Blood cells are produced inside bone in specialized connective tissue called **bone marrow**. The cells in blood are **erythrocytes, leukocytes,** and **platelets**.

2. Blood contains the potential fibers of **fibrinogen**, which become actual fibers (**fibrin**) during clotting. Clotted blood is a semisolid connective tissue that rapidly stops bleeding and provides the cells necessary for wound healing.

3. The extracellular amorphous ground substance of blood is the fluid and proteins in blood plasma.

B. Blood functions

1. Blood carries oxygen and nutrients to cells throughout the body and carries cell waste products to the kidneys and lungs.

2. Blood helps maintain homeostasis in the body.

3. Blood is a readily available specimen that can be withdrawn from a patient. Blood chemistry and blood cytology reveal much about a patient's health.

C. Bone marrow is a highly specialized reticular connective tissue that has hematogenic and osteogenic potential.

1. In the fetus, bone marrow is an actively hematopoietic tissue.

2. Adult bone marrow is less hematopoietic because it regresses from hematopoietic **red marrow** to resting, fat-storing **yellow marrow**.

II. ERYTHROCYTES

A. Distribution

1. **Erythrocytes,** or **red blood cells,** are the most prevalent cells in peripheral blood.
   a. The peripheral blood of an individual contains 25,000,000,000,000 (**twenty-five trillion**) erythrocytes, and the spleen and bone marrow contain many more.
   b. Each cubic millimeter of blood contains approximately 5×10^6 red cells. The total peripheral blood volume is approximately 5 L.

2. Red cells comprise about 45% of the total blood volume. When a sample of blood is centrifuged to sediment the red cells, the **hematocrit** is about 45 (i.e., 45% of the sample is packed cells). Hematocrit depends on a patient's health status and can vary greatly.

B. Structure

1. Red cells are biconcave discs that measure 7–8 μm in diameter and about 2 μm thick at the edge (Figure 11-1).

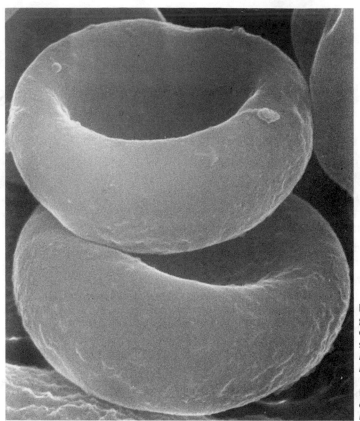

Figure 11-1. Scanning electron micrograph of two normal human erythrocytes illustrating their biconcave disc shape. (Reprinted from Johnson KE: *Histology: Microscopic Anatomy and Embryology.* New York, John Wiley, 1982, p 108. Micrograph courtesy of Dr. E.N. Albert, Department of Anatomy, George Washington University Medical Center.)

2. Although red cells have a minute surface area (approximately 130 μm^2), the aggregate surface area of all red cells in the body's 5 L of blood is about 2000 times greater than the surface area of skin. The immense capacity of blood to bind and transport gas is partially due to the tremendous surface area of red cells.

3. Although bounded by a plasma membrane, red cells are somewhat unusual in that they possess **no other organelles**.

4. About one-third of the erythrocyte mass is hemoglobin, a protein composed of four globin chains and heme (an iron-containing porphyrin with a high oxygen-binding capacity).
  a. Hemoglobin has a molecular weight of 68 kD. Hemoglobin A, the predominant hemoglobin in adult blood, contains two pairs of α-**globin chains** and two pairs of β-**globin chains** $(\alpha_2 \beta_2)$.
  b. Oxygen binds to the iron in the heme when the iron is a reduced ferrous iron.
  c. Red cell cytoplasm contains **hemoglobin reductase,** an enzyme that maintains hemoglobin in a reduced state.

C. Ultrastructure

1. Red cells are surrounded by a typical plasma membrane but are otherwise **largely devoid of intracellular organelles**.
  a. In adults, red cells lack a nucleus. Primitive fetal red cells have a heterochromatinized nucleus, which is probably inactive.
  b. Red cells do not have mitochondria, endoplasmic reticulum, or Golgi apparatus. Mature red cells also lack ribosomes, although peripheral blood occasionally contains immature red cells called **reticulocytes,** which may contain aggregated reticular clusters of ribosomes.

2. Red cells have a **few intracellular membrane-bound vesicular elements;** however, most of the cytoplasm has a faintly granular appearance caused by hemoglobin molecules.

3. Red cells contain glycolytic enzymes and **spectrin,** an intracellular cytoskeletal protein associated with the plasma membrane.

4. The erythrocyte membrane contains numerous **intramembranous particles** in the plasmalemma P-face, which are visible in freeze-fracture-etch specimens.

D. Functions

1. **Red cells transport oxygen and carbon dioxide** and perform most of their work within capillaries. Often the capillary diameter is considerably smaller than the diameter of the red cells (e.g., in the capillary bed in the lungs). Consequently, red cells in the microvasculature are subjected to considerable stress. Damaged red cells are removed from the circulation and destroyed.
   a. Red cells have a life span of approximately 120 days.
   b. Red cell destruction occurs in the spleen, liver, and bone marrow.

2. **Role of hemoglobin in gas transport**
   a. Red cells transport oxygen from pulmonary alveoli to peripheral tissues and carbon dioxide from peripheral tissues to pulmonary alveoli.
   b. In alveoli, oxygenated blood carries about 96 mm Hg of oxygen. In peripheral tissues, oxygen tension (P_{O_2}) decreases to as little as 40 mm Hg.
   c. The carbon dioxide tension (P_{CO_2}) in peripheral venous blood is about 50 mm Hg, which is much greater than the P_{CO_2} in alveoli. Consequently, carbon dioxide diffuses through the plasma into alveoli.

E. Abnormal erythrocytes

1. In some pathologic states, red cells have pronounced **morphologic abnormalities**. For example, some hemolytic diseases cause an abundance of reticulocytes.

2. Normally shaped red cells can be abnormally small (**microcytic**) or abnormally large (**macrocytic**).

3. Red cells may exhibit gross **abnormalities in shape,** such as those which occur in **sickle cell anemia,** a disease caused by a genetic alteration in the hemoglobin β-chain.
   a. A single amino acid substitution in the hemoglobin causes the abnormal shape and, consequently, the excessive destruction of red blood cells.
   b. Nearly 10% of all blacks in the United States are heterozygous for the sickle gene. Children of two heterozygous parents are affected by the gene according to Mendel's law: 25% are normal wild type, 50% are heterozygous carriers, and 25% have frank sickle cell anemia.

III. LEUKOCYTES

A. Leukocyte classes. Peripheral blood contains two kinds of leukocytes: granulocytes and agranulocytes.

1. **Granulocytes** include:
   a. **Neutrophils**
   b. **Eosinophils**
   c. **Basophils**

2. **Agranulocytes** include:
   a. **Lymphocytes**
   b. **Monocytes**

B. Neutrophils

1. **Distribution**
   a. Neutrophils, or polymorphonuclear leukocytes (PMNs), are the most common leukocytes in normal human peripheral blood, comprising 40%–60% of all leukocytes in the blood.
   b. A cubic millimeter of blood contains about 4500 neutrophils.

2. **Structure** (Figure 11-2). Neutrophils are 12–15 μm in diameter. The nucleus has three to five lobes, is largely heterochromatin, and contains no nucleolus. The cytoplasm is moderately acidophilic and contains two types of granules.

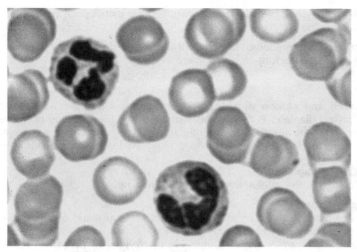

Figure 11-2. Light micrograph of two neutrophils among erythrocytes in a normal human peripheral blood smear. (Reprinted from Johnson KE: *Histology: Microscopic Anatomy and Embryology.* New York, John Wiley, 1982, p 111.)

3. **Ultrastructure** (Figure 11-3)

   a. The neutrophil plasma membrane is unremarkable. A nuclear envelope surrounds the nucleus, which has several prominent lobes containing a large amount of dense, peripheral granular heterochromatin and a small amount of filamentous euchromatin.

   b. The cytoplasm contains many glycogen granules, a few mitochondria, a small Golgi apparatus, and little endoplasmic reticulum.

   c. The cytoplasm also contains two varieties of membrane-bound **granules**. These granules are modified lysosomes and have a bacteriocidal function.

   (1) **Azurophilic granules (primary**, or **type A)** stain with azure dye and are diagnostic for neutrophils. These large (0.4 μm) electron-dense granules comprise about 20% of the granule population and are visible in the light microscope.

   (2) **Specific granules (secondary**, or **type B)** are smaller (0.2 μm) and may contain crystalloids. They comprise 80% of the granule population and are not visible in the light microscope.

   d. Neutrophils have many cortical microfilaments and some cytoplasmic microtubules.

4. **Functions.** The primary function of neutrophils is the phagocytosis and destruction of bacteria. Bacteria can be phagocytosed after **opsonization** (see Chapter 9 II B 2 d).

Specific granule

Azurophilic granule

Glycogen granules

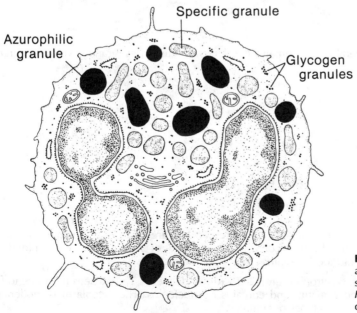

Figure 11-3. Diagram of a neutrophil as it appears in the electron microscope. (Reprinted from Lentz TL: *Cell Fine Structure.* Philadelphia, WB Saunders, 1971, p 21.)

a. Bacteria release chemoattractants that cause neutrophil **chemotaxis** and promote the adhesion of neutrophils to bacteria. **Complement components** are neutrophil chemoattractants released during inflammation.

b. Neutrophils first adhere to bacteria and then engulf them in a membrane-bound **phagosome**. Phagosomes fuse with secondary granules and then primary granules to form **phagolysosomes**.

c. Azurophilic granules contain **myeloperoxidase,** an enzyme that produces bacteriocidal molecular oxygen from hydrogen peroxide (H_2O_2).

d. Specific granules contain the protein **lactoferrin,** which binds the ferric ions required for bacterial multiplication. Lactoferrin also initiates a negative feedback loop that prevents neutrophil production in the bone marrow.

C. Eosinophils

1. **Distribution.** Eosinophils comprise about 1% of all leukocytes in blood; a cubic millimeter contains about 200.

2. **Structure** (Figure 11-4)
   a. Eosinophils are motile phagocytic cells, which have diameters similar to neutrophils yet appear quite distinct in the light microscope.
   b. The nucleus has two or three lobes, which contain a striking array of large (0.6–1 μm) red or orange **eosinophilic granules**.

3. **Ultrastructure** (Figure 11-5)
   a. Eosinophils have a typical cell membrane and contain scattered mitochondria, a Golgi apparatus near the nucleus, numerous eosinophilic granules (modified lysosomes) and some nonspecific, electron-dense granules.
   b. Eosinophilic granules have prominent crystalloids.
   c. Microfilaments are prominent in the eosinophil cortex.

4. **Functions**
   a. Eosinophils kill parasitic larvae as they enter peripheral blood or the lamina propria of the gut.
      (1) Poor sanitation can cause schistosomiasis or ascariasis in humans.
      (2) Patients infected with parasite larvae have elevated eosinophil counts. In extreme cases, 90% of peripheral blood leukocytes are eosinophils.
   b. When parasites invade the body, **mast cells** draw eosinophils to the infection site.
      (1) Eosinophils are attracted to inflammation sites by **eosinophil chemotactic factor** (released by mast cells) and by **lymphokines** (released by T lymphocytes).
      (2) Eosinophils help regulate mast cell response to inflammation by releasing an enzyme that degrades the histamines released by mast cells at inflammation sites.

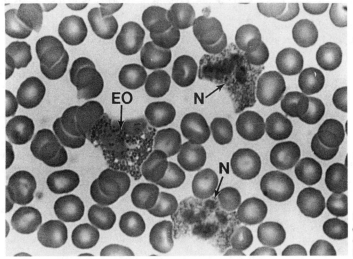

Figure 11-4. Light micrograph of an eosinophil (*EO*) and two neutrophils (*N*) among erythrocytes in a normal human peripheral blood smear.

Crystalloids

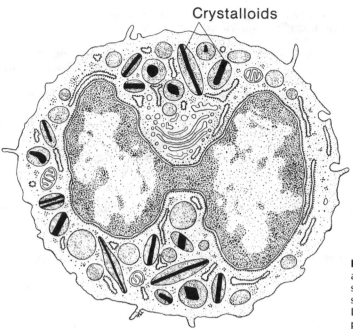

Figure 11-5. Diagram of an eosinophil as it appears in the electron microscope, showing crystalloids within the specific granules. (Reprinted from Lentz TL: *Cell Fine Structure.* Philadelphia, WB Saunders, 1971, p 27.)

   c. Eosinophilic granule crystalloids have a dominant component called the **major basic protein,** which has a poorly understood antiparasitic function.
     (1) Complement components draw eosinophils to parasites.
     (2) Eosinophils adhere to the parasites and degranulate, releasing major basic protein.
   d. Eosinophilic granules contain lysosomal enzymes that destroy dead parasites.

D. Basophils

   1. Distribution. Basophils are the rarest leukocytes. A cubic millimeter of blood contains about 5 basophils.

   2. Structure
     a. Basophils are about the size of neutrophils and eosinophils and contain a nucleus with two or three lobes.
     b. A mass of small (0.5 μm) basophilic specific granules often obscures the nucleus. These granules stain magenta (metachromatically) with the dye mixtures used to stain blood smears.
     c. Basophil granules are membrane-bound and contain crystalline regions, which suggests that they are modified lysosomes.
     d. Basophil granules contain **histamine** (a potent acute vasodilator), **heparin** (a glycosaminoglycan anticoagulant), and **slow reacting substance** (a slow-acting vasodilator).

   3. Functions
     a. Basophils mediate the inflammatory response and secrete **eosinophil chemotactic factor**.
     b. Basophils and mast cells have similar structures and functions; however, they are distinct cell types.
     c. In response to certain antigens, basophils stimulate the formation of **immunoglobulin E** (IgE)—a class of antibodies.
     (1) IgE binds to basophil and mast cell surfaces but has no immediate effect.
     (2) Subsequent exposure to the same antigen can cause a basophil and mast cell response restricted to specific organs (e.g., bronchial asthma in the lungs or a severe and systemic response such as anaphylactic shock brought on by a bee sting).

E. Lymphocytes

   1. Distribution
     a. Lymphocytes are the most common agranulocytes. A cubic millimeter of blood contains about 2500.

    b. Lymphocytes are not confined to peripheral blood. They also exist in connective tissue lamina propria, lymph nodes, the spleen and tonsils, and bone marrow.

2. Structure
    a. Lymphocyte diameter varies from 5–8 μm in small lymphocytes to 15 μm in large lymphocytes (Figure 11-6).
    b. Lymphocytes have a round, densely stained nucleus, which occupies most of the cell volume, and a thin shell of cytoplasm around the nucleus. As lymphocyte diameter increases, the cell's cytoplasm volume increases faster than the nuclear volume.
    c. Lymphocytes do not contain specific granules.
    d. Lymphocyte chromatin stains unevenly, exhibiting many densely stained blocks of heterochromatin.

3. Ultrastructure
    a. A plasma membrane surrounds each lymphocyte. The plasma membrane P-face contains many intramembranous particles.
    b. Lymphocytes have cell-surface immunoglobulins and membrane-bound receptors for specific immunoglobulins.
    c. Lymphocytes contain a heterochromatic nucleus surrounded by a porous nuclear envelope, numerous lysosomes, a Golgi apparatus, and abundant smooth endoplasmic reticulum.
    d. Blast transformation (see Chapter 12 VII B 2) converts lymphocytes into **plasma cells**. During this phase, lymphocytes proliferate, develop more rough endoplasmic reticulum, and begin secreting immunoglobulins, and the Golgi apparatus becomes enriched.

4. Function. Lymphocytes are key cells in the immune system (see Chapter 12 VII B for information about lymphocytic function).

F. Monocytes

    1. Distribution. A cubic millimeter of blood contains about 300 monocytes.

    2. Structure
        a. Monocytes, the largest leukocytes, have diameters of 12–18 μm.
        b. Monocytes have an agranular cytoplasm and a rounded nucleus that has indentations on one side. They do not contain specific granules.
        c. In contrast to lymphocyte chromatin, monocyte chromatin stains uniformly, revealing a delicate network (Figure 11-7).

    3. Ultrastructure. Monocytes contain many lysosomes and have a prominent Golgi apparatus.

    4. Function. Monocytes are the direct precursors to **macrophages** (see Chapter 9 II B 2).
        a. Alveolar macrophages are monocytes that leave the closed circulation and enter the lungs to remove inspired debris.

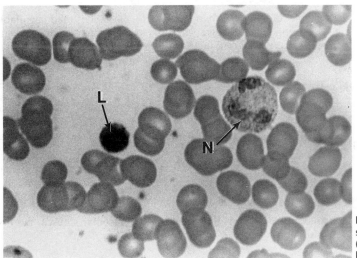

Figure 11-6. Light micrograph of a small lymphocyte (*L*) and a neutrophil (*N*) among erythrocytes in a normal human peripheral blood smear.

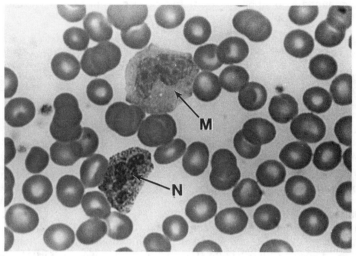

Figure 11-7. Light micrograph of a monocyte (*M*) and a neutrophil (*N*) among erythrocytes in a normal human peripheral blood smear.

    b. Monocytes that enter connective tissue to perform phagocytic functions are sometimes called **histiocytes.**

IV. PLATELETS

A. Structure and distribution

    1. Platelets are 2–4 μm in diameter. A cubic millimeter of blood contains 200,000–400,000.

    2. Platelets have a central **granulomere,** which stains purple in blood smears, and a peripheral **hyalomere,** which stains faintly. They may be clustered in small or large masses.

B. Ultrastructure

    1. Platelets are cell fragments. They are surrounded by a plasma membrane that has a thick glycocalyx and deep surface invaginations that lead to canaliculi.

    2. The peripheral cytoplasm of the platelet is rich in microfilaments and microtubles. These cytoskeletal elements are responsible for maintaining its lenticular shape.

    3. Platelets contain **dense core granules** with serotonin in them and **alpha granules,** which contain clotting factors and various lysosomal enzymes.

C. Function. Platelets are important for blood clotting. When exposed to substances made available when blood vessels are cut (e.g., collagen), platelets aggregate rapidly, adhering to each other and to fibrin. Masses of aggregated platelets and fibrin are the basis for clots.

V. BONE MARROW HISTOLOGY

A. Location and blood supply

    1. Bone marrow exists in interstices between bone spicules in the shafts of bones and the trabecular networks of cancellous bone.

    2. Marrow compartments are lined by osteogenic endosteum and contain anastomosing networks of vascular channels.

    3. Blood vessels penetrate the bone to reach the marrow compartment.

    4. Within bone, arteries branch into arterioles that carry blood to the **vascular sinuses,** the first elements of the venous drainage of the marrow.

    5. Bone marrow circulation is closed.

   6. Hematopoietic compartments between the endosteum and the vascular sinuses contain irregularly arranged cords of hematopoietic cells or fat cells.

   7. In red marrow, hematopoietic cells outnumber fat cells. In yellow marrow, fat cells outnumber hematopoietic cells.

B. Vascular sinuses are lined with a continuous endothelium that has a typical basement membrane.

   1. Reticular cells and **reticular fibers** surrounding the endothelium comprise the vascular sinus adventitia and a stromal framework for the hematopoietic compartment. Reticular cells can be converted to fat cells and may be phagocytic.

   2. New blood cells pass through the vascular sinus endothelium into the circulation. Experts in bone marrow structure believe they pass through transient **transcellular pores** in endothelial cell cytoplasm rather than between endothelial cells.

VI. HEMATOPOIESIS

A. Hematopoiesis in human development

   1. Hematopoiesis begins early in embryonic development.
      a. During the second week of gestation, angiogenic cell clusters appear in the yolk sac.
      b. In the sixth week, hematopoiesis begins in the liver.

   2. Hematopoiesis in bone marrow begins during the second month of gestation. From this time on, most hematopoiesis occurs in bone marrow.

B. Hematopoiesis theories

   1. Historically, histologists have disagreed about the origin of blood cells.
      a. Some believe that blood cells are **monophyletic**—that all types of blood cells develop from one type of stem cell.
      b. Others propose that they are **polyphyletic**—that a unique type of stem cell exists for each type of blood cell.

   2. Evidence derived from experiments on irradiated mice appears to support the monophyletic theory. This evidence strongly suggests that a lymphocyte-like cell called a **colony-forming unit–spleen (CFU-S)** has the potential to differentiate, under appropriate conditions, into each type of blood cell.
      a. In these experiments, mice subjected to radiation doses sufficient to kill all bone marrow stem cells died from a total depletion of blood cells.
      b. When unirradiated cells were transplanted from bone marrow of syngeneic mice, the host survived and developed new colonies of hematopoietic tissue in the spleen and bone marrow.
      c. Donor cells that received sublethal doses of radiation before transplantation sustained chromosomal damage that caused characteristic alterations in the karyotype.
         (1) Subsequent karyotype analysis of splenic nodules taken from hosts revealed that colony cells had the same aberrant karyotype, suggesting that they were derived from one type of precursor cell.
         (2) Often, colonies contained megakaryocytes, monocytes, and erythrocyte precursors that had the same chromosome markers. This observation is consistent with the monophyletic hypothesis.

   3. CFU-S cells seem to commit to the formation of specific cell lines long before the characteristics of the differentiated cell lines develop (e.g., hemoglobin for red cells, surface immunoglobulins for lymphocytes, azurophilic granules for granulocytes). The CFU-S cells probably form distinct cell lines for the production of red cells, granulocytes, lymphocytes, and monocytes and macrophages (Figure 11-8).

C. Erythropoiesis

   1. Origin of erythrocytes. Erythrocytes develop from the CFU-S cells in bone marrow.
      a. CFU-S cells commit to erythrocyte formation when they differentiate into **burst-forming unit–erythroid (BFU-E) cells.**

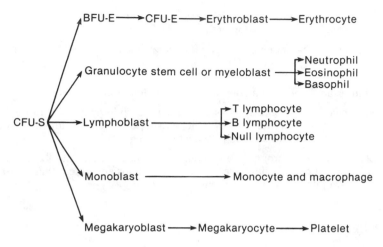

Figure 11-8. Diagram of colony-forming unit–spleen (*CFU-S*) cell differentiation into the different formed elements of the blood. *BFU-E* = burst-forming unit–erythroid; *CFU-E* = colony-forming unit–erythroid.

    b. BFU-E cells differentiate into **colony-forming unit–erythroid (CFU-E) cells,** which are sensitive to **erythropoietin,** a 70-kD glycoprotein produced in the renal juxtaglomerular apparatus.

  2. Erythropoiesis stages
    a. Erythropoietin causes CFU-E cells to differentiate into **erythroblasts**. These cells contain a prominent nucleus with a large nucleolus, many ribosomes and mitochondria, and a well-developed Golgi apparatus.
    b. Ribosomes accumulate in the cytoplasm, increasing erythroblast basophilia due to the staining properties of RNA. **Basophilic erythroblasts** are large (15 μm) free cells that exhibit striking cytoplasmic basophilia.
    c. Cytoplasmic ribosomes bind hemoglobin mRNA and begin synthesizing hemoglobin. Hemoglobin reduces the cell's basophilia, and the cell becomes a **polychromatophilic erythroblast**.
    d. When hemoglobin dominates the cytoplasm, which now stains pink, the cell becomes an **orthochromatic erythroblast** (normoblast).
    e. As hemoglobin accumulates, the nucleus condenses as it undergoes a heterochromatic involution, and the cell shrinks.
    f. Eventually, the cell sheds the nucleus and most mitochondria and polyribosomes to form an erythrocyte that is 7–8 μm in diameter.
    g. When red cells are destroyed, hemoglobin degrades into bilirubins and other materials, which are excreted in bile, and iron, which is transported by the serum glycoprotein **transferrin** to bone marrow, where it is used to synthesize new hemoglobin.

D. Granulopoiesis and agranulopoiesis

  1. Origin of granulocytes. Granulocytes develop from CFU-S cells.
    a. The first recognizable granulocyte precursor is the **myeloblast**. This small cell (10 μm in diameter) has a large euchromatic nucleus with several nucleoli and **no granules** in its basophilic cytoplasm.
    b. The cell becomes a **promyelocyte** when the cytoplasm accumulates a few **azurophilic (nonspecific) granules,** the nucleus accumulates heterochromatin, and slight indentations occur in the nucleus.

  2. Promyelocyte differentiation
    a. Promyelocyte differentiation begins as **specific granules** (neutrophilic, eosinophilic, or basophilic) accumulate in the cytoplasm and the nucleus continues to condense and lobulate. These cells are called **neutrophilic, eosinophilic,** or **basophilic myelocytes,** depending on the type of specific granules present.
    b. A **metamyelocyte** is a cell that has accumulated many specific granules but has yet to complete the process of nuclear condensation and lobulation.

(1) For example, a **neutrophilic metamyelocyte** has a full complement of neutrophilic specific granules and a relatively uncondensed nucleus with a single indentation, whereas a neutrophil has a heterochromatic nucleus with three to five distinct lobes.

(2) It takes about 2 weeks for the committed myeloblast to become a neutrophil.

c. The number of myelocytes and metamyelocytes in bone marrow is about 10 times the number of granulocytes circulating in blood.

3. Agranulocytes

    a. Monopoiesis

        (1) **Monocytes** develop from the same CFU-S cells as granulocytes but have a different developmental process that includes **monoblast** and **promonocyte** stages.

        (2) Definitive monocytes pass briefly through the peripheral blood and become free **macrophages**.

    b. Lymphopoiesis. Lymphocytes develop from **lymphoblasts,** which are derived from CFU-S cells. Lymphoblasts produce three classes of lymphocytes.

        (1) **T lymphocytes (T cells)** begin the differentiation process in the thymus and then seed the thymus, spleen, lymph nodes, and the lamina propria.

        (2) **B lymphocytes (B cells)** arise from lymphoblasts in bone marrow and seed the spleen, lymph nodes, lamina propria, and thymus. Plasma cells differentiate from B cells.

        (3) **Null lymphocytes (null cells)** are a population of cytotoxic lymphocytes that lack traditional T and B cell markers. (See Chapter 12 VII B for more information about null cells and lymphocyte differentiation.)

E. Thrombopoiesis

    1. Megakaryocytes are platelet precursors that develop from **megakaryoblasts** (differentiated CFU-S cells).

        a. Megakaryocytes are enormous cells (diameter of 100 μm or more) and contain a complex, multilobulated, polyploid nucleus. They exist only in bone marrow.

        b. Megakaryocyte cytoplasm has a perinuclear zone typical of cells that synthesize cytoplasmic membranes and granules.

        (1) The cytoplasm contains abundant rough endoplasmic reticulum and a well-developed Golgi apparatus.

        (2) The cell also has a prominent peripheral cytoplasmic component that is rich in granules and filled with a complex network of smooth endoplasmic reticulum.

2. Platelets

    a. In the vascular channels of bone marrow, megakaryocytes slowly release small cytoplasmic fragments called **platelets** into the bloodstream.

    b. Megakaryocyte fragmentation occurs when the cell surface membrane fuses with smooth endoplasmic reticulum membranes.

    c. Platelets remain in the bloodstream for 7–10 days and are continuously replaced by new platelets from bone marrow.

STUDY QUESTIONS

Directions: Each question below contains five suggested answers. Choose the **one best** response to each question.

1. All of the following statements concerning eosinophils are true EXCEPT

(A) peripheral blood contains fewer eosinophils than neutrophils
(B) patients with schistosomiasis have elevated eosinophil counts
(C) peripheral blood contains fewer eosinophils than lymphocytes
(D) peripheral blood contains fewer eosinophils than basophils
(E) eosinophils are capable of phagocytosis

2. All of the following statements concerning platelets are true EXCEPT

(A) they aggregate when exposed to collagen
(B) they are megakaryocyte fragments
(C) they contain a small, dense nucleus
(D) they have a plasma membrane
(E) peripheral blood contains fewer platelets than erythrocytes

Directions: The groups of questions below consist of lettered choices followed by several numbered items. For each numbered item select the **one** lettered choice with which it is **most** closely associated. Each lettered choice may be used once, more than once, or not at all.

Questions 3–7

Match each description of a peripheral blood component with the letter that corresponds to that component in the micrograph below.

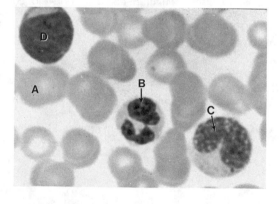

3. Produced in bone marrow and destroyed in the spleen; contains hemoglobin

4. Bacterial phagocyte that contains azurophilic granules

5. Derived from bone marrow precursors; can become a macrophage

6. Granulocyte that can become a phagocyte in the gastrointestinal lamina propria

7. Agranulocyte that can become a phagocyte inside pulmonary alveoli

Questions 8–13

For each description of blood cell production, select the process or processes with which it is most closely associated.

(A) Erythropoiesis
(B) Granulopoiesis
(C) Both
(D) Neither

8. Stimulated by a glycoprotein hormone from the kidneys

9. Active in fetal bone marrow

10. Initiated in the yolk sac during the first month gestation

11. Active in the spleen of a normal adult

12. A reduction in cytoplasmic RNA and extrusion of the nucleus

13. An accumulation of modified lysosomes in the cytoplasm

ANSWERS AND EXPLANATIONS

1. The answer is D. [*III B 1, C 1, 2, 4, D 1*] Eosinophils are motile, phagocytic cells that have two or three nuclear lobes and red or orange granules (eosinophilic granules) in the cytoplasm. In peripheral blood, eosinophils are more numerous than basophils and less numerous than neutrophils and lymphocytes. Eosinophils kill parasite larvae that enter the peripheral blood or the lamina propria of the gut. Consequently, patients with parasitic infections (e.g., schistosomiasis) have elevated levels of eosinophils in their blood.

2. The answer is C. [*IV A, B*] Platelets are fragments of megakaryocytes that are 2–4 μm in diameter. They have a plasma membrane but do not contain a nucleus. Platelets are involved in blood clotting. When stimulated by collagen, platelets adhere to fibrin and each other, forming the basis for clots. Peripheral blood contains far more erythrocytes than platelets. A cubic millimeter of peripheral blood contains $2–4 \times 10^2$ platelets but approximately 5×10^6 erythrocytes.

3–7. The answers are: 3-A, 4-B, 5-C, 6-B, 7-C. [*II A, B, D; III B 3, 4, F*] This is a light micrograph of a peripheral blood film. Red cells (A) contain hemoglobin, an oxygen-binding heme, and account for about 45% of the blood volume. Red cells are produced in bone marrow, exist in the bloodstream for approximately 120 days, and are destroyed in the spleen, bone marrow, and liver.

Neutrophils (B) can become phagocytes in many areas of the body, including lamina propria (the loose irregular connective tissue domain beneath the mucosal epithelium of many visceral organs).

Monocytes (C and D) are the largest leukocytes. They contain many lysosomes and have a prominent Golgi apparatus. Monocytes are the immediate precursors of macrophages. Alveolar macrophages are monocytes that enter the lungs to destroy inspired debris.

8–13. The answers are: 8-A, 9-C, 10-A, 11-D, 12-A, 13-B. [*VI A, C, D 1–2*] Erythropoiesis and granulopoiesis occur in fetal bone marrow; erythropoiesis begins in the yolk sac during the first month of gestation. In the early stages of erythropoiesis, erythropoietin (a glycoprotein hormone produced by the renal juxtaglomerular apparatus) causes CFU-E (colony-forming units–erythroid) cells to differentiate into erythroblasts. As hemoglobin production in the erythroblast increases, cytoplasmic RNA decreases. In the final stages of erythropoiesis, the erythroblast sheds its nucleus and most mitochondria as it becomes a true erythrocyte. Nuclear extrusion does not occur in granulopoiesis.

During granulopoiesis, granulocytes become promyelocytes as they accumulate azurophilic (nonspecific) granules. Promyelocytes become differentiated as they accumulate specific granules (neutrophils, eosinophils, or basophils). These specific granules are modified lysosomes.

Blood cell formation does not occur in the spleen of normal adults.

12

The Immune System

I. INTRODUCTION

A. Immune system characteristics

1. The immune system is an elaborate system that protects the body against invasion by pathogenic organisms, malignant transformation of cells, and inadvertent introduction of foreign substances into the body.

2. Biomedical scientists are making impressive advances in understanding the functional aspects of the immune system. Today, for example, physicians use interferon and interleukin 2 (IL-2) therapeutically to reduce cancer mortality. In the future, physicians probably will have various immunological reagents at their disposal to combat viral disease, malignancy, autoimmune disease, and possibly aging.

B. Immune system components

1. **Immunoglobulins** are antibody molecules in the blood. They have a central role in the immune system (see VII).

2. **Lymphoid tissue**
   a. **Solitary lymphocytes** or small aggregates of lymphocytes are the simplest form of lymphoid tissue and exist in the loose connective tissue of the body.
   b. **Lymphoid nodules** are larger collections of lymphocytes.
   c. **Lymph nodes** are collections of lymphoid nodules connected to the lymphatic circulation by lymphatic vessels and are distributed throughout the body. They are abundant in the neck, inguinal region, and axillary region. Lymph nodes are filters for lymphatic circulation.
   d. **Tonsils** are large collections of lymphoid nodules associated with cryptic infoldings of the pharyngeal stratified squamous epithelium.
   e. **The thymus** is a lobulated gland in the neck and the primary site of T lymphocyte differentiation during the early years of life.
   f. **The spleen** is a reservoir for erythrocytes and a complex filter for the systemic circulation.

C. The complement system

1. The complement system is a group of serum proteins (complement components) that are activated by circumstances such as antigenic stimulation. Once activated, complement components can cause the lysis of foreign cells and the destruction of other invaders.

2. The complement system has an important role in controlling antigen-antibody reactions. Its complex interlocking cascade of serum proteins is activated by antigenic stimulation and can cause direct lysis of many kinds of microorganisms and cells.

3. The complement system also has an important role in the inflammatory response.

II. DIFFUSE LYMPHOID TISSUE

A. **Solitary lymphocytes** are cellular elements of the immune system that exist in almost all compartments of the body. In healthy individuals, they are widely scattered in the connective tissue domains of the body.

1. The **lamina propria** is commonly infiltrated with solitary lymphocytes.
    a. Lymphocytes can leave the lamina propria beneath epithelial layers, cross the epithelial basement membrane, and enter the epithelium.
    b. In the lamina propria and epithelium, lymphocytes can express their immunologic functions.

2. Lymphocytes are present in the lumen of some organs. For example, lymphocytes are common in the vagina.

3. Lymphocytes can enter certain secretion products. For example, they are an abundant cellular component of the milk produced by lactating mammary glands.

B. Lymphoid nodules

1. Lymphocytes in the lamina propria can aggregate into lymphoid nodules ranging in size from several micrometers to many millimeters.

2. Lymphoid nodules are particularly prominent in the lamina propria of the trachea and bronchi and in the gastrointestinal, urinary, and reproductive systems (Figure 12-1).

3. Most lymphoid nodules are served by a rich plexus of lymphatic capillaries; however, they are not encapsulated nor do they have the supply of larger lymphatic vessels characteristic of lymph nodes.

III. LYMPH NODES

A. Histology

1. **Divisions.** Lymph nodes are abundant in the axillary and inguinal regions and in the neck. They are surrounded by a thin capsule and are divided into a **cortex** packed with lymphocytes and a **medulla** with looser, cord-like arrangements of lymphocytes and sinuses (Figure 12-2).

2. **Nodules.** Lymph nodes contain nodules of dense aggregates of lymphocytes.
    a. **Primary nodules** are homogeneous collections of small, densely packed lymphocytes. Immature B cells are most commonly present in primary nodules.
    b. **Secondary nodules** (follicles) have peripheral dense regions of small lymphocytes and **germinal centers** containing larger lymphocytes and some macrophages. Differentiating B cells, macrophages, and T cells are present in germinal centers.

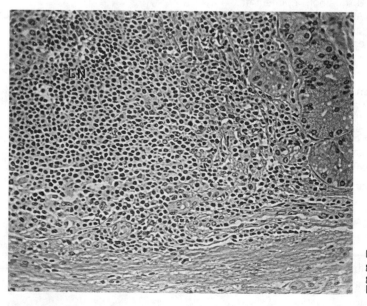

Figure 12-1. Light micrograph of the gastric mucosa, showing a small aggregation of lymphocytes forming a lymphoid nodule (*LN*).

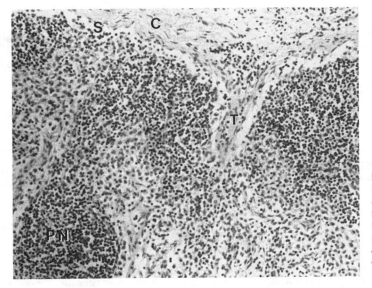

Figure 12-2. Light micrograph of a lymph node, showing the capsule (*C*) and an associated trabecula (*T*). The subcapsular sinus (*S*) and a primary nodule (*PN*) also are visible. (Reprinted from Johnson KE: *Histology: Microscopic Anatomy and Fmbryology.* New York, John Wiley, 1982, p 132.)

3. Internodular cortical lymphocytes are present between the cortical nodules, and the **tertiary cortex** is situated between the cortical nodules and the medulla. Tertiary cortex lymphocytes are primarily T cells.

4. A mesh of extracellular **reticular fibers** supports lymph node cells.

B. Functions

1. Lymph nodes filter particulate material and soluble antigens. Lymph percolates through the tortuous sinuses slowly, maximizing the time for phagocytic activity and immune surveillance.

2. Lymph nodes are sites for the clonal expansion of specific populations of B cells and T cells.

3. During immune response, macrophage-lymphocyte interaction is intense in lymph nodes.

C. Lymphatic circulation

1. **Lymph** is blood filtrate produced in capillaries.
   a. It contains some of the proteins of whole blood but not all of the cells. Also, it contains monocytes and an abundance of lymphocytes.
   b. As lymph is expressed from capillaries, it percolates through tissues and enters lymphatic capillaries, which are widely distributed in the body.

2. **Lymphatic capillaries**
   a. Lymphatic capillaries may be as large as 100 μm in diameter and are composed of flattened endothelial cells joined at their margins by adhesive junctions. Often, adjacent edges of endothelial cells overlap.
   b. Lymphatic capillaries have a sparse basement membrane, an irregular luminal border, many mitochondria, a well-developed Golgi apparatus, and numerous pinocytotic vesicles in their cytoplasm.
   c. Most lymphatic capillaries are continuous capillaries, although some are fenestrated capillaries (e.g., capillaries in the lacteals of intestinal villi). They usually have a larger diameter than cardiovascular system capillaries.
   d. Lymphatic capillaries are considerably more leaky than most continuous capillaries. Macromolecules, bacteria, and lymphocytes from surrounding connective tissue pass across lymphatic capillaries (in both directions) with less restriction than typically occurs in continuous capillaries.

3. **Larger lymphatic vessels**
   a. Lymphatic capillaries join larger **collecting vessels,** which share many of the structural characteristics of veins (e.g., valves).
      (1) Collecting vessel walls are composed of poorly demarcated layers.

(2) The layering is less distinct than the layering in the walls of veins, and lymphatic vessel walls are thinner than the walls of veins of comparable size.

b. Collecting vessels drain the lymphatic capillaries, pass through lymph nodes, and ultimately drain into the largest lymphatic vessel in the body: the **thoracic duct**.

  (1) The thoracic duct drains the lower limbs, most abdominal and thoracic viscera, and the left side of the head and neck.

  (2) It courses through the diaphragm and mediastinum and usually enters the circulatory system at the junction of the left subclavian vein and the left internal jugular vein.

c. Another large lymphatic vessel drains the right side of the head and neck, right upper limb, right side of the heart, right lung, and the right side of the liver. This vessel enters the circulatory system near the junction of the right subclavian vein and the right internal jugular veins.

d. The thoracic duct and other large lymphatic vessels have an indistinct layered structure that usually includes some elastic tissue in addition to the mural tissues present in all collecting lymphatics.

4. Lymphatic circulation in lymph nodes

  a. Lymph drains into the convex surface of the node through **afferent lymphatic vessels,** which communicate with **subcapsular (cortical) sinuses** (Figure 12-3).

  b. Cortical sinuses run between cortical nodules, through the tertiary cortex, and connect to **medullary sinuses,** which follow trabeculae to the central part of the node.

  c. Medullary sinuses in the central part of the node are formed from an anastomosing network of channels that drain into hilar **efferent lymphatic vessels**.

  d. Blood vessels enter and leave lymph nodes at the **hilus**. These blood vessels are unremarkable except for postcapillary venules in and around the tertiary cortex. These postcapillary venules have cuboidal or columnar endothelial cells that pass T cells and B cells **between their lateral boundaries**.

IV. TONSILS AND PEYER'S PATCHES

A. Tonsils are lymphoid organs near the pharynx. They consist of an invaginated epithelium, which forms crypts, and a thick lamina propria infiltrated with lymphocytes often arranged in primary and secondary nodules.

  1. The **pharyngeal tonsil** (adenoid tonsil) is located in the pseudostratified ciliated columnar epithelium of the nasopharynx.

  2. The **palatine tonsils** are located in the stratified squamous epithelium of the oropharynx (Figure 12-4).

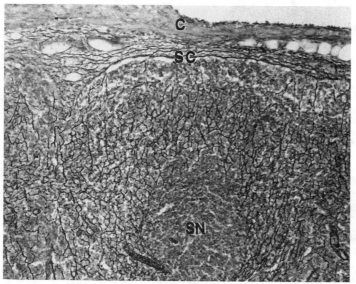

Figure 12-3. Light micrograph of a lymph node. Staining makes the reticular fibers appear black. The center of a large secondary nodule (*SN*) is visible with the capsule (*C*) and subcapsular sinuses (*SC*). (Reprinted from Johnson KE: *Histology: Microscopic Anatomy and Embryology.* New York, John Wiley, 1982, p 133.)

Figure 12-4. Light micrograph of a portion of the palatine tonsil, showing large expanses of lymphoid infiltration, a primary nodule (*PN*), and the stratified squamous epithelium (*SSE*) typical of the palatine tonsil.

3. The **lingual tonsils** are located in the stratified squamous epithelium covering the tongue.

B. Peyer's patches are lymphoid aggregates in the gastrointestinal tract.

1. Peyer's patches can be found anywhere in the wall of the gastrointestinal tract but are particularly prominent in the wall of the **ileum,** the small intestinal segment that enters the colon. Here, they are macroscopically visible.

2. Structures similar to Peyer's patches are prominent in the **vermiform appendix**.

3. All gut-associated lymphoid tissue **(GALT)** consists of many lymphocytes surrounding primary and secondary nodules (Figure 12-5).

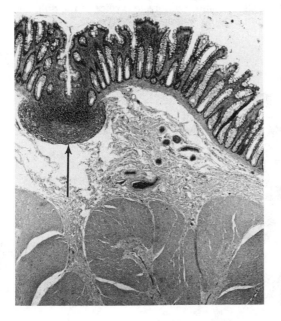

Figure 12-5. Light micrograph of a lymphoid nodule (*arrow*) called a Peyer's patch in the colon. Peyer's patches also are abundant in the ileum of the small intestine.

V. THE SPLEEN

A. Gross appearance

1. The spleen is a visceral organ lying between the gastric fundus and the diaphragm. It usually has the appearance of a curved wedge of tissue with its convex surface lying against the diaphragm.

2. The spleen is 12 cm long, 7 cm wide, and 4 cm thick and weighs 100–150 g in a healthy adult.

B. Histology

1. The spleen is encapsulated in a dense irregular **connective tissue capsule** covered by mesothelium. Connective tissue **trabeculae** project from the capsule and form a complex network of connective tissue elements in the parenchyma of the organ.

2. **Red and white splenic pulp** occupies the compartments between trabeculae. **Red pulp** is rich in erythrocytes and tortuous vascular channels. **White pulp** is composed of collections of lymphocytes similar to the primary and secondary nodules of lymph nodes.
   a. The **periarterial lymphatic sheaths (PALS)** are dense aggregates of small lymphocytes in cylindrical clusters around the **central arteries** in white pulp (Figure 12-6).
   b. In addition, **secondary nodules** with well-formed **germinal centers** contribute to the white pulp.
   c. The PALS and marginal zone around secondary follicles contain T cells; the secondary nodules are rich in B cells.

C. Splenic vasculature

1. The **splenic arteries** enter the spleen at the hilus and immediately branch into **trabecular arteries**.

2. **Central arteries,** the vessels that pass through the PALS, branch from trabecular arteries.

3. **Follicular arteries** branch from central arteries, supply secondary nodules in the white pulp, and carry blood to smaller arterioles.
   a. Some of these arterioles empty into a reticular mesh of irregularly arranged sinuses called the **open splenic circulation**.
   b. Other arterioles empty into splenic sinuses that are lined by elongated cells shaped like barrel staves (the so-called **closed splenic sinuses**).

4. Blood from the open reticular mesh and the splenic sinuses drains into **veins of the pulp** and the **trabecular veins**. Blood leaves the spleen through the **splenic veins**.

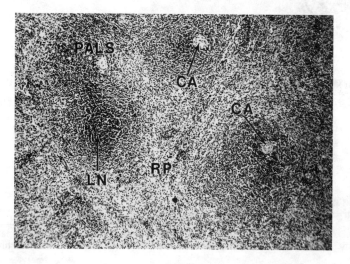

Figure 12-6. Light micrograph of the spleen, showing central arteries (*CA*), the periarterial lymphatic sheath (*PALS*), and a lymphoid nodule (*LN*) of the white pulp. Red pulp (*RP*) surrounds the white pulp.

D. The splenic reticulum. Reticular cells and reticular fibers similar to those in lymph nodes exist throughout much of the parenchyma of the spleen. These cells and fibers form a scaffolding that supports both red and white pulp. Reticular cells and fibers around central arteries are arranged in concentric layers (as opposed to a sponge mesh) to support the PALS.

E. Splenic sinuses and cords. Red pulp contains a complex mixture of **splenic sinuses** and a reticular mesh of **splenic cords**.

    1. Splenic sinuses are peculiar vascular channels 40 μm in diameter, which are lined by tapered cells elongated **parallel** to the long axis of the sinus (Figure 12-7).

        a. These cells contain many contractile microfilaments and lack attachment specializations between them.

        b. The microfilaments probably contract to open gaps between cells, just as pressing down on the top of a wooden barrel creates gaps between the staves.

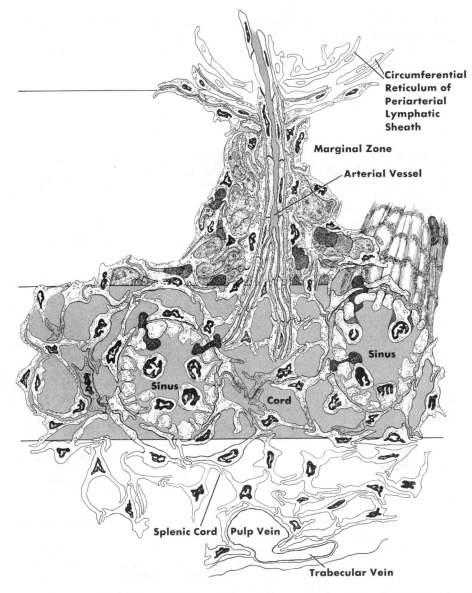

Figure 12-7. Illustration of the splenic circulation and the arrangement of sinusoidal cells. (Reprinted from Weiss L: *The Cells and Tissues of the Immune System.* Englewood Cliffs, Prentice-Hall, 1972, p 60.)

2. **Splenic cords.** The splenic sinuses are surrounded by a fenestrated basement membrane and by adventitial reticular cells that comprise part of the splenic cords. Blood cells in the splenic cords pass through the wall of a splenic sinus between the lining cells, into the lumen of the splenic sinus, and then out of the spleen through the splenic veins.

F. Functions

1. The spleen is a complex **filter**. In the cords and sinuses, aged and damaged erythrocytes are examined, removed from the circulation, and destroyed.
   a. As erythrocytes age, they become less plastic due, in part, to age-dependent changes in the chemical composition of the plasma membrane.
   b. Young, healthy cells can make their way through the phagocyte-laden cords into sinuses and out of the spleen. Aging cells are retarded in the cords and eventually are engulfed and destroyed.
   c. Splenic macrophages and phagocytic reticular cells also clear particulate material from the blood.

2. The white pulp has an **immune surveillance** function, much like the function of lymph nodes. After entering the white pulp, B cells concentrate in secondary nodules and T cells concentrate in the PALS. Thus, the spleen traps and processes antigens in blood, and lymph nodes trap and process antigens in lymph.

VI. THE THYMUS

A. Gross appearance

1. The thymus is an irregularly shaped organ located in the mediastinum near the great vessels of the heart. The **size of the thymus varies considerably throughout life**.
   a. At birth, it weighs about 15 g. It continues to grow during childhood and puberty until it weighs 40 g.
   b. After puberty, the thymus undergoes a gradual regression and fatty degeneration. By age 40, it weighs about 10 g.

2. The thymus has **several large irregular lobes** that are divided into smaller (1 mm in diameter) lobules.

3. A dense irregular connective tissue capsule covers the entire gland, and trabecular branches of the capsule surround each lobule.

B. Histology

1. Each lobule has a **cortex** rich in lymphocytes and a **medulla** rich in reticular epithelial cells (Figure 12-8).
   a. Reticular epithelial cells in the medulla form a cellular reticular network similar to the networks formed by reticular cells in lymph nodes, bone marrow, and the spleen; however, the reticular network of the thymic medulla lacks the numerous reticular fibers present in the networks of these other organs.
   b. Instead, the tips of reticular epithelial cells in the thymus are joined to each other by **desmosomes,** forming a mesh that is infiltrated by numerous lymphocytes and macrophages.

2. The medulla contains concentric whorls of degenerating epithelial cells, which are called **Hassall's corpuscles** or thymic corpuscles (Figure 12-9).

C. Functions

1. The thymus is a key organ in the immune system because it is here that lymphocytes derived from bone marrow proliferate and differentiate into T cells.
   a. Cells destined to become T cells are bone marrow stem cells.
   b. Stem cells from the bone marrow mature into T cells and undergo clonal expansion in the thymus, and then seed distant sites in lymph nodes, the spleen, and elsewhere in the body.

2. Lymphocyte differentiation is mediated by a **lymphokine** called **thymosin,** which is secreted by reticular epithelial cells.

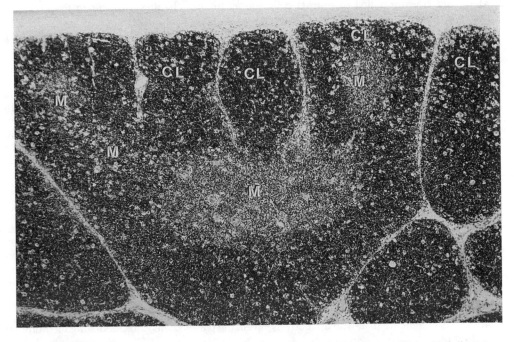

Figure 12-8. Light micrograph of the thymus, showing cortical lobules (*CL*) and the medulla (*M*).

VII. BASIC CELLULAR IMMUNOLOGY

A. The immune response

1. **Foreign proteins or polysaccharides** (antigens) that enter the circulatory system evoke an immune response in the host organism.
   a. A **humoral immune response** is the synthesis of large quantities of antibodies (immunoglobulins) specifically directed against the foreign antigen.
   b. A **cellular immune response** is the differentiation of cytotoxic cells specifically for the foreign antigens.

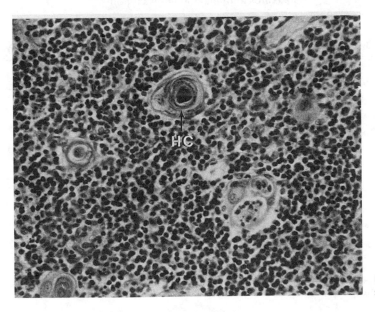

Figure 12-9. Light micrograph of Hassal's corpuscles (*HC*) in the thymus.

2. **Malignant transformation of normal cells** causes striking changes to occur in the chemical characteristics of the cell surface.
   a. The immune system recognizes these tumor-specific antigens as foreign and initiates a cellular immune response, which may destroy the malignant cells.
   b. If the system fails to combat the malignant cells, the tumor will continue to grow and may spread to secondary sites (metastasis).

3. **Histocompatibility antigens** are glycoproteins that exist on the surface of all cells. The structure of these antigens is determined by a complex family of genes called the **major histocompatibility complex (MHC)**.
   a. Parents and their children are genetically heterogenous with respect to the MHC. Thus, tissues grafted from one family member to another are almost as likely to be rejected as grafts between unrelated individuals.
   b. Drugs such as **cyclosporine** suppress the cellular immune response and have facilitated successful transplantation of hearts, livers, kidneys, and other organs between unrelated people.

B. **Lymphocytes.** Although lymphocytes are morphologically uncomplicated, these free cellular elements of the immune system are functionally heterogenous. Studies in cellular immunology have revealed three broad functional classes of lymphocytes: **B lymphocytes** (B cells), **T lymphocytes** (T cells), and **null cells**. These lymphocytes are derived from undifferentiated bone marrow precursors (the same CFU-S cells that form lymphoblasts). Table 12-1 summarizes the types of lymphocytes, their characteristics, and their functions.

1. **B cell characteristics and distribution**
   a. B cells can differentiate into antibody-synthesizing plasma cells, thus mediating the humoral immune response.
   b. B cells leave the bone marrow, diffuse throughout the body, and then differentiate due to poorly understood inductive influences.
      (1) B cells are so-named because scientists observed that, in birds, CFU-S cells differentiate into B cells in an organ called the **bursa of Fabricius**.
      (2) No structure equivalent to the bursa of Fabricius has been identified in humans; however, the name remains.
   c. B cells comprise about 30% of the circulating lymphocytes and are primarily responsible for humoral immunity, which is the production of specific serum immunoglobulins for various environmental antigens (see VIII). B cells produce antibodies in two distinct ways.
      (1) In the first and most simple case, antigens bind to the B cell surface, probably due to immunoglobulin M (IgM) and immunoglobulin D (IgD) on the lymphocyte surface. Then, B cells undergo clonal proliferation and a terminal differentiation into **plasma cells,** which are highly specialized for immunoglobulin secretion.

Table 12-1. Lymphocyte Subpopulations

Cell Type	Distinctive Characteristics	Functional Role in Immune System
B cells	High levels of surface immunoglobulins, form EAC rosettes	Differentiate into plasma cells, which secrete immunoglobulins and memory B cells
T cells	Low levels of surface immunoglobulins, form E rosettes	Cell-mediated immune response, second line of defense
Helper T cells	. . .	Stimulate B cell differentiation
Suppressor T cells	. . .	Suppress B cell differentiation
Killer T cells	. . .	Lyse foreign cells
Natural killer (NK) cells	Large lymphocytes with azurophilic cytoplasmic granules	Cell-mediated immune response, first line of defense
Null cells	Lack B and T cell markers, no surface immunoglobulins and no rosette formation	Reserve population of undifferentiated lymphocytes (?)

EAC = antibody-and-complement coated erythrocytes; E = erythrocytes.

(2) In the second, more complex case, antigens are bound to the surface of **helper T cells**. The antigens are released by the helper T cells and become bound to **macrophages**. Finally, the B cells interact specifically with stimulated macrophages and, subsequently, undergo clonal proliferation and plasma cell differentiation.

d. Certain B cells persist in the form of **memory B cells** after initial exposure to an antigen. These cells are capable of rapid clonal expansion if exposure to the antigen occurs again, even years after the initial exposure.

e. B cells are most heavily concentrated in the germinal centers of lymphoid tissue aggregates, such as those in lymphoid nodules and nodes and in the spleen. For example, B cells comprise approximately 50% of the lymphocytes in the spleen. B cells are concentrated in the cortical aggregates of lymph nodes and the primary and secondary nodules of splenic white pulp.

f. Approximately 75% of the lymphocytes in bone marrow are B cells.

2. B cell differentiation. Plasma cells are common cellular elements of connective tissues in the body, especially connective tissues in the lamina propria of visceral organs, where they synthesize antibodies.

a. B cells stimulated by an antigen begin to proliferate extensively to produce clones of cells committed to plasma cell differentiation. This process is called the **blast transformation**.

b. As B cells differentiate into plasma cells, their cytoplasm undergoes specific changes that prepare the cell for **glycoprotein synthesis** (immunoglobulins are glycosylated proteins).

(1) Cytoplasmic basophilia increases dramatically as B cells become plasma cells due to a sudden increase in the amount of rough endoplasmic reticulum.

(2) Immunoglobulin genes are transcribed, and mRNA for immunoglobulin G (IgG) synthesis leaves the nucleus through nuclear pores in the nuclear envelope and binds to ribosomes of the rough endoplasmic reticulum.

(3) Next, IgG mRNA is translated and newly synthesized peptide H and L chains enter the rough endoplasmic reticulum cisternae.

(4) Membrane-bound vesicles containing incomplete IgG bud from the rough endoplasmic reticulum and move to the Golgi membranes, which recognize unglycosylated IgG chains and begin adding specific sequences of carbohydrate residues to the IgG.

(5) Glycosylated IgG buds from the Golgi apparatus in membrane-delimited secretion vacuoles, which migrate to and fuse with the cell surface. Then, IgG is released from the cell and enters the blood plasma.

(6) In plasma, IgG circulates freely throughout the body and readily enters connective tissues where appropriate.

c. Specific antigens promote the production of specific antibodies, which are catabolized after the antigenic stimulus is destroyed.

d. Specific memory B cells remain in the body for years and can undergo clonal expansion and differentiation into a new specific population of plasma cells whenever the body is exposed to the antigen that evoked the original humoral immune response.

3. T cell characteristics and distribution

a. T cells can differentiate into cytotoxic cells, thus mediating the cellular immune response.

b. T cells comprise about 70% of circulating lymphocytes and are responsible for the complex phenomenon of cellular immunity.

c. T cells are the predominant type of lymphocyte in the deep tertiary cortex of lymph nodes and are present in the PALS in the spleen.

d. Unlike B cells, T cells have small amounts of surface immunoglobulins; however, they can bind antigens to their surface, possibly because their cell surface contains products of **immune response (Ir) genes,** which can bind antigens specifically.

4. T cell functions

a. Lymphokine production. During antigenic stimulation, T cells undergo clonal proliferation and then differentiation to produce **lymphokines** [e.g., **macrophage inhibitory factor (MIF)**], chemotactic factors for other formed elements of peripheral blood, and other nonspecific cytotoxic substances.

b. Killer T cells are T cells that can contact and kill specific abnormal cells. They have the primary function of mediating the lysis of foreign or abnormal cells within the body. They accomplish this task using two basically distinct mechanisms.

(1) In the first mechanism, cell-bound histocompatibility antigens on T cell target cells activate small lymphocytes, causing them to proliferate and then differentiate into a population of specific cytolytic T cells.

(2) Killer T cells have **Fc receptors** for the Fc fragment of IgG antibodies bound to the surface of the target cells. In the second mechanism, killer T cells recognize the surface antibodies of the target cell and subsequently cause their lysis.

c. **Helper T cells and suppressor T cells** modulate the humoral immune response of B cells and the production of immunoglobulins by plasma cells.

    (1) Helper T cells and macrophages are required to stimulate B cell differentiation and plasma cell function.

    (2) Suppressor T cells provide an antagonistic effect that inhibits B cell differentiation and plasma cell function.

d. **Nature of T cell response.** In effect, many T cell functions are stimulated by specific stimuli; however, the response of the T cell per se is localized and relatively nonspecific.

    (1) For example, when stimulated, certain T cells secrete MIF, which has a strong **local** (as opposed to systemic) effect.

    (2) Macrophages in the immediate vicinity of T cells secreting MIF are immobilized and remain stationary for extended periods of time and, thus, have a greater opportunity to engulf and destroy the stimulus that evoked MIF production.

5. **Natural killer (NK) cells** are a recently discovered population of cells that mediate the lysis of foreign cells.

    a. NK cells are larger (12–15 μm diameter) than small lymphocytes (7–9 μm diameter) and are closer to the size of a neutrophil.

    b. NK cells have a lobulated nucleus, a substantial amount of granular cytoplasm, and azurophilic granules in their cytoplasm.

    c. NK cells are the first line of defense against foreign cells. They are differentiated, cytolytic entities that can act immediately against foreign cells, destroying them rapidly.

        (1) The proliferation and differentiation of killer T cells from small lymphocytes requires some time. During this time, an abnormal tumor cell might proliferate faster than the killer T cells designed to combat it.

        (2) As a tumor develops, NK cells probably attack it and contain the development of tumor cells, while cytotoxic and cytolytic killer T cells are in the process of developing into a second line of defense against the malignant tumor cells.

    d. NK cells also are efficient cell killers because they can kill more than one cell. An NK cell that has lysed one target cell can leave that cell and proceed to a new target cell.

6. **Null cells** are lymphocytes that lack B cell and T cell characteristics but can differentiate into lymphocyte killer cells and can become hematopoietic stem cells. Null cells probably are an undifferentiated lymphocytic population whose functional significance currently is not well understood.

VIII. IMMUNOGLOBULIN MOLECULAR BIOLOGY

A. General characteristics

1. Antibodies, or **immunoglobulins,** are specific protein molecules. A high concentration of immunoglobulins circulate freely in blood and lymph. They also are bound to the surface of lymphocytes.

2. The **five broad classes of immunoglobulins** are designated by the capital letters A, D, E, G, and M and are abbreviated as IgA, IgD, IgE, IgG, and IgM. (Table 12-2 summarizes the characteristics of the five immunoglobulin subclasses.)

B. Immunoglobulin G (IgG) is the most prevalent antibody in blood and lymph.

1. IgG is composed of four polypeptide chains linked together in an -S-S- group. This complex is a glycoprotein with a molecular weight of 150 kD.

2. When IgG is digested with a proteolytic enzyme such as papain, it splits into three macromolecules: two **Fab fragments** (antigen binding fragments) and one **Fc fragment** (crystallizable fragment). The Fc portion of the molecule is the segment that is recognized by a cell's Fc receptors.

Table 12-2. Immunoglobulin Subclasses

Subclass	Percentage of Total Serum Ig	Molecular Weight	Function
IgG	75%	150-kD monomer	Circulating antibodies
IgA			
Serum	15%	150-kD monomer or 315-kD dimer with J polypeptide	Antibodies of bodily secretions, colostrum
Secretions	NA	400-kD dimer with J polypeptide and secretory component	. . .
IgM	10%	900-kD pentamer with J polypeptide	Activates complement cascade, B cell surface markers
IgD	Trace	180-kD monomer	B cell marker, mediates B cell differentiation (?)
IgE	Trace	190-kD monomer	Mast cell activation in allergic reactions

Ig = immunoglobulin; NA = not applicable.

3. IgGs comprise about 75% of all immunoglobulins in normal adult serum and are the "workhorses" of the immune system.
   a. IgG complexes with specific circulating antigens to activate the complement system and cause opsonization.
   b. IgG binds to the Fc receptor of macrophages and activates their cytotoxic functions.

C. Immunoglobulin A (IgA)

1. IgAs are present in serum as monomers or dimers linked by the J polypeptide and constitute about 15% of serum immunoglobulins.

2. IgAs are the dominant antibodies in glandular secretions. Sweat, tears, mucus, vaginal and prostatic secretions, and milk contain high concentrations of IgA.
   a. Lymphocytes in the connective tissue or lamina propria of these glands synthesize IgA dimers. The dimers are bound to glandular epithelial cells by a specific surface protein called the **secretory component**. Then, IgA dimers, with the secretory component attached, are transported through the cell.
   b. The secretory component binds to the domain of the IgA dimer that is sensitive to proteolysis, thus forming a **secretory IgA (sIgA)** molecule.
   c. In the gastrointestinal tract, the secretory component resists destruction of the sIgA molecule by enteric proteases, thereby increasing the half-life of this antibody.

3. sIgAs are the first line of defense in the immune system; they probably prevent potentially pathogenic microorganisms from entering the body.

D. Immunoglobulin M (IgM)

1. IgMs comprise about 10% of the serum immunoglobulins and exist as 900-kD pentamers in blood.

2. IgM is a minor component of the B cell surface and serves as a marker for this lymphocyte subclass.

3. IgM also is a very potent activator of the complement cascade.

E. Immunoglobulin D (IgD) is a monomeric molecule. It is the predominant immunoglobulin on the B cell surface, and trace amounts are present in serum. Its exact function is not known, but recent evidence suggests that IgDs may modulate B cell differentiation.

F. Immunoglobulin E (IgE)

1. IgE is a monomeric molecule present in trace amounts in serum. It binds with high affinity to specific Fc receptors on the surface of mast cells.

2. IgE synthesis is evoked by exposure to certain antigens called **allergens** (e.g., ragweed pollen, bee venom). These IgEs then bind to mast cells.

3. When the host is exposed to the allergen a second time, IgEs bound to mast cells recognize the allergen and immediately degranulate, either locally (producing asthmatic attacks, swelling, and rashes) or systemically (producing **anaphylactic shock**).

STUDY QUESTIONS

Directions: Each question below contains five suggested answers. Choose the **one best** response to each question.

1. Each of the following statements concerning the spleen is true EXCEPT

(A) its connective tissue capsule has a mesothelial coating
(B) its sinuses are continuous capillaries
(C) the white pulp contains T cells and B cells
(D) T cells are concentrated in the PALS
(E) B cells are concentrated in lymphoid nodules

2. Each of the following statements concerning the palatine tonsils is true EXCEPT

(A) they lack secondary nodules
(B) they contain primary nodules
(C) they contain numerous B cells
(D) they are associated with an epithelium different from the epithelium of the pharyngeal tonsils
(E) they are associated with stratified squamous epithelium

3. Each of the following statements concerning lymph nodes is true EXCEPT

(A) T cells are present in secondary nodules
(B) T cells predominate in the deep tertiary cortex
(C) secondary nodules lack macrophages
(D) reticular fibers surround subcapsular sinuses
(E) secondary nodules contain reticular fibers

4. Each of the following statements concerning B cells is true EXCEPT

(A) they can differentiate into plasma cells
(B) they must interact with macrophages before they produce antibody
(C) they are derived from bone marrow
(D) they are abundant in secondary nodules
(E) they are less common than T cells in the PALS in the spleen

Directions: The groups of questions below consist of lettered choices followed by several numbered items. For each numbered item select the **one** lettered choice with which it is **most** closely associated. Each lettered choice may be used once, more than once, or not at all.

Questions 5–9

Match each characteristic below with most appropriate immunoglobulin subclass.

(A) IgG
(B) IgA
(C) IgM
(D) IgD
(E) IgE

5. A prominent antibacterial component of breast milk and sweat that is synthesized by plasma cells

6. A pentameric immunoglobulin and a prominent component of the B cell surface

7. Predominant immunoglobulin subclass in plasma; comprises the bulk of circulating antibodies

8. Involved in mast cell function

9. In its dimeric form, it is associated with a J polypeptide and a secretory component that protects it from enteric proteolysis

Questions 10–14

Match each description below with the appropriate lettered site or structure in the following micrograph.

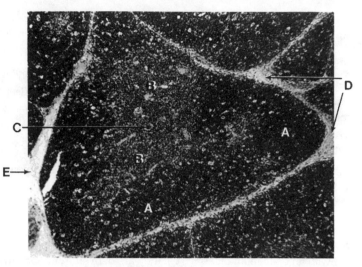

Reprinted from Johnson KE: *Histology: Microscopic Anatomy and Embryology.* New York, John Wiley, 1982, p 137.

10. Cluster of degenerated thymic reticular epithelial cells

11. Area rich in reticular epithelial cells

12. Area rich in T cells

13. Area rich in cells derived from bone marrow

14. Area rich in cells derived from pharyngeal pouches and joined by desmosomes

ANSWERS AND EXPLANATIONS

1. The answer is B. [*V A–C, F*] The spleen is a peritoneal visceral organ. It has a dense connective tissue capsule, which is surrounded by mesothelial cells of the gastric mesentery. The white pulp contains B cells in primary and secondary nodules and T cells in the periarterial lymphatic sheaths (PALS). The splenic sinuses in the red pulp are highly modified discontinuous capillaries that allow red blood cells to pass freely from the circulation into the red pulp.

2. The answer is A. [*IV A*] The palatine tonsils are cryptic structures in the oral cavity, which are associated with a stratified squamous epithelium derived from the endoderm. The pharyngeal tonsil is associated with pseudostratified columnar epithelium. All tonsils contain primary and secondary lymphoid nodules, help control the bacterial flora in the pharynx, and are rich in B cells.

3. The answer is C. [*III A 2 b*] Lymph nodes are filters for antigens in lymph. They contain B cells in primary and secondary nodules and T cells in the deep tertiary cortex. Secondary nodules contain germinal centers, which are sites where B cells are activated and differentiated into immunoglobulin-secreting plasma cells. Thus, secondary nodules contain B cells, helper T cells, suppressor T cells, and macrophages. Reticular fibers exist throughout lymph nodes, including around the subcapsular sinuses and in nodules.

4. The answer is B. [*VII B 1, 2*] B cells are derived from bone marrow stem cells and commonly differentiate into antibody-synthesizing plasma cells after interacting with helper T cells and macrophages. However, memory B cells can differentiate into plasma cells without interacting with macrophages. B cells are abundant in secondary lymph nodules but are outnumbered by T cells in the PALS.

5–9. The answers are: 5-B, 6-C, 7-A, 8-E, 9-B. [*VIII B–F; Table 12-2*] Immunoglobulins are antibody molecules secreted by plasma cells. IgG is the predominant antibody subclass in blood and lymph. IgA exists as a dimer in bodily secretions such as milk and sweat. It contains a J polypeptide, which links monomers, and a secretory component, which helps epithelial cells in the glandular epithelium recognize dimeric IgA and renders the IgA less susceptible to proteolytic degradation. IgM is a pentameric immunoglobulin and a prominent B cell surface marker. IgD is present on the surface of many developing lymphocytes. IgE is present on the surface of mast cells and basophils and is involved in allergic reactions.

10–14. The answers are: 10-C, 11-B, 12-A, 13-A, 14-B. [*VI B, C*] This is a micrograph of a lobule of the thymus. Hassall's corpuscles (*C*) are clusters of degenerated thymic reticular epithelial cells. The medulla (*B*) of thymic lobules is rich in reticular epithelial cells, which are modified epithelial cells derived from the third and fourth pharyngeal pouches and joined by desmosomes. T cells are abundant in the cortex (*A*) of thymic lobules. T cells are derived from bone marrow and then differentiate in the thymus. In the thymus, the boundaries of lobules (*D* and *E*) usually are quite evident.

13
Muscular Tissue

I. MICROFILAMENT-BASED MOTILITY

A. General features. Motility in eukaryotic systems requires the cytoplasmic accumulation of the motility proteins **actin** and **myosin**. All motile cells in the body contain substantial amounts of these proteins. Specialized cells that move rapidly, repeatedly, or forcefully have an abundance of actin and myosin arranged in specialized contractile arrays.

B. Molecular biology of motility proteins

1. Actin, a 42-kD molecular weight protein widely distributed in nature and prevalent in muscular tissue, forms the microfilaments that are observed frequently in the cortex of many cell types.

    a. Actin exists as a globular protein called **G-actin** (about 5.5 nm in diameter), which can polymerize into filaments called **F-actin**.

    b. F-actin is composed of two beaded chains of G-actin subunits wound together in a loose spiral with a 36-nm periodicity. **Thin filaments** in muscle tissue are long strands of F-actin helices and their associated regulatory proteins tropomyosin, troponin, and α-actinin (Figure 13-1).

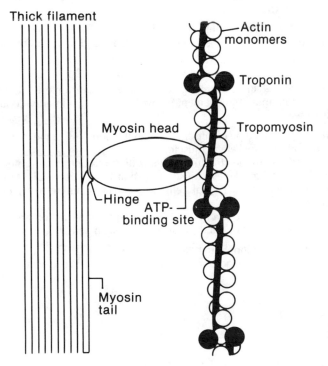

Figure 13-1. Diagram showing how myosin and the thin filaments of muscle tissue interact. (Reprinted from Thorpe NO: *Cell Biology*. New York, John Wiley, 1984, p 615.)

(1) **Tropomyosin** molecules are composed of two intertwined α-helices and are about 40 nm long. They lie end-to-end in the groove of the actin helix and cover the myosin binding site on the thin filament when intracellular calcium is low. Each tropomyosin molecule has a troponin complex bound to one end, and it covers seven G-actin subunits.

(2) **Troponin** is a complex of three globular subunits. Troponin complexes are regularly spaced along the thin filament. **Troponin T (TnT),** the first subunit, has a tropomyosin binding site. **Troponin C (TnC),** the second subunit, has a calcium ion binding site. **Troponin I (TnI),** the third subunit, has a site that inhibits actin-myosin interaction.

(3) α-**Actinin** binds to the ends of thin filaments. It exists where thin filaments insert into the Z lines of sarcomeres (see III C) and may help bind microfilaments to the plasma membrane in moving cells.

2. **Myosin** is a 470-kD molecular weight polypeptide that is widely distributed and not restricted to muscle tissue. Proteolytic cleavage divides myosin into **light meromyosin (LMM)** and **heavy meromyosin (HMM)** fragments. HMM is the head of the molecule and LMM is the long tail.

   a. **Thick filaments** in muscle fibers are groups of myosin molecules. Strong lateral interaction between the relatively straight LMM chains forms rod-like, multimolecular aggregates.

   b. HMM has an **S-1 fragment** and an S-2 fragment. S-1 fragments project radially from the axis of the myosin aggregate, forming cross-bridges between thick and thin filaments (see III D 2).

      (1) When intracellular calcium levels are low, S-1 fragments cannot bind to the S-1 binding site in the actin of thin filaments.

      (2) When calcium levels are high, calcium binds to TnC, causing a conformational change in tropomyosin that exposes the S-1 binding site, thus allowing actin-myosin interaction.

C. **Contractile force** is generated by the interaction of actin and myosin, which produces relative movement between these proteins (Figure 13-2).

   1. Actin and myosin are linked to organelles, the cytoskeleton, and the cell surface. Consequently, relative movement of these proteins generates weak contractile force at the expense of **adenosine triphosphate (ATP).**

   2. Although meager at the cellular level, contractile forces are substantial enough to promote intracellular movement of organelles and cellular locomotion. In large aggregates of highly specialized motile tissues, the combined contractile activities of many cells can generate impressive forces.

   3. Cells that are capable of autonomous motility, but not specialized for it, move much more slowly than skeletal muscle cells, which are specialized for rapid and forceful contraction.

D. **Types of muscular tissue**

   1. **Smooth muscle cells** are distributed throughout the cardiovascular, gastrointestinal, urogenital, and respiratory systems. Smooth muscle movement is involuntary and primarily controlled by the autonomic nervous system.

   2. **Skeletal muscle** is the most abundant type of muscle tissue in the body. All of the major muscles used to move bones and tendons are skeletal muscle.

      a. Skeletal muscle is **striated** due to the structural organization of contractile proteins within individual muscle cells. Almost all skeletal muscles have at least partial voluntary control.

      b. Skeletal muscle cells, or **fibers,** are syncytial cells (i.e., cells containing many nuclei and formed by fusion of uninuclear cells).

      c. Skeletal muscles move bones relative to one another. Voluntary contraction and relaxation of skeletal muscles facilitate movement and purposeful physical activity.

   3. **Cardiac muscle** tissue is striated and individual cardiac muscle fibers are uninuclear. Cardiac muscle exists only in the heart and in portions of the aorta and venae cavae immediately adjacent to the heart.

II. SMOOTH MUSCLE

A. Function

1. Smooth muscle contraction is controlled by the autonomic nervous system.

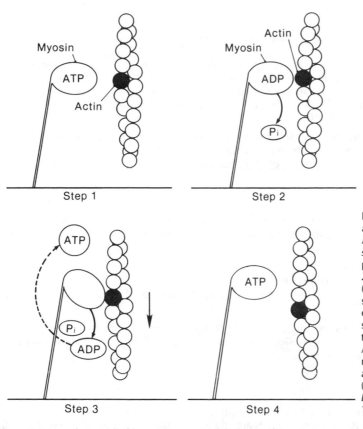

Figure 13-2. Diagram showing how a myosin molecule interacts with ATP and actin during contraction. In *step 1*, ATP binds to the myosin and prevents its binding to actin. In *step 2*, ATP is hydrolyzed to ADP and P_i (inorganic phosphate), and the myosin binds to actin. In *step 3*, there is a conformational change in the myosin and ADP is released. This ADP is then regenerated to ATP. In *step 4*, ATP binds to myosin again and this reduces the strength of the myosin-actin complex so that it dissociates. (Reprinted from Thorpe NO: *Cell Biology*. New York, John Wiley, 1984, p 615.)

2. Smooth muscle fibers are essential for controlling the size and motility of the lumina in the cardiovascular, gastrointestinal, urogenital, and respiratory systems.
   a. Smooth muscle fibers in the walls of muscular arteries help control the distribution of blood in the body.
   b. Rhythmic peristalsis of the muscle layers in the gut tube (Figure 13-3) facilitates food transportation and digestion.
   c. Smooth muscle in the male and female reproductive tracts propels gametes toward each other for fertilization.

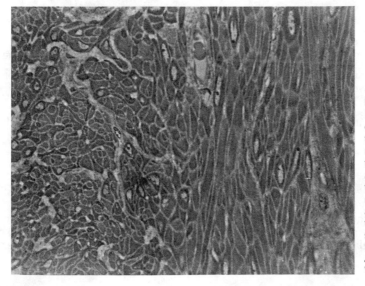

Figure 13-3. Light micrograph of the layers of smooth muscle in the wall of the small intestine. On the right, the smooth muscle fibers are arranged roughly concentrically with the lumen and usually are described as having a circular arrangement. On the left, they are arranged longitudinally, running parallel to the long axis of the gut tube. Notice the centrally placed nuclei and absence of striations. (Reprinted from Johnson KE: *Histology: Microscopic Anatomy and Embryology*. New York, John Wiley, 1982, p 77.)

B. Histology

1. Light microscopic investigation of smooth muscle masses reveals large numbers of cells, each containing a single central nucleus.
   a. Smooth muscle fibers are elongated, tapering cells that are 10–20 μm in diameter and range in length from 20 μm in blood vessels to 1 mm or more in the uterine wall during pregnancy.
   b. Occasionally, nuclei are not visible in sections cut perpendicular to the long axis of smooth muscle fibers because cells pack together and the fat portions of one cell often mesh with the thin portions of adjacent cells.

2. Special stains (e.g., periodic acid-Schiff reaction) reveal a basement membrane surrounding individual smooth muscle cells. These membranes contain collagen Types III and V. (For more information about collagen, see Chapter 7 III C).

C. Ultrastructure

1. Electron microscopy reveals that the smooth muscle cell perinuclear region contains abundant mitochondria and a modest Golgi apparatus, and that filamentous structures dominate the cell periphery.

2. Thick filaments (containing myosin) and thin filaments (containing actin) are visible in thin sections cut perpendicular to the long axis of smooth muscle cells. These filaments are less formally arranged in smooth muscle cells than they are in striated (skeletal, cardiac) muscle cells.

3. Often, smooth muscle cells are joined by **gap junctions** (sites where ion flux occurs along a low resistance pathway) to form multicellular, functionally integrated tissues. (For more information about gap junctions, see Chapter 8, III C 2 e.)

D. Contraction mechanism

1. The smooth muscle cell contraction mechanism is a topic of considerable research interest. Most researchers believe that thin filaments are relatively stable and are inserted into the plasma membrane by α-actinin molecules.

2. In contrast, thick, myosin-rich filaments are relatively labile except immediately preceding contraction (e.g., after sympathetic stimulation), when myosin molecules may form large polymers that participate in a sliding filament mechanism that is fundamentally similar to that found in striated muscle tissue (see III D).

3. The difference between smooth and striated muscle may lie in the relative stability of the thin and thick lattices within each type of muscle fiber as well as differences in the packing of thin and thick filaments.

E. Innervation

1. Smooth muscle is innervated by relatively few sympathetic and parasympathetic nerve fibers.

2. Smooth muscle masses usually have an antagonistic system of activation and inhibition. In some tissues, cholinergic (sympathetic) fibers stimulate motility and adrenergic (parasympathetic) fibers depress motility. In other tissues, these roles are reversed.

3. Smooth muscle cells also have autonomous motility, which is modulated by the muscle's innervation.

III. SKELETAL MUSCLE

A. Function. Most skeletal muscle is involved in the voluntary propulsion of bones and ligaments. Higher brain centers have primary control of these muscles, and their contractions are initiated by spinal cord motor (ventral root) neurons. Some skeletal muscle contractions are involuntary (e.g., reflex arcs, eye blinking).

B. Histology. Light microscopic examination of sections cut parallel to the long axis of individual muscle cells reveals that gross muscles (e.g., biceps) have clear striations (Figure 13-4), reflecting the highly ordered arrangement of actin-rich thin filaments and myosin-rich thick filaments. Electron microscopy reveals the structural basis of striations (Figure 13-5).

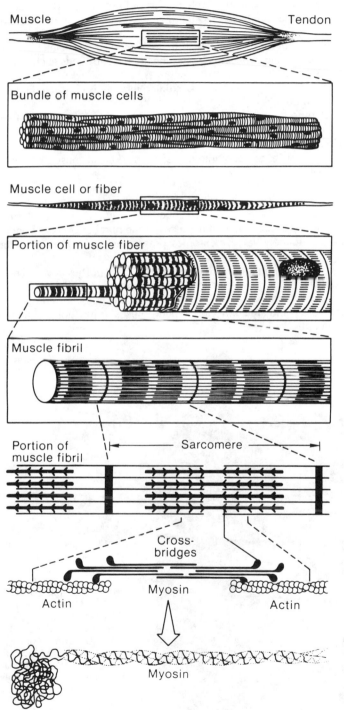

Figure 13-4. Diagram of the organization of skeletal muscle tissue from gross muscles to the molecular level. (Reprinted from Thorpe NO: *Cell Biology*. New York, John Wiley, 1984, p 610.)

1. **Structural organization**
    a. **Skeletal muscle fibers** are syncytial cells formed by fusion of multiple uninucleated myoblasts during embryonic development. Nuclei, mitochondria, and glycogen granules are displaced to the cell periphery by **myofibrils** (i.e., masses of regularly arranged bundles of thin and thick filaments). Gross muscle subdivisions called **fascicles** contain numerous muscle fibers.
    b. **Connective tissue** components of skeletal muscle consist of fibroblasts and collagen fibers.
       (1) The **epimysium** is a connective tissue capsule that surrounds each gross muscle (e.g., biceps).
       (2) The **perimysium** is a connective tissue sheath that covers each fascicle.
       (3) The **endomysium** is a delicate connective tissue sheath that surrounds each muscle cell and binds it to adjacent cells.
    c. **Sarcomeres** are the fundamental structural units of striated muscle. Each striated myofibril is composed of numerous sarcomeres.
    d. **Blood supply.** Skeletal muscles contain numerous continuous capillaries that deliver a generous supply of blood.
    e. **Myotendinous junction.** Skeletal muscles often end on tendons, which attach them to bones. The myotendinous junction is a complex interdigitation of skeletal muscle fibers, Type I collagen, and tendon fibroblasts.

2. **Structural variations.** The myofibrillar composition of gross muscles varies. Histochemical reactions for mitochondrial enzymes reveal three basic fiber types: red, white, and intermediate. The difference in fibril color is partially due to differences in the number of mitochondria and the amount of pigmented mitochondrial cytochromes present in each type. Most muscles contain variable mixtures of red, white, and intermediate fibers.
    a. **Red fibers** have numerous mitochondria with densely packed cristae. They contract relatively slowly and are served by neuromuscular junctions that have few folds and synaptic vesicles.
    b. **White fibers** have fewer mitochondria than red fibers and the mitochondria have fewer cristae. They contract rapidly and are connected to neuromuscular junctions with elaborate folds and numerous synaptic vesicles.

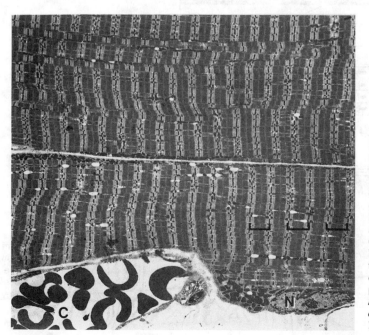

Figure 13-5. Electron micrograph of skeletal muscle. A myocyte nucleus (*N*) is visible in the lower right corner and a small capillary (*C*) containing erythrocytes is visible in the lower left corner. The numerous sarcomeres (in brackets) that comprise muscle fibers are clearly visible. (Prepared by Dr. C. Sexton, Department of Radiology, Yale University, and provided courtesy of Dr. M. K. Reedy, Department of Anatomy, Duke University. Reprinted from Johnson KE: *Histology: Microscopic Anatomy and Embryology.* New York, John Wiley, 1982, p 78.)

    c. **Intermediate fibers,** as their name implies, fall between red and white fibers in the number of mitochondria and mitochondrial cristae they contain and in the ultrastructure of their neuromuscular junctions.

    d. Recent studies indicate that red and white fibers contain myosins that are closely related but that are chemically and immunologically distinct.

C. Sarcomere ultrastructure (Figure 13-6). Each sarcomere is composed of a regular array of myosin-rich thick filaments and actin-rich thin filaments.

    1. Each sarcomere extends from **Z line** to **Z line**. Thin filaments insert at Z lines (areas rich in α-actinin) and project toward the sarcomere center.

    2. The **A band** is an electron-dense region representing the length of the thick filament. The **M line** is a specialized region at the sarcomere center. Here, thick filaments have radially arranged connections to each other.

    3. Some overlap of thick and thin filaments occurs along the length of the A band, except in the **H band** (i.e., the middle of the A band), where no overlap occurs.

    4. The **I band** is a sarcomere region containing only thin filaments. Each I band spans two sarcomeres; thus, the midpoint of each I band is the Z line.

D. Contraction mechanism

    1. **Regulation of muscle contraction.** Contractile force in all muscle is generated by a change in the relative position of thin and thick filaments, which is regulated by the concentration of intracellular calcium.

        a. ATP hydrolysis supplies the energy for cell contraction. Under appropriate conditions, the actin-myosin complex has ATPase activity.

        b. ATP hydrolysis and the ensuing release of energy causes conformational changes in motile tissue proteins that result in purposeful movement (see Figure 13-2).

    2. **Sliding filaments and swinging cross-bridges.** The length of a sarcomere varies according to the contractile status of the cell. As individual sarcomeres change length, the length of the entire gross muscle changes. When a muscle contracts, thick and thin filaments slide past one another.

        a. As a contraction and relaxation cycle begins, calcium concentration around the myofibrils increases suddenly, causing a conformational change in the troponin molecule that exposes the S-1 (cross-bridge) binding site of actin, and a myosin-actin complex forms.

        b. Then, another conformational change occurs and the S-1 fragment swings like an oar in an oarlock while still attached to the thin filament, causing the thin filament to slide relative to the thick filament. When this occurs at millions of cross-bridges, the sarcomere grows shorter. (The A band remains the same length, but the H and I bands shorten.)

        c. After ATP hydrolysis and the swinging of cross-bridges, the calcium concentration falls rapidly, dissociating the myosin-actin complex, and the contraction stops.

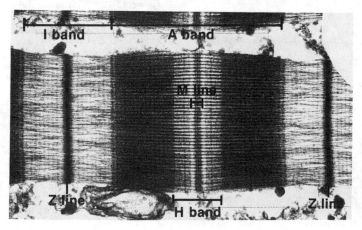

Figure 13-6. A high resolution electron micrograph of skeletal muscle, showing individual sarcomeres extending from Z line to Z line. (Courtesy of Dr. M. K. Reedy, Department of Anatomy, Duke University. Reprinted from Johnson KE: *Histology: Microscopic Anatomy and Embryology.* New York, John Wiley, 1982, p 79.)

3. ATP is hydrolysed to adenosine diphosphate (ADP), which subsequently is phosphorylated to form ATP. ATP hydrolysis is active during muscle contraction. ATP synthesis occurs in mitochondria. The plethora of mitochondria in skeletal and cardiac muscle are required for rapid regeneration of ATP.

E. Membrane system. Striated muscle cells have a plasmalemma called the **sarcolemma** and a highly modified smooth endoplasmic reticulum called **sarcoplasmic reticulum,** which is integrally involved in regulating the calcium concentration around myofibrils (Figure 13-7).

1. Sarcoplasmic reticulum forms an anastomosing network of interconnected cisternae (flat, membrane-delimited sacs) that communicate with dilated **terminal cisternae**.

2. The sarcolemma contains numerous **T tubules** that penetrate the mass of myofibrils. Each sarcomere is served by two T tubules that invaginate from the sarcolemma near the A band and I band junction.

3. T tubules contact the terminal cisternae. One T tubule and two cisternae form a **triad**. In human skeletal muscle, triads are located near the junctions of A and I bands. The remainder of the I band is covered by mitochondria aligned somewhat perpendicular to the myofibril long axis.

4. Sarcoplasmic reticulum is rich in **calsequestrin,** a protein that binds calcium.
   a. The sarcoplasmic reticulum of a resting muscle contains a high concentration of calcium. When a nervous impulse to initiate contraction reaches the neuromuscular junction, sarcolemma depolarization occurs.
   b. A wave of depolarization moves along the sarcolemma and T tubules similar to the way an action potential is propagated along a nerve fiber. Consequently, sarcolemma depolarization causes depolarization in the sarcoplasmic reticulum.
   c. When the terminal cisternae membranes become permeable to calcium, calcium ions flow into the region around the myofibrils and initiate actin-myosin complex formation and thus contraction.
   d. Sarcoplasmic reticulum contains an ATPase that actively pumps calcium back into the sarcoplasmic reticulum, where calsequestrin binds it, thus completing the contraction cycle.

F. Innervation. Skeletal muscle fibers interact with motor neurons at the neuromuscular junction. Nerve fibers contain synaptic vesicles filled with **acetylcholine,** which initiates the depolarization wave when it is released into the synaptic cleft.

1. **The neuromuscular junction** (Figure 13-8)
   a. Muscle cells have scattered surface invaginations called **primary synaptic clefts** where axons for motor neurons to specific muscle regions terminate.
      (1) The sarcolemma has numerous deep folds in the primary synaptic cleft called **secondary synaptic clefts** or **junctional folds,** which increase the area of contact between the nerve and muscle.
      (2) Nerve and muscle cell membranes are parallel and are separated by a synaptic cleft that contains the basement membrane that surrounds the muscle fiber.

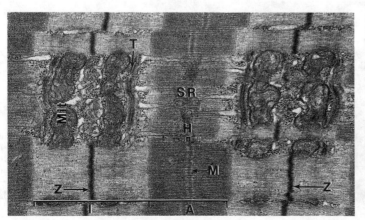

Figure 13-7. A high resolution electron micrograph of a sarcomere in skeletal muscle. The sarcomere extends from Z line to Z line and is centered on the narrow M band. This sarcomere is contracted. The H band (*H*) is narrow due to extensive interdigitation between thin filaments of actin in the I band (*I*) and thick filaments of myosin in the A band (*A*). Mitochondria (*Mit*) lie over the I band and close to the T tubules (*T*) and sarcoplasmic reticulum (*SR*). (Courtesy of Dr. H.A. Padykula, Department of Anatomy, University of Massachusetts. Reprinted, with changes, from Weiss L, Greep R: *Histology.* New York, McGraw-Hill, 1977, p 267.)

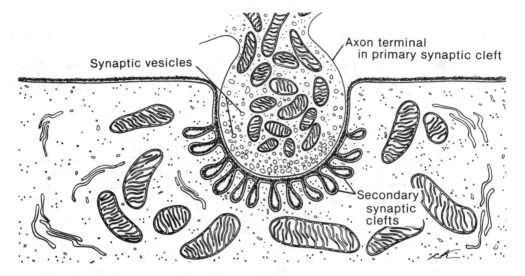

Figure 13-8. Diagram of the neuromuscular junction. (Reprinted from Bloom W, Fawcett DW: *A Textbook of Histology*. Philadelphia, WB Saunders, 1968, p 287.)

  b. Myocyte cytoplasm near the neuromuscular junction contains an abundance of mitochondria, rough endoplasmic reticulum, and free ribosomes, which may be used to synthesize acetylcholine receptors in the muscle cell membrane.
  c. An axon innervating a neuromuscular junction lacks myelin. Instead, the associated Schwann cell forms a protective cap over the junction.

2. Synaptic vesicles
  a. When a nerve stimulates a skeletal muscle fiber, the synaptic vesicle releases acetylcholine and the vesicle's membrane fuses with the sarcolemma, becoming an integral component of the muscle membrane.
  b. After releasing acetylcholine, membrane patches move from the central neuromuscular junction to the axon periphery, where they are removed from the cell surface and integrated into the cell as pits and vesicles coated with the protein **clathrin.** Eventually, coated vesicles are integrated into the membrane system at the neuromuscular junction and are used again as synaptic vesicles.
  c. Thus, some of the synaptic vesicular membrane is recycled in the cell rather than being destroyed by primary lysosomes, which typically occurs when material is returned to the cell by endocytosis. Membrane recycling is more efficient than constant de novo synthesis of synaptic membrane.

G. Regeneration

  1. The nuclei in skeletal muscle fibers cannot divide. Instead, muscle regeneration and hypertrophy are facilitated by **satellite cells,** small uninuclear cells enclosed in the basement membrane of skeletal muscle fibers.

  2. When a muscle is injured or stimulated to grow by repeated strenuous exercise, satellite cells divide, producing new cells to replace damaged cells.

  3. If strenuous exercise continues, one daughter cell from satellite cell mitosis fuses with existing muscle fibers, causing a hypertrophic increase in muscle mass. The other daughter cell remains a stem cell.

IV. CARDIAC MUSCLE

  A. Function. Cardiac muscle contracts rhythmically and continuously, pumping blood through the circulatory system. This provides cells with a constant supply of oxygen and nutrients and removes cell waste products.

B. Histology (Figure 13-9)

1. Cardiac muscle tissue exists only in the myocardium and proximal portions of the aorta and venae cavae.

2. Cardiac myocytes have one central nucleus and are joined to each other by **intercalated disks**.

3. Cardiac muscle is striated, like skeletal muscle, and sarcomeres are arranged similarly in cardiac and skeletal muscle. Cardiac muscle cells may have branches.

C. Ultrastructure

1. Electron microscopy reveals other differences between cardiac and skeletal muscle.
   a. Cardiac muscle has larger T tubules and a less defined sarcoplasmic reticulum. Each T tubule is associated with one terminal cisterna, forming **diads** instead of triads.
   b. Cardiac muscle contains many more mitochondria, which run parallel to the myofibril long axis rather than perpendicular to it in the I bands.

2. Electron microscopy reveals that intercalated disks are complex interdigitations between adjacent cells. The disks have three types of junctional specializations.
   a. **Fasciae adherentes,** numerous filamentous masses on the inner aspects of apposed cells, are anchoring sites for the sarcomere nearest the end of the cells.
   b. **Maculae adherentes** in the intercalated disk bind cells to prevent separation during contraction.
   c. Numerous **gap junctions** exist between cells adjacent to the intercalated disks. Gap junctions provide direct ionic (electrical) communications between cells and help coordinate contraction in many cells.

D. Innervation. Cardiac muscle is innervated by branches of the sympathetic and parasympathetic nervous systems. A complex set of modified cardiac muscle fibers called the **Purkinje fibers** forms an intracardiac impulse conduction system. (For more information about cardiac muscle innervation see Chapter 15 II B 7.)

E. Regeneration. Cardiac muscle tissue has no regenerative capacity. During myocardial infarction, cardiac muscle tissue dies and is replaced by fibroblast-rich scar tissue.

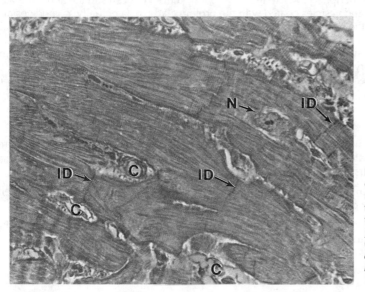

Figure 13-9. Light micrograph of cardiac muscle tissue showing the nucleus (*N*) of one cell and, characteristically, many capillaries (*C*). Intercalated disks (*ID*) are intercellular junctions between muscle cells. (Reprinted from Johnson KE: *Histology: Microscopic Anatomy and Embryology.* New York, John Wiley, 1982, p 82.)

STUDY QUESTIONS

Directions: Each question below contains five suggested answers. Choose the **one best** response to each question.

1. Which of the following conditions exists in a muscle fiber that is stretched beyond the point where it can generate tension when stimulated?

(A) The H band is narrow because thin filaments overlap thick filaments extensively

(B) The H band is broad because thin filaments do not overlap the thick filaments

(C) The H band is narrow because there is very little overlapping of thick and thin filaments

(D) The I band is shorter because thick and thin filaments overlap extensively

(E) The M line is small because there is very little overlapping of thick and thin filaments

2. Each of the following statements concerning skeletal muscle cell characteristics is true EXCEPT

(A) they are syncytial cells with peripheral nuclei

(B) they are joined to each other by intercalated disks

(C) they contain many elongated mitochondria

(D) they have membrane triads at the junction of the A and I bands

(E) they contain an elaborate sarcoplasmic reticulum

3. Each of the following statements concerning cardiac muscle tissue is true EXCEPT

(A) each cardiac muscle cell contains a single central nucleus and many sarcomeres

(B) cardiac muscle tissue exists in the proximal aorta

(C) cardiac muscle cells are joined by intercalated disks

(D) cardiac muscle tissue contains gap junctions

(E) cardiac muscle tissue regenerates by satellite cell mitosis

4. Each of the following statements concerning smooth muscle cells is true EXCEPT

(A) they do not contain sarcomeres

(B) they contain stable thick filaments

(C) they contain many mitochondria

(D) they have gap junctions

(E) they do not contain T tubules

5. Thin filaments have all of the following components EXCEPT

(A) myosin

(B) troponin C

(C) F-actin

(D) troponin I

(E) tropomyosin

Directions: The groups of questions below consist of lettered choices followed by several numbered items. For each numbered item select the **one** lettered choice with which it is **most** closely associated. Each lettered choice may be used once, more than once, or not at all.

Questions 6–10

Match each location listed below with the regulatory protein or proteins contained there.

(A) Troponin
(B) Tropomyosin
(C) Both
(D) Neither

6. Thick filaments

7. Thin filaments

8. I band

9. H band

10. A band

Questions 11–13

Match each characteristic listed below with the appropriate filaments.

(A) Thick filaments
(B) Thin filaments
(C) Both
(D) Neither

11. Present in the A band

12. Insert into the Z line

13. Permanently attached to cross-bridges

ANSWERS AND EXPLANATIONS

1. The answer is B. [*III C, D 2*] A sarcomere cannot generate tension when stimulated to contract unless thick and thin filaments overlap. Thick and thin filaments overlap in most of the A band but not in the center of the A band (i.e., the H band). When a muscle is stretched beyond the point of thick and thin filament overlap, the H band becomes broad. Without overlap, S-1 myosin heads in the thick filament cannot interact with the actin-containing thin filaments. As a sarcomere stretches, Z lines move further apart. Eventually, the thin filaments no longer interdigitate, and the sarcomere can no longer generate tension.

2. The answer is B. [*III B 1 a, E*] Intercalated disks are the intercellular junctions in cardiac muscle tissue. Skeletal muscle cells are multinucleated syncytial cells formed by myoblast fusion. Cell nuclei are located around the periphery of a central cytoplasmic domain, which is filled with sarcomeres. Cells contain abundant elongated mitochondria, which produce the ATP necessary for muscle contraction. Skeletal muscle cells have an elaborate sarcoplasmic reticulum with terminal cisternae. Triads, formed by one T tubule and two sarcoplasmic reticulum terminal cisternae, are located at the junction of the A and I bands in skeletal muscle sarcomeres.

3. The answer is E. [*IV B, C, E*] Cardiac muscle is abundant in the heart and proximal sections of the aorta and venae cavae. Each cardiac muscle cell contains a single nucleus, located in the center of the cell, and many sarcomeres. Cells are joined to each other by intercalated disks. Extensive gap junctions exist between cells, near the intercalated disks. In cardiac muscle, gap junctions coordinate the contraction of groups of cardiac muscle cells. Cardiac muscle tissue has no regenerative capacity. Satellite cells lie within the basement membrane of the skeletal muscle fibers and facilitate skeletal muscle regeneration and hypertrophy.

4. The answer is B. [*II B, C, D 2*] Smooth muscle is abundant in the gastrointestinal and reproductive systems. Smooth muscle contraction controls the peristaltic motility in the organs of these systems. Smooth muscle cells do not contain sarcomeres or T tubules, but they do contain many thin filaments, labile thick filaments, and numerous mitochondria, which produce the ATP for muscle contraction. Gap junctions join cells and coordinate muscle contraction.

5. The answer is A. [*I B 1*] All types of muscle tissue contain thin filaments, which consist of F-actin cables and the regulatory proteins tropomyosin and troponin. Troponin is a complex of troponin C, troponin I, and troponin I. Tropomyosin undergoes conformational changes to cover or expose the myosin binding site on actin-containing thin filaments. Tropomyosin and troponin regulate the binding of the myosin in thick filaments to the actin in thin filaments.

6–10. The answers are: 6-D, 7-C, 8-C, 9-D, 10-C. [*I B 1 b; III C*] Troponin and tropomyosin are regulatory proteins associated with the actin in thin filaments; they are not present in thick filaments. When calcium binds to troponin, it causes changes in tropomyosin so that the myosin binding site of actin is exposed. The I and A bands contain thin filaments and, therefore, contain both troponin and tropomyosin. The H band does not contain thin filaments and, therefore, does not contain either protein.

11–13. The answers are: 11-C, 12-B, 13-A. [*III C, D 2*] Thick filaments consist of myosin molecules packed together so that their S-1 fragments project toward thin filaments, forming cross-bridges. Cross-bridges are part of myosin and are permanently attached to thick filaments; they transiently attach to thin filaments during contraction.

Thin filaments contain actin, troponin, tropomyosin, and α-actinin. They insert into Z lines by way of α-actinin.

Both thin and thick filaments are present and overlap in the A band.

I. INTRODUCTION

A. Nervous tissue characteristics

1. Nervous tissue integrates and coordinates the functions of other tissues in the body.

2. Nervous tissue is composed of diverse types of neuronal and glial cells that are derived from embryonic neuroepithelium. During tissue development, some epithelial features are lost. For example, the basement membrane that surrounds the nervous system becomes highly vascularized. This is not characteristic of other epithelia.

B. Nervous system components

1. **Divisions.** The nervous system is actually two structurally and functionally interconnected systems: the **central nervous system (CNS),** which includes the brain and spinal cord, and the **peripheral nervous system (PNS),** which includes the nerves and ganglia scattered throughout the peripheral portions of the body.

2. **Cells**
   a. **Neurons** are the basic cellular elements of the nervous system. Neuron structure varies tremendously.
      (1) Motor neurons, which conduct motor impulses from the spinal cord to skeletal muscles, have cell bodies located in the ventral (anterior) horns of the spinal cord and axonal terminations on muscle fibers a meter or more away.
      (2) Other CNS neurons have a highly complex dendritic arborization; still others are small interconnecting cells.
   b. **Glial cells,** such as **astrocytes** and **Schwann cells,** perform subsidiary functions unrelated to communication.

3. **Synapses** are the sites of anatomic and functional interaction between neurons.

II. NEURONS

A. Histology. Motor neurons in the anterior horn of the spinal cord of adults exemplify the structure and function of specialized neurons (Figure 14-1).

1. Each motor neuron has a **cell body,** one **axon** (usually), and a variable number of **dendrites,** which vary in morphology. The axon and dendrites are long cytoplasmic processes that radiate from the cell body. The cytoplasm around the cell body is called the **perikaryon.**

2. Each motor neuron has a **large, euchromatinized nucleus** containing a **prominent nucleolus.** Protein synthesis is ongoing in the perikaryon, and newly synthesized proteins are transported radially, especially toward the axon. The nucleus and associated organelles, most notably the rough endoplasmic reticulum, are unusually prominent in the cell body.

3. Motor neurons secrete **synaptic vesicles,** membrane-delimited packets of neurotransmitters such as **acetylcholine,** at the neuromuscular junction.

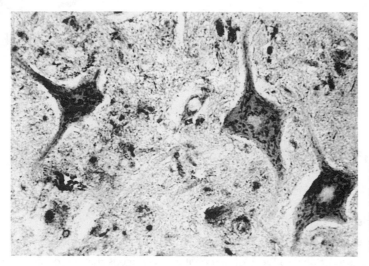

Figure 14-1. Light micrograph of motor neurons in the spinal cord. The perikaryon of each of these large cells is rich in darkly stained Nissl substance. (Reprinted from Johnson KE: *Histology: Microscopic Anatomy and Embryology.* New York, John Wiley, 1982, p 90.)

B. Ultrastructure (Figure 14-2)

1. Light microscopic examination reveals many clusters of rough endoplasmic reticulum and free cytoplasmic ribosomes (called **Nissl substance**) in the perikaryon. These clusters cause cytoplasmic basophilia.

2. The perikaryon also contains numerous mitochondria interspersed throughout the Nissl substance. Because the cell is actively producing protein for secretion, the cell has a prominent Golgi complex.

3. Neurons also contain an abundance of cytoplasmic microtubules called **neurotubules** and microfilaments called **neurofilaments**. Neurotubules are 27 nm in diameter and neurofilaments are 10 nm in diameter.

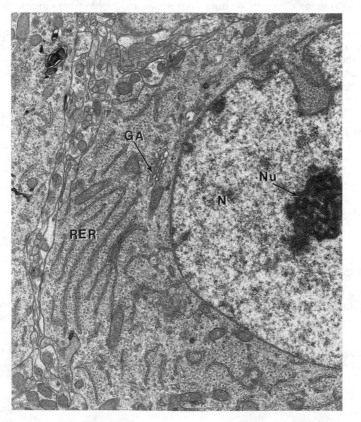

Figure 14-2. Electron micrograph of a motor neuron, showing a euchromatic nucleus (*N*) containing a prominent nucleolus (*Nu*), rough endoplasmic reticulum (*RER*), and a Golgi apparatus (*GA*). (Micrograph courtesy of Dr. E.N. Albert, Department of Anatomy, George Washington University.)

4. Axons do not contain an abundance of Nissl substance or Golgi complexes. Instead, axons and dendrites are heavily laden with neurotubules and neurofilaments. Dendrites also contain scattered small mitochondria and usually are covered by numerous synapses.

5. Researchers believe that neurotubules and neurofilaments are involved in the rapid **anterograde** (toward the neuromuscular junction) and slower **retrograde** (toward the cell body) transport of cytoplasm that occurs in the motor neuron.

C. Myelination

1. Schwann cells, a type of glial cell, myelinate motor neuron axons in the PNS.

   a. Schwann cell length is limited. Therefore, a long motor neuron axon is covered by thousands of Schwann cells. At the **node of Ranvier**—the point where one myelinating Schwann cell ends and another begins—the axon is exposed to the fluid and ions surrounding it.

   b. Unmyelinated nerve fibers in the PNS are associated with Schwann cells in a different way than myelinated nerve fibers. In this case, axons from several unmyelinated fibers are embedded in each Schwann cell.

2. Myelin is produced by the layers of Schwann cell membrane that form during nervous system development. Presumably, layers form as the Schwann cell or its **mesaxon** rotates.

   a. During nervous system development, the mesaxon membrane grows and wraps around the axon, fusing with itself, thus creating many layers of membrane fusion sites similar to those formed at a zonula occludens.

   b. The thick coat of fused membrane layers insulates the axon but allows ions from the axon to flow into the extracellular space around the axon at gaps between Schwann cells.

3. Nerve impulse conduction

   a. Because the insulation layers are discontinuous, nerve impulses, which really are ionic fluxes, pass along the axon by **saltatory conduction** (i.e., by skipping from one node to another).

   b. Nerve conduction velocity is much greater in myelinated nerve fibers than in unmyelinated fibers. In myelinated fibers, conduction velocity is directly proportional to the thickness of the myelin sheath.

4. Schmidt-Lanterman clefts, which are conical defects in the myelin sheath, move up and down the Schwann cell carrying Schwann cell cytoplasm and its nutrients to the living membranes of the myelin sheath.

5. The neuromuscular junction is the point where a motor neuron contacts the muscle fiber bundle it innervates. Here, Schwann cells do not myelinate the axon, but remain close to it.

III. NEUROGLIA are the nervous system cells that perform a variety of supportive functions ancillary to impulse conduction. Neuroglia produce the myelin that insulates and functionally isolates neurons.

A. Ependymal cells are the epithelial cells that line the neural canal and brain ventricles. In the **choroid plexus** (see V F 2), ependymal cells produce **cerebrospinal fluid (CSF),** a blood filtrate. Ependymal cells form a leaky cellular barrier between the brain parenchyma and the CSF.

B. Astrocytes are present around blood vessels in the brain parenchyma (Figure 14-3). Two morphologic varieties exist: **fibrous astrocytes** (in white matter) and **protoplasmic astrocytes** (in grey matter). Astrocytes also help form the **glia limitans** at the external boundary of the CNS.

C. Myelin-producing glia

1. Oligodendroglia produce myelin in the CNS.

2. Schwann cells produce myelin in the PNS.

D. Microglia are part of the mononuclear phagocyte system. They are derived from bone marrow monocytes and, thus, differ from other neuroglia, which are derived from the neuroepithelium. Microglia help remove effete elements of the nervous system in health and disease.

E. Satellite cells are present in PNS ganglia. They surround neurons and may functionally isolate neurons in locations such as dorsal root ganglia and sympathetic ganglia (Figure 14-4).

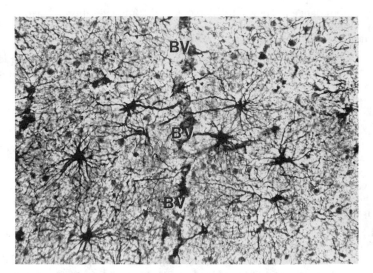

Figure 14-3. Light micrograph of the fibrous astrocytes associated with central nervous system blood vessels (*BV*).

IV. SYNAPSES

A. Types

1. **Axodendritic** synapses join an axon to a dendrite.

2. **Axosomatic** synapses join an axon to a nerve cell body.

3. **Axoaxonic** synapses join one axon to another axon.

4. **Dendrodendritic** synapses join one dendrite to another dendrite.

B. Morphology

1. Cytoplasmic continuity between neurons does not exist at any type of synapse. Neurons are separated by the neurolemma surrounding each neuronal element or process.

2. Near the synaptic ending, nerve axons divide into many unmyelinated, fine branches called **telodendria**.

3. Typically, axonal processes exhibit terminal swellings called **terminal boutons** at synapses.

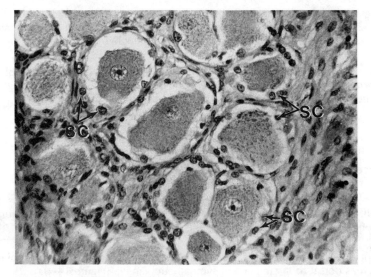

Figure 14-4. Light micrograph of satellite cells (*SC*) around sensory neurons in a dorsal root ganglion.

C. Ultrastructure

1. Neurofilaments, which are abundant in the rest of the axon, abruptly terminate at the synapse.

2. Electron microscopy reveals that terminal boutons are laden with mitochondria and numerous small (40–65 nm in diameter) membrane-delimited **synaptic vesicles**. Synaptic vesicles contain **neurotransmitters** such as **acetylcholine, norepinephrine, dopamine,** and **serotonin**.

3. Presynaptic and postsynaptic membranes are closely apposed, but are separated by a definite 20-nm **synaptic cleft**.

4. When an action potential reaches the presynaptic membrane, it stimulates the release of the vesicles containing the neurotransmitter, which in turn initiates a wave of depolarization at the postsynaptic membrane of the next neuron in the chain.

5. In this manner, nerve impulses pass from one nerve cell to the next, even though individual cells are not in direct anatomic communication.

D. Functional cell biology. Synaptic vesicles probably originate in the Golgi apparatus membranes of the neuronal cell body.

1. Since the perikaryon is relatively devoid of synaptic vesicles, neurotransmitters and synaptic vesicle precursor membranes may be synthesized in the perikaryon and then transported toward the synapse by anterograde axoplasmic flow. Histochemical studies have revealed that the enzymes for neurotransmitter synthesis are present in neuron perikaryon. Final assembly of synaptic vesicles may occur in the bouton itself, where scattered profiles of smooth endoplasmic reticulum have been observed.

2. When a nerve is stimulated, an action potential travels along the nerve to the synapse, causing synaptic vesicle membranes to fuse with the presynaptic side of the axon membrane. Then, the neurotransmitter enters the synaptic cleft and stimulates the postsynaptic membrane.

3. Evidence indicates that the presynaptic membrane is then endocytosed in the form of **clathrin-coated pits** and **coated vesicles**.
   a. **Horseradish peroxidase (HRP)** is a high molecular weight protein molecule with enzymatic activity that can be demonstrated histochemically. Soon after HRP is percolated into the synaptic cleft, coated vesicles contain HRP activity. Later, HRP can be found in smooth endoplasmic reticulum vesicles in the bouton.
   b. These results suggest that there is a rapid turnover and recycling of portions of the presynaptic membrane.

4. Repeated stimulation of a nerve will eventually deplete the supply of synaptic vesicles, leading to nerve fatigue and the cessation of synaptic function.

V. CENTRAL NERVOUS SYSTEM

A. General characteristics

1. The CNS includes the brain and spinal cord. It contains both neurons and glial cells. The human CNS contains more than **10 billion neurons** and approximately **50 billion glial cells**.

2. In some sections of the CNS, neuronal cell bodies form macroscopic clusters called **nuclei,** and large bundles of myelinated nerve fibers run together in **tracts**.
   a. Many functional nuclei in the brain contain vast collections of nerve cell bodies, axons, dendrites, and glial cells, which form a complex tangle of processes called the **neuropil**.
   b. The neuropil contains about half of all neuronal cytoplasm in the CNS. Nerve cell bodies contain the remainder.

3. Functional variations from one CNS region to another are partially due to variations in the kind of neurons present in each region and to variations in the organization of communicating processes in the surrounding neuropil.

B. Neuronal variation

1. **Morphologic variation**
   a. **Shape and size.** Neurons vary tremendously in shape and size.
      (1) Some neurons have smaller diameters than erythrocytes. Others are 100 μm or more in diameter.

    (2) Many of the smaller neurons are Golgi type II cells that perform an integrative function. Granule cells in the cerebellar granular layer are another type of small neuron with a few short dendrites.
  b. Polarity. Neurons are classified as multipolar, bipolar, or unipolar.
    (1) Multipolar neurons account for the majority of CNS neurons.
    (2) Bipolar neurons are relatively uncommon except in sensory epithelia, where they are abundant. Bipolar neurons are relatively common, however, in embryos.
      (a) Many of these bipolar neurons later undergo a process that brings the axon and dendrite close together, resulting in a cell that appears unipolar but really is a modified bipolar cell called a **pseudounipolar neuron**.
      (b) These cells are commonly found in ganglia associated with satellite cells.
    (3) Unipolar cells are rare in humans.

2. Regional variation
  a. Pyramidal cells exist in the cerebral cortex.
    (1) These cells have a roughly pyramidal shaped cell body and a long slender axon projecting from their base (Figure 14-5).
    (2) They have numerous dendrites with many long branches and regions where hundreds of synaptic terminals from other cortical neurons make contact.
  b. Purkinje cells exist in the cerebellar cortex, at the boundary between the molecular and granular layers of the cortical grey matter.
    (1) Each cell has a rounded cell body, a slender axon that projects through the granular layer into the white matter, and three to five main dendrites that divide into an elaborate dendritic arborization.
    (2) The cerebellar cortex contains several hundred thousand Purkinje cells. If, as some researchers estimate, each Purkinje cell dendritic arborization contains several hundred thousand synaptic boutons, then the total number of synapses is truly astounding.
  c. Golgi type I neurons have long axons that begin in the CNS and terminate a great distance away in the CNS or peripherally. Anterior horn motor neurons and cerebral pyramidal cells are examples of Golgi type I neurons.
  d. Golgi type II neurons have short axons that begin and end in restricted regions within the CNS. Bipolar cells in the retina and olfactory epithelium are examples of Golgi type II neurons.

C. Grey matter and white matter

1. The brain and spinal cord contain **grey matter,** which is relatively rich in neuronal cell bodies and glia, and **white matter,** which is relatively rich in neuronal axons, dendrites, glia, and myelin.

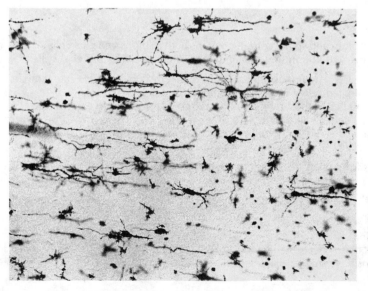

Figure 14-5. Light micrograph of the pyramidal neurons common in a Golgi preparation of the cerebral cortex of a monkey. (Micrograph courtesy of Dr. E.N. Albert, Department of Anatomy, George Washington University.)

2. In the spinal cord, the centrally located grey matter is divided into **anterior horns** (containing motor neurons), **intermediate horns** (containing autonomic neurons), and **posterior horns** (containing sensory neurons). Most grey matter in the spinal cord is surrounded by white matter.

3. In the brain, the relationship between grey and white matter is slightly different, due to the presence of superimposed cortical grey matter.

D. The spinal cord

1. The central canal is a small hole in the middle of the spinal cord lined by ependymal cells. The central canal contains CSF and communicates with the ventricles of the brain.

2. Horns of the spinal cord. Transverse sections of the spinal cord reveal three horns surrounding the central canal.
   a. The **anterior horns** (ventral horns) are paired structures containing the cell bodies of motor neurons.
   b. The **posterior horns** (dorsal horns) are paired structures containing the cell bodies of sensory neurons.
   c. The **intermediate horns** are paired structures containing the cell bodies of autonomic neurons.

3. Meninges. The spinal cord is surrounded by a layer of meninges that are identical to the meninges of the brain (see V F 1).

E. The brain has a second layer of grey matter (the **cortex**) that is superimposed peripherally over the white matter.

1. The cerebrum
   a. A series of grey matter **nuclei,** containing numerous neuronal cell bodies, exists deep in the cerebrum.
      (1) Large bundles of fiber tracts, consisting largely of neuronal axons, connect the nuclei.
      (2) The nuclei are surrounded by a layer of white matter composed of axons and myelin sheaths derived from oligodendroglial cells. Numerous axons project both away from and toward the deep nuclei.
   b. The **cerebral cortex** is a layer of grey matter surrounding the white matter. It is composed of a peripheral layer of pyramidal neurons and associated interneurons and glia.

2. The cerebellum (Figure 14-6)
   a. The cerebellar portion of the brain also contains an inner layer of grey matter, an intermediate layer of white matter, and a cortical (outer) layer of grey matter.
   b. The **cerebellar cortex** consists of a granular layer, a Purkinje cell layer, and a molecular layer.
      (1) The cerebellar cortex is folded into sections called **folia,** which enclose central cores of white matter. Adjacent to the white matter is a **granular layer,** rich in neurons and glia, which has a highly cellular appearance.
      (2) Outside the granular layer is a **Purkinje cell layer** containing the cell bodies of Purkinje cells and **basket cells.** Basket cells surround the cell body of Purkinje cells.
      (3) The complex dendritic arborizations of the Purkinje cells project away from the cell body and into the outermost **molecular layer** of the cerebellar cortex. The molecular layer also contains glial cells and nerve fibers.

F. CNS support tissues and structures

1. Meninges surround the brain and spinal cord. They are derived partially from mesoderm and partially from the neural crest. Although they are not nervous tissue, it is appropriate to discuss them here.
   a. The **dura mater,** the outermost layer of meninges, is a tough, thick, connective tissue capsule around the entire CNS. It contains blood vessels and, in some locations, large venous sinuses.
      (1) In the spinal canal, the dura is separated from the periosteum of the vertebrae. Within the skull, the dura is loosely associated with the skull bones and their periosteum.
      (2) The inner aspect of the dura is lined by a mesothelium and faces a thin serous cavity called the **subdural space.**

Figure 14-6. Light micrograph of the cerebellar cortex folia. Purkinje cells (*PC*) have numerous dendrites in the molecular layer (*ML*) and in the nuclei at the outer edge of the granular layer (*GL*). White matter (*WM*) has nerve cell processes that carry impulses to and from the granular layer.

   b. The **arachnoid mater,** the intermediate layer of the meninges, lies within the dura mater. It is an avascular layer encased in a thin capsule of connective tissue cells and fibers and has anastomosing connective tissue trabeculae that project from the dura toward the surface of the brain and spinal cord.
   c. The **pia mater,** the innermost layer of the meninges, carries the main blood supply for the CNS. This layer is so intimately linked to the arachnoid mater that many histologists refer to them jointly as the **pia-arachnoid layer**.
   d. The **subarachnoid space** is a fairly extensive area between the arachnoid mater and pia mater, which is filled with CSF. It is quite wide along the length of the spinal cord and over most of the brain and, in several locations over the brain, enlarges to form **cisternae**. The largest of these, the **cisterna magna,** lies just posterior to the cerebellum and contains three holes where the fourth ventricle of the brain communicates with the subarachnoid space.
2. Choroid plexus. Near the cisterna magna, the walls of the brain thin until they are no more than a complex anastomosing network of capillaries from the pia-arachnoid layer coated with a continuous epithelial layer of convoluted ependymal cells. These capillaries and ependymal cells are the **choroid plexus**.
   a. Locations. The closed loops of capillaries that form the choroid plexus project into the brain ventricles. The body contains four patches of choroid plexus tissue.
      (1) Two exist in the walls of the **lateral ventricles** within the cerebral hemispheres.
      (2) Two others are found in the roof of the **third and fourth ventricles**.
   b. Histology
      (1) The specialized pial capillaries of the choroid plexus are fenestrated and modified to produce CSF, a blood filtrate important for the metabolism and protection of the brain parenchyma.
      (2) Each choroid plexus cuboidal epithelial cell has a robust apical junctional complex projecting toward the ventricular lumen and a prominent apical microvillous brush border.
      (3) Choroid plexus epithelial cells contain numerous prominent mitochondria and rest on a thick basement membrane, which they share with capillary endothelial cells.
      (4) Choroid plexus epithelial cells have the fine structure typical of cells involved in the selective transport of ions. Their fine structure is similar to that of cuboidal cells in renal proximal convoluted tubules.

3. **Cerebrospinal fluid (CSF)** is a blood filtrate with an ionic and macromolecular composition markedly different from serum. It contains far less protein but much more sodium, potassium, and chloride. CSF fills the subarachnoid space, the ventricles of the brain, and the central canal of the spinal cord.

   a. Most CSF is produced at the choroid plexus and drains from the CNS back into the systemic circulation at the **arachnoid villi** (i.e., projections of the subarachnoid space into the enlarged subdural spaces that run along the superior midline of the brain).

   b. A dual system of selective filters produces CSF.

      (1) The choroid plexus contains fenestrated capillaries that selectively pass some ions and blood proteins while retaining other proteins and all cellular elements.

      (2) Epithelial cells of the choroid plexus selectively transport molecules from the blood filtrate produced at the fenestrated capillaries.

   c. CSF contains some lymphocytes and may function similarly to lymphatic fluid in the rest of the body. (The nervous system does not contain lymphatic vessels.)

   d. The adult CNS contains about 100 ml of CSF, which is produced at the rate of 200 ml each day.

4. **CNS capillaries**

   a. CNS tissue has a high metabolic rate and, therefore, consumes substantial amounts of oxygen. Consequently, **the brain requires an abundant supply of blood**. The entire brain parenchyma is laced with capillaries that communicate with arterioles that branch from large blood vessels in the subarachnoid space.

   b. Capillaries in the brain parenchyma are continuous capillaries; they do not have transcellular fenestrations.

   c. Endothelial cells in these capillaries are tightly joined together by junctional complexes that have robust, occluding apical junctions. (Chapter 8 III C contains a more complete discussion of junctional complexes.)

5. **The blood-brain barrier** prevents many macromolecules in the capillaries from entering the brain parenchyma. It also prevents certain drugs and their metabolites from entering the parenchyma.

   a. The tight junctions formed by the zonulae occludentes that join endothelial cells in the choroid plexus capillaries are the anatomic basis for the blood-brain barrier.

   b. One histologic demonstration of the existence of the blood-brain barrier uses HRP (see IV D 3 a) to trace leaky or tight blood vessels.

      (1) When HRP is injected into the systemic circulation, it fills brain capillaries rapidly but cannot cross the capillary endothelium to enter brain parenchyma, except at circumventricular organs such as the infundibulum and the area postrema.

      (2) Conversely, when HRP is injected into CSF, it diffuses rapidly through the interstices of the neuropil because ependymal cells do not have tight junctions.

VI. PERIPHERAL NERVES

A. **Histology.** The large nerves studied in the gross lab actually are bundles of many motor and sensory nerves that run between a part of the body and the spinal cord or brain. A **nerve fiber** is composed of an axon and its myelinating Schwann cells.

B. **Connective tissue sheaths**

1. In a gross nerve, each axon is surrounded by a delicate connective tissue sheath called the **endoneurium**. Endoneurium is not nervous tissue. (Only neurons and glia are properly classified as nervous tissue.) The endoneurium surrounds the Schwann cell basement membrane with a sheath of collagen fibers and encapsulating, flattened fibroblasts.

   a. Clusters of axons, Schwann cells, and endoneurial fibroblasts are gathered into bundles called **fascicles** (Figure 14-7).

   b. Fascicles are encapsulated by a connective tissue sheath of collagen fibers and fibroblasts called the **perineurium**.

2. In a gross nerve, bundles of fascicles are surrounded by a connective tissue capsule of collagen fibers, adipocytes, and blood vessels called the **epineurium**.

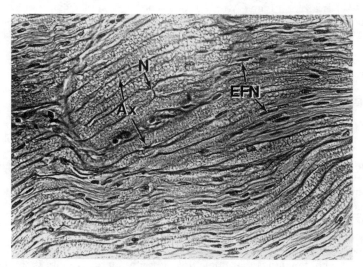

Figure 14-7. Light micrograph of a peripheral nerve fascicle. Visible are a node of Ranvier (*N*), an endoneurial fibroblast nuclei (*EFN*), and axons (*Ax*).

3. Capsules become smaller as the diameter of the nerve trunk decreases as the gross nerve leaves the CNS and divides into progressively smaller nerve branches.

C. Peripheral nerve endings. Most peripheral nerve endings perform a sensory function. They can be simple free nerve endings or more complex encapsulated nerve endings. (Chapter 13 discusses the connection between nerves and skeletal muscle at neuromuscular junctions.)

1. **Free nerve endings** are distributed throughout the body. They exist in skin, muscles, tendons, bones, and many visceral organs. They consist of small nerve fibers with a few branches and are involved in the sensation of pain.

2. **Encapsulated nerve endings**
   a. **Golgi tendon organs** are similar to muscle spindles and occur at muscle-tendon junctions.
   b. **Lamellar (Pacinian) corpuscles** (in skin, mucous membranes, and the cornea) and **genital corpuscles** (in external genitalia and nipples) are deep pressure receptors that have extremely thick capsules composed of multiple layers of concentrically arranged connective tissue cells that are continuous with the endoneurial cells surrounding the sensory fiber.
   c. **Meissner's corpuscles** are touch receptors located in dermal papillae in the skin of the soles, palms, and fingertips. They consist of a nerve ending associated with a roughly pear-shaped collection of connective tissue cells.

STUDY QUESTIONS

Directions: The groups of questions below consist of lettered choices followed by several numbered items. For each numbered item select the **one** lettered choice with which it is **most** closely associated. Each lettered choice may be used once, more than once, or not at all.

Questions 1–5

Match each description below with the most appropriate neuron structure.

(A) Nissl substance
(B) Neurotubules
(C) Synaptic vesicles
(D) Axons
(E) Dendrites

1. Cellular processes with many synapses that conduct nervous impulses toward the neuronal cell body

2. Large cytoplasmic aggregates of rough endoplasmic reticulum that stain intensely with basic dyes

3. Contains neurotransmitters such as acetylcholine and serotonin

4. Abundant in axons and involved in anterograde bulk cytoplasmic transport

5. Myelinated by Schwann cells and conducts nervous impulses away from the cell body

Questions 6–10

Match each description below with the most appropriate nervous system cells.

(A) Neurons
(B) Glial cells
(C) Both
(D) Neither

6. Cells that produce myelin

7. Cells that often have numerous processes radiating from the cell body

8. Cells that contain Nissl substance

9. Cells that are abundant in the spinal cord

10. Cells that line the neural canal and ventricles of the brain

Questions 11–15

Match each nerve fiber component described below with the appropriate lettered structure in the following electron micrograph.

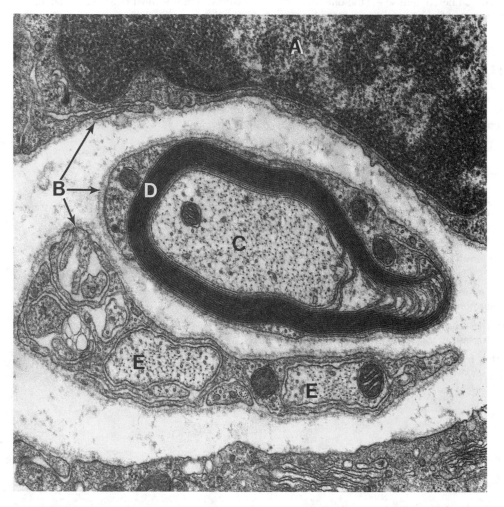

Micrograph courtesy of Dr. J.M. Rosenstein, Department of Anatomy, George Washington University.

11. Axon of a myelinated nerve

12. Schwann cell nucleus

13. Axon of an unmyelinated nerve

14. Schwann cell basement membrane

15. Layers of Schwann cell membrane

ANSWERS AND EXPLANATIONS

1–5. The answers are: 1-E, 2-A, 3-C, 4-B, 5-D. [*II A–C*] Nissl substance is the name given to the large aggregates of rough endoplasmic reticulum and free cytoplasmic ribosomes in the perikaryon of large motor neurons. These clusters cause cytoplasmic basophilia.

Neurotubules are the microtubules in neurons. They are especially abundant in axons, where they are involved in rapid axoplasmic flow away from the cell body (anterograde flow).

Synaptic vesicles are membrane-delimited sacs containing neurotransmitters such as acetylcholine and serotonin.

Axons are cellular processes that conduct nerve impulses away from the cell body. Motor neuron axons are long processes that are myelinated by numerous Schwann cells in the PNS.

Dendrites conduct nerve impulses toward the cell body. Their most striking structural characteristic is that they commonly are covered with many synapses.

6–10. The answers are: 6-B, 7-C, 8-A, 9-C, 10-B. [*I B 2; II B, C; III*] Neurons are the main functional cells in the nervous system. They conduct nerve impulses and communicate with each other through synapses. Motor neurons, located in the spinal cord anterior horns, are large cells that contain Nissl substance.

Glial cells perform ancillary, noncommunicative supportive functions in the nervous system. Myelin is produced by glial cells such as oligodendroglia in the CNS and Schwann cells in the PNS. Ependymal cells are glial cells that line the neural canal and ventricles of the brain; ependymal cells in the choroid plexus produce CSF.

Both neurons and glial cells often have a complex morphology that includes numerous radiating processes. Both cell types are abundant throughout the nervous system.

11–15. The answers are: 11-C, 12-A, 13-E, 14-B, 15-D. [*II C*] This is an electron micrograph of a myelinated nerve fiber. In the PNS, myelin (*D*) is produced by Schwann cells (nucleus labeled *A*), a type of glial cell. The axon (*C*) is surrounded by a myelin sheath. Each Schwann cell is surrounded by a basement membrane (*B*). Unmyelinated nerve fiber axons (*E*) often are embedded in Schwann cell cytoplasm without the formation of a thick myelin sheath.

15
The Cardiovascular System

I. INTRODUCTION. The cardiovascular system, which includes the heart and circulatory system, conducts blood through the body, ensuring that each cell receives a rich supply of oxygen and nutrients.

A. Components. The cardiovascular system consists of a four-chambered heart and a closed circulatory system of arteries and veins. Cardiac contraction and relaxation propels blood through the circulatory system, which includes the pulmonary circulation and the systemic circulation.

1. **Pulmonary circulation** carries blood from the heart to the lungs and then back to the heart.
   a. **Pulmonary arteries** carry deoxygenated blood to the lungs.
   b. **Pulmonary capillaries** in the lungs are specialized for gas exchange. Here, blood gives up carbon dioxide and takes on oxygen.
   c. **Pulmonary veins** return oxygenated blood to the heart.

2. **Systemic circulation** supplies blood to all other parts of the body.
   a. **Elastic arteries.** Oxygenated blood leaves the heart and enters the aorta, the largest elastic artery. The aorta branches into a complex network of muscular arteries that distribute blood to major subdivisions of the body.
   b. **Arterioles** are smaller arteries that branch from small muscular arteries and carry blood to a rich anastomosing network of capillaries.
   c. **Capillaries** are the smallest branches of the cardiovascular system and are extremely abundant and attenuated in peripheral tissues. Virtually all cells receive and expel metabolites, which diffuse through capillary walls.
   d. **Venules** are small vessels that receive drainage from the capillary network and carry it to veins.
   e. **Veins** receive blood from venules and carry it back to the heart. Near the heart, veins empty into the superior and inferior **venae cavae,** the largest veins in the body.

3. **Heart chambers.** The heart contains two atria and two ventricles.
   a. The right atrium receives venous drainage from the superior and inferior venae cavae. When the right atrium contracts, blood enters the right ventricle. When the right ventricle contracts, blood enters the pulmonary arteries (pulmonary circulation).
   b. After gas exchange occurs in the lungs, oxygenated blood flows back to the left atrium of the heart through pulmonary veins. When the left atrium contracts, blood enters the left ventricle. When the left ventricle contracts, blood is injected under high pressure into the aorta (systemic circulation).
   c. The period of ventricular contraction is called **systole.** The period of ventricular dilation is called **diastole.**

B. Morphology. The cardiovascular system is a specialized, functionally complex system for transporting erythrocytes and leukocytes in the peripheral blood. It exhibits considerable morphologic variation to perform its many functions.

1. For example, the walls of the heart chambers contain a myriad of cardiac muscle fibers that contract in an organized rhythm to propel blood.

2. Walls of muscular arteries and arterioles contain an abundance of smooth muscle fibers. These fibers contract and relax to change the luminal diameter, thereby maintaining peripheral vascular tone and regulating the flow of blood to organs during changing physiologic states.

C. Tissue organization

1. **Endothelium** is a simple squamous epithelium associated with a basement membrane. The entire cardiovascular system is lined by endothelium.

   a. In the heart, endocardial endothelium lines the chambers and coats all structures projecting into these chambers (e.g., ventricular trabecular muscles, valves).

   b. In major vessels, endothelium exists at the lumina and is continuous with the cardiac endothelium (Figure 15-1). Capillaries consist solely of endothelium and basement membranes and, perhaps, a small number of adventitial connective tissue fibers.

2. **Tunicae.** The walls of the heart and most blood vessels are composed of concentric layers of tissue (Figure 15-2).

   a. The **tunica intima** is the innermost tissue layer. In blood vessels, it is adjacent to the lumen and includes the endothelium and its basement membrane. In the heart, the tunica intima is called the **endocardium**.

   b. The **tunica media** is the middle tissue layer. In blood vessels, it is composed of smooth muscle cells, elastic fibers, collagen fibers, and some other connective tissue cells. In the heart, it contains cardiac muscle fibers instead of smooth muscle fibers. The tunica media in the heart is called the **myocardium**.

   c. The **tunica adventitia,** the outermost tissue layer, is composed of loose connective tissue and may contain adipose tissue. In the heart, the tunica adventitia is called the **epicardium**.

      (1) Often, connective tissue elements surrounding blood vessels attach the adventitial connective tissue to other parts of the body, resulting in an indistinct boundary between the tunica adventitia and surrounding tissues.

      (2) In contrast, the heart wall has a distinct outer boundary because the heart lies in the pericardial cavity—a closed serous cavity in the chest lined by a **mesothelium** (a simple squamous epithelium)—and is surrounded by a visceral layer called the **pericardium**.

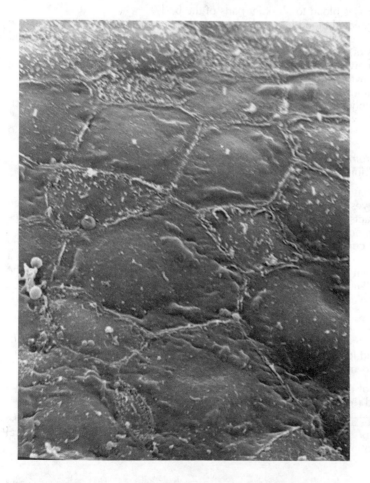

Figure 15-1. Scanning electron micrograph of the simple squamous epithelium present in the luminal surface of a blood vessel.

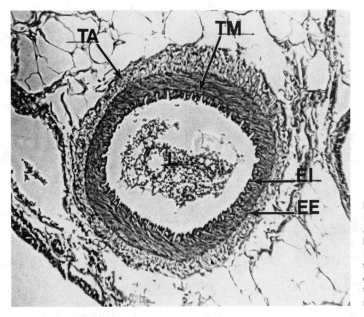

Figure 15-2. Light micrograph of the layers in the wall of a muscular artery. The lumen (L) is surrounded by a thin tunica intima. The elastica interna (EI) is the boundary between the tunica intima and the tunica media (TM). The elastica externa (EE) is the boundary between the tunica media and the tunica adventitia (TA).

3. **Blood vessel composition** varies tremendously; the exact composition of a blood vessel depends on its function. Some vessels do not contain all layers (e.g., capillaries consist solely of an endothelium and a basement membrane).

II. THE HEART

A. Endocardium (Figure 15-3)

1. Endocardium contains a smooth continuous endothelial layer that covers all inner surfaces of the heart, including the valves. Cells rest on a basement membrane and are tightly joined to each other by occluding tight junctions.

2. Beneath the endothelium is a **subendothelial component** of the endocardium. This component contains collagen fibers, elastic fibers, an amorphous ground substance containing proteoglycans and glycoproteins, fibroblasts, and scattered smooth muscle cells.

B. Myocardium (see Figure 15-3) is composed primarily of striated muscle cells called **cardiac muscle cells,** or **myocytes.**

1. Each cardiac muscle cell contains one central, elongated **nucleus** that includes some peripheral heterochromatin and a large amount of central euchromatin.

2. **Sarcomeres** dominate the cytoplasm and have the same banding patterns as skeletal muscle sarcomeres (see Chapter 13 III C). They also have thick and thin filaments like skeletal muscle.

3. **Cardiac muscle cells differ from skeletal muscle cells.**
   a. Cardiac muscle cells contain less sarcoplasmic reticulum than skeletal muscle cells, and terminal cisternae join broader, flatter T tubules. T tubules and cisternae form **diads** rather than the triads present in skeletal muscle.
   b. Cardiac muscle cells contain more mitochondria (35% of the cytoplasmic volume as compared to 2% in skeletal muscle cells).
   c. Cardiac muscle cells are joined by complex intercellular junctions called **intercalated disks** (Figure 15-4). Skeletal muscle cells exist as an anatomic syncytium.
      (1) Intercalated disks have a broad transverse contact region called the **fascia adherens**. Here, a small cleft exists between cells, similar to the cleft in the zonula adherens of epithelial junctional complexes.

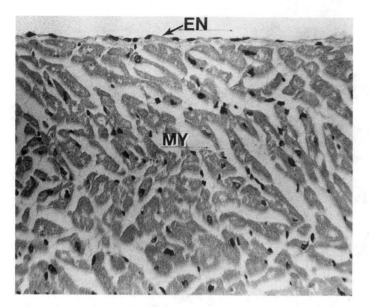

Figure 15-3. Light micrograph of the endocardium (*EN*) and myocardium (*MY*) in the heart.

(2) The fascia adherens is an adhesion site for adjacent cells and an intracellular anchoring site for the actin-rich thin filaments of sarcomeres; desmosomes with 10-nm tonofilaments are part of the fasciae adherentes and may help anchor cells together.

(3) Each intercalated disk has an extensive array of gap junctions in the longitudinal sarcolemma near the fascia adherens.

(4) Gap junctions couple cells and allow the free passage of ions and small molecules from cell to cell through their aqueous channels. This creates a low resistance pathway that links groups of cardiac muscle cells. Contraction-initiating impulses from the cardiac conducting system travel along these pathways to cause sudden, coordinated contraction of numerous myocytes.

4. Myocytes in the atria differ from myocytes in the ventricles.
   a. Atrial cells are smaller and lack the numerous T tubules present in ventricular cells.
   b. Atrial cell capacitance is lower and the conduction velocity of contraction-initiating impulses is greater because atrial cells lack T tubule surface membranes.
   c. Atrial cells have more gap junctions, which may improve conduction velocity.

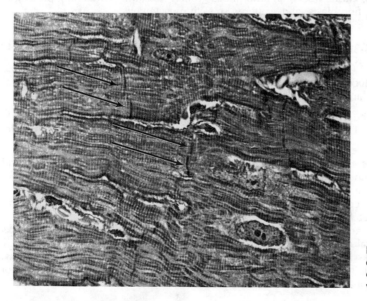

Figure 15-4. Light micrograph of cardiac muscular tissue, showing intercalated disks (*arrows*) joining individual cells.

 d. Atrial cells have a more prominent Golgi apparatus and dense cytoplasmic vesicles, which may contain catecholamine-like substances.

 5. **The cardiac skeleton** is a dense fibrous connective tissue mass that supports myocardial cells.
   a. The cardiac skeleton includes the **trigona fibrosa,** which is the membranous part of the interventricular septum, and the **annuli fibrosi,** which surround the atrioventricular and arterial foramina.
   b. Some parts of the cardiac skeleton contain **chondroid,** which is connective tissue that can have the appearance of fibrocartilage and may contain chondrocyte-like cells.
   c. In older individuals, the cardiac skeleton may be calcified to some extent.

 6. **The heart valves** consist of chondroid and are attached to the cardiac skeleton. Thin projections called **chordae tendineae** attach the valves to papillary muscles in the ventricles. Endocardial endothelium completely covers the valves, chordae tendineae, and papillary muscles.

 7. **The conducting system** of the myocardium initiates and coordinates muscle contraction.
   a. This system consists of the **sinoatrial (SA) node, atrioventricular (AV) node, AV bundle,** and **Purkinje fibers**.
   b. All parts of the conducting system are composed of modified cardiac muscle cells. For example, Purkinje fibers are larger than surrounding ventricular myocytes, have fewer myofibrils, and the myofibrils are disarrayed. The cytoplasm of Purkinje fibers appears pale and contains many glycogen granules, unlike surrounding ventricular myocytes.

C. Epicardium, the outermost layer of the heart, consists of a loose connective tissue of fibroblasts, collagen fibers, and adipose tissue. It is coated by the visceral pericardium and a simple squamous mesothelium.

III. ARTERIAL CIRCULATION

A. Large elastic arteries such as the aorta, pulmonary arteries, common carotid arteries, and common iliac arteries connect the heart and large muscular arteries. These vessels have a large lumen and a relatively thin wall (about 10% of their diameter), and they contain many medial elastic fibers (Figure 15-5).

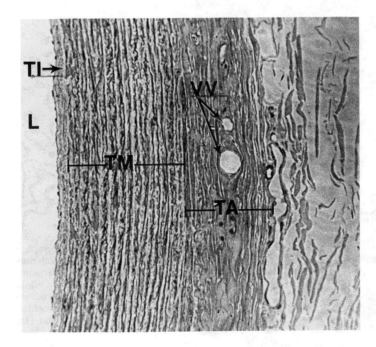

Figure 15-5. Light micrograph of the wall of the aorta, a large elastic artery. The tunica intima (*TI*) is a thin layer at the lumen (*L*). The tunica media (*TM*) contains many layers of elastic laminae. The tunica adventitia (*TA*) contains vasa vasorum (*VV*).

1. The tunica intima has an unremarkable endothelium of cells joined by extensive zonulae occludentes and numerous gap junctions, which lies on a thin basement membrane. The intima contains elastic fibers, collagen fibers (Types I, III, and IV), smooth muscle cells, and metachromatic amorphous ground substance. In the aorta, the intima constitutes approximately 25% of the artery wall. An **elastica interna** boundary separates the intima from the media.

2. The tunica media of elastic arteries contains 50 or more concentric fenestrated elastic laminae. Interlaminar bridges of elastic tissue may connect laminae, and some collagen fibers and smooth muscle cells are interspersed between laminae (Figure 15-6). A poorly defined **elastica externa** boundary separates the media from the adventitia.

3. The tunica adventitia is thin and consists of scattered collagen fibers and small elastic fibers. Small blood vessels called **vasa vasorum,** small lymphatics, and nerve fibers penetrate into the media from the adventitia.

B. Muscular arteries are 1.0–0.03 cm in diameter. They have a relatively thick wall (about 25% of their diameter) and contain more smooth muscle and fewer elastic fibers in their media than elastic arteries. Otherwise, they are structurally similar to large elastic arteries.

1. Muscular arteries have continuous endothelial epithelium overlying a thin, continuous basement membrane. Endothelial cells are joined by occluding junctions and large gap junctions.

2. Collagen fibers and smooth muscle cells are present in the subendothelial compartment of the intima.

3. All muscular arteries have prominent elastica interna and elastica externa that form the inner and outer boundaries of the tunica media.

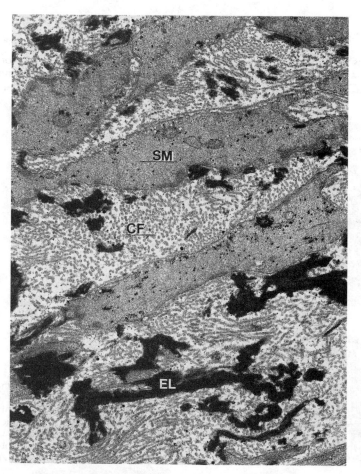

Figure 15-6. Electron micrograph of the media of the aorta. Smooth muscle cells (*SM*) are mixed with numerous collagen fibers (*CF*) and elastic laminae (*EL*). (Micrograph courtesy of Dr. E.N. Albert, Department of Anatomy, George Washington University.)

4. The media consists principally of smooth muscle cells surrounded by a basement membrane. Large muscular arteries have up to 50 concentric layers of smooth muscle cells interspersed with thin fenestrated elastic laminae.

5. The elastica externa boundary separates the media from the adventitia. In small muscular arteries, the elastica externa is sparse or absent (Figure 15-7).

6. The adventitia of muscular arteries is relatively thick compared to the vessel diameter and contains many collagen fibers, fibroblasts, adipocytes, and a few smooth muscle cells. Vasa vasorum, lymphatic vessels, and nerve fibers in the adventitia project into the media.

C. Arterioles branch from muscular arteries and are the smallest arteries in the body.

1. Many of the larger arterioles have an elastica interna and one or two medial layers of smooth muscle cells.

2. The adventitia merges into the connective tissue of the surrounding organ.

3. Many arterioles have precapillary sphincters, which open and close every few seconds, near the entrance to capillary beds.

4. The smallest arterioles do not have an elastica interna, and their media may contain only scattered, modified smooth muscle cells called **pericytes**.

IV. CAPILLARIES

A. Continuous capillaries are the most common and widespread variety of capillaries.

1. **Locations.** Continuous capillaries are present in many connective tissues, the dermis of skin, the brain, skeletal muscle, bones, lungs, and some exocrine glandular organs such as the pancreas and salivary glands. They also exist in parts of the testes and ovaries.

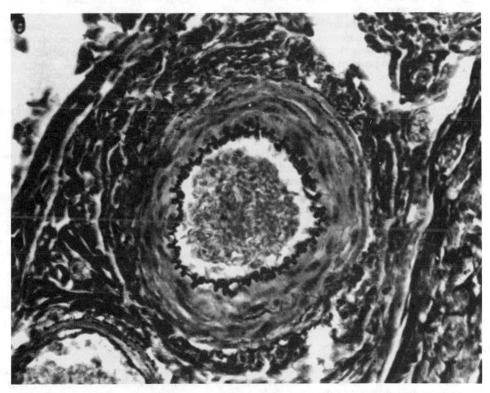

Figure 15-7. Light micrograph of a small muscular artery. This blood vessel is clogged with clumped erythrocytes. The elastica interna shows very clearly because the elastic fibers have been specially stained. The tunica media is primarily smooth muscle cells, and the adventitia contains a mixture of collagenous and elastic fibers. (Reprinted from Johnson KE : *Histology: Miscroscopic Anatomy and Embryology.* New York, John Wiley, 1982, p 159.)

2. Histology

    a. Continuous capillaries consist of a continuous endothelial luminal epithelial layer where endothelial cells are joined by extensive occluding junctions and gap junctions.

    b. These capillaries prevent the leakage of macromolecules in either direction between the blood and organ parenchyma. They have a thin, but continuous basement membrane and contain numerous pinocytic and endocytic vesicles, which control the passage of materials into and out of the blood.

    c. Continuous capillary walls are extremely attenuated, which also facilitates diffusion of materials into and out of the blood. For example, pulmonary capillaries may have a total mural thickness of as little as 100 nm.

B. Fenestrated capillaries

    1. Locations. Fenestrated capillaries are present in endocrine tissues, the renal glomerulus and around renal tubules, the choroid plexus, the mucosa of the gastrointestinal tract, the ciliary body of the eye, and the stria vascularis in the membranous labyrinth of the ear.

        a. In endocrine tissue, the fenestrations probably are important in the transport of hormones into the blood.

        b. In the kidneys, fenestrated capillaries are important for blood filtration and urine production.

        c. In some locations, fenestrated capillaries help produce important blood filtrates (e.g., cerebrospinal fluid in the choroid plexus, aqueous humor in the ciliary body, and endolymph in the stria vascularis).

    2. Histology. Fenestrated capillaries have extremely attenuated walls and a thin, continuous basement membrane. They contain endothelial cells traversed by transcellular pores called **fenestrae**.

        a. Fenestrae are about 70 nm in diameter and are bounded by plasma membranes.

        b. Many fenestrae are covered by a diaphragm, which appears as a single thin line of electron-dense material in the electron microscope. The composition and function of diaphragms is not known.

        c. Fenestrated renal capillaries do not have diaphragms across their fenestrae and have an unusually thick basement membrane, which is intimately apposed to the podocyte basement membrane. These two thick basement membranes constitute a selective barrier that allows some serum proteins to pass into the glomerular filtrate while barring others (see Chapter 26 II C 2 a).

C. Discontinuous capillaries form an incomplete endothelial barrier between vascular channels and organ parenchyma. Large intercellular gaps exist between endothelial cells, and the basement membrane is either discontinuous or nonexistent.

    1. Discontinuous capillaries are abundant in the liver sinusoids, where they allow proteins synthesized in parenchymal cells to enter blood circulating through the sinusoids. Some liver sinusoid littoral cells are differentiated into the phagocytic **Kupffer cells** of the mononuclear phagocyte system. Sinusoids in other parts of the body do not contain a discontinuous endothelium. For example, bone marrow sinusoids are lined by continuous capillaries, and sinusoids in the adrenal cortex and other endocrine glands are lined by fenestrated capillaries.

    2. Highly modified discontinuous capillaries are present in splenic red pulp (see Chapter 12 V D).

    3. Lymphatic capillaries are discontinuous capillaries with extremely attenuated or nonexistent basement membranes.

V. VENOUS CIRCULATION

A. Venules

    1. The postcapillary **pericytic venules** are leaky blood vessels. Their endothelial intercellular junctions are not as intimate as those formed in continuous capillaries and arterioles, and tracer proteins leak passively from them.

        a. Endothelial cells in pericytic venules have receptors that have a high affinity for vasoactive substances such as **histamines**. These venules become leaky when stimulated by histamine (e.g., during inflammation).

    b. Pericytic venules have a thin basement membrane associated with their endothelium and a few pericytes that contain numerous microfilaments.

  2. Pericytic venules empty into **muscular venules,** slightly larger vascular channels that have a layer or two of smooth muscle cells in their media. In tissue sections, muscular venules usually parallel small arterioles.

    a. Muscular venules have a continuous endothelium that does not leak because cells are joined by prominent zonulae occludentes and gap junctions.

    b. The media contains several layers of smooth muscle cells; the substantial but thin adventitial connective tissue layer merges indistinguishably into the surrounding connective tissue of the organ.

B. Veins

  1. Small veins receive venous drainage from muscular venules. Their diameter is larger than venules because their media contains more layers of smooth muscle cells and also contains scattered elastic fibers. The elastic fibers in veins are not arranged as orderly as elastic fibers in arteries, and thus the elastica interna and elastica externa are not prominent.

  2. Venous valves. As veins increase in diameter, they acquire pairs of valves. These are especially prominent in the veins of the lower extremities where they prevent reflux of venous blood.

  3. Large veins. In some larger veins, longitudinal muscle fibers are prominent in the media (veins in the limbs) or in the adventitia (veins in the abdominal cavity). Large veins also have vasa vasorum, nerves, and lymphatic vessels in the adventitia that penetrate part of the way through to the media.

STUDY QUESTIONS

Directions: Each question below contains five suggested answers. Choose the **one best** response to each question.

1. Each of the following statements concerning muscular arteries is true EXCEPT

(A) they contain more elastic fibers than smooth muscle cells
(B) the media contains smooth muscle cells
(C) they have an endothelium at the lumen
(D) they have an elastica externa
(E) the media contains elastic fibers

2. Each of the following statements concerning cardiac muscle cells is true EXCEPT

(A) they have less sarcoplasmic reticulum than skeletal muscle cells
(B) they are abundant in the walls of the ventricles and atria
(C) they are abundant in the epicardium
(D) they contain T tubules
(E) gap junctions facilitate ionic communication between cardiac muscle cells

3. Each of the following statements concerning large veins is true EXCEPT

(A) they have valves
(B) they have well-defined elastica interna and externa
(C) the adventitia may contain smooth muscle
(D) the media contains elastic fibers
(E) the media contains smooth muscle

4. Each of the following statements concerning continuous capillaries is true EXCEPT

(A) they have a basement membrane
(B) they have occluding junctions between cells
(C) they are abundant in splenic red pulp
(D) they have gap junctions
(E) they are abundant in skeletal muscle

5. Which of the following statements best characterizes venules?

(A) They carry blood to capillary networks
(B) Their walls are thicker than arteriole walls
(C) Their adventitia is distinct from surrounding connective tissues
(D) Their media contains many layers of smooth muscle
(E) They have histamine receptors and leak fluids during inflammation

6. Many blood vessels have elastic fibers in their tunica media. Which of the following blood vessels contains the most elastic fibers?

(A) The radial artery
(B) The renal artery
(C) The aorta
(D) The inferior vena cava
(E) The superior vena cava

7. Which of the following blood vessels has the thickest wall relative to its diameter?

(A) The inferior vena cava
(B) An arteriole
(C) The saphenous vein
(D) A venule
(E) The ulnar artery

Directions: The groups of questions below consist of lettered choices followed by several numbered items. For each numbered item select the **one** lettered choice with which it is **most** closely associated. Each lettered choice may be used once, more than once, or not at all.

Questions 8–12

Match each description below with the appropriate lettered structure in the following micrograph.

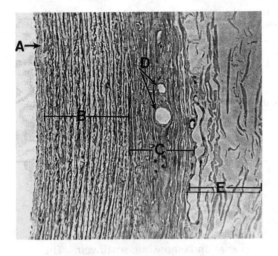

8. Layer containing many concentric elastic laminae and smooth muscle fibers

9. Layer that does not contain smooth muscle fibers or an endothelial layer

10. Layer that does not contain elastic laminae but has an endothelial layer

11. The tunica media

12. Supplies blood for cells in the vessel wall

Questions 13–17

Match each characteristic listed below with the most appropriate type of capillary.

(A) Continuous
(B) Fenestrated
(C) Discontinuous
(D) Lymphatic

13. Abundant in skeletal muscle

14. Predominant in areas producing blood filtrates such as urine

15. Line liver sinusoids

16. Present in the choroid plexus

17. Form the blood-brain barrier

ANSWERS AND EXPLANATIONS

1. The answer is A. [*III B*] Most of the named arteries in the body (e.g. ulnar artery) are muscular arteries. The media of muscular arteries contains an abundance of both smooth muscle cells and elastic fibers; however, it contains more smooth muscle cells than elastic fibers. In contrast, the media of large elastic arteries such as the aorta contains more elastic fibers than smooth muscle cells. Like all blood vessels, muscular arteries have a luminal endothelium. They also have a well-defined elastica interna and externa.

2. The answer is C. [*II B, C*] Cardiac muscle cells are present throughout the myocardium and the proximal roots of the venae cavae and pulmonary arteries; they are not present in the epicardium. Like skeletal muscle cells, cardiac muscle cells contain prominent sarcomeres, sarcoplasmic reticulum, and T tubules; however, they have less sarcoplasmic reticulum than skeletal muscle cells. Cardiac muscle cells are joined by intercalated disks containing numerous gap junctions with aqueous channels for intercellular ionic communication.

3. The answer is B. [*V B*] Many large veins contain valves that prevent reflux of blood. They have a thick media that is rich in smooth muscle cells and elastic fibers; however, they do not have a well-defined elastica interna and externa due to the disorganization of elastic fibers. In some veins, such as veins in the abdominal cavity, the adventitia is rich in smooth muscle.

4. The answer is C. [*IV C 2*] The body contains three types of capillaries: continuous, fenestrated, and discontinuous. Continuous capillaries are the most common type. They are surrounded by a basement membrane and have tight occluding junctions and gap junctions between cells. They are abundant in brain parenchyma, connective tissues, and skeletal muscle. Fenestrated capillaries are present in endocrine tissues. Splenic red pulp contains highly modified discontinuous capillaries.

5. The answer is E. [*V A 1 a*] Venules receive blood from capillaries and empty into small veins. Their walls are thinner than arteriole walls because their media, which consists of a few smooth muscle cells or pericytes, is thinner. They have a thin adventitia with fibroblasts that merges with surrounding connective tissues. Some venules become leaky when stimulated by histamine, especially during inflammation.

6. The answer is C. [*III A*] The media of each vessel listed contains some elastic fibers and smooth muscle; however, the media of large elastic arteries consists of many concentric elastic laminae. The aorta is the largest elastic artery in the body and, therefore, contains the greatest number of elastic fibers. The left ventricle ejects blood under pressure high enough to distend the aorta. The elastic laminae in the media of the aorta expand to absorb some of this pressure and then return the pressure to the system as they relax. This smooths the flow of blood and is energy efficient.

7. The answer is E. [*III B*] All large blood vessels have a thick media containing many smooth muscle cells. Among the vessels listed, the ulnar artery—a large muscular artery—has the thickest media relative to its diameter because blood pressure is highest in muscular arteries. The inferior and superior venae cavae are large veins that have relatively thin walls compared to arteries of the same size, because blood pressure is lower in veins than in arteries. Capillaries and venules have an extremely thin media, which contains only a few smooth muscle cells or a thin layer of pericytes.

8–12. The answers are: 8-B, 9-C, 10-A, 11-B, 12-D. [*I C; III A*] This is a micrograph of the wall of the aorta, the largest elastic artery in the body. The tunica intima (A) is a thin layer at the lumen and contains the endothelium. It is the innermost layer of a blood vessel, lying between the lumen and the media, and is homologous to the endocardium. In large vessels such as the aorta, the intima consists of the endothelium and its basement membrane and a thin connective tissue domain containing smooth muscle cells, fibroblasts, and collagen fibers. The tunica media (B) is a thick layer dominated by numerous concentric layers of elastic fibers interspersed with a few smooth muscle cells and fibroblasts. The tunica adventitia (C) surrounds the tunica media and joins the blood vessel to surrounding connective tissues (E). The adventitia contains no smooth muscle and merges into surrounding connective tissue. The vasa vasorum (D) supply blood to the many living cells that comprise the vessel wall.

13–17. The answers are: 13-A, 14-B, 15-C, 16-B, 17-A. [*IV*] Continuous capillaries are present in skeletal muscle, the lungs, and the brain. They form a continuous epithelial barrier that restricts diffusion of materials from the blood into the tissues. Bone marrow has continuous capillaries, with large pores that open from time to time, allowing passage of cells from the hematopoietic compartment into the circulation.

Fenestrated capillaries are present in renal glomeruli. The fenestrations are holes through the walls of capillary endothelial cells.

Discontinuous capillaries are present in the liver and spleen. Gaps large enough to allow passage of cells exist between the endothelial cells of discontinuous capillaries.

Cerebrospinal fluid, a blood filtrate containing some of the proteins found in whole blood, is produced in the choroid plexus. Fenestrated capillaries in the choroid plexus allow a select subset of blood proteins to enter cerebrospinal fluid.

Continuous capillaries in the brain form the blood-brain barrier by restricting the flow of some substances into the brain parenchyma.

I. INTRODUCTION

A. Components. The respiratory system is composed of the nasal passages; the olfactory epithelium; a conducting airway, which includes the nasopharynx, larynx, trachea, bronchi, and bronchioles; and the respiratory portions of the lungs.

B. Functions. The respiratory system modifies inspired air. Air is cleansed, warmed or cooled, and moistened. The system also facilitates olfaction, respiratory gas exchange, and sound production.

C. Specialization for respiration

1. **Alveolar epithelium.** Gas exchange occurs in the lungs through the air-liquid interface of the thin alveolar epithelium. Three specializations in this epithelial lining facilitate gas exchange. Epithelial cells are very thin, they are closely associated with a rich plexus of capillaries, and some cells are specialized to secrete surfactant.

2. **Surfactant** forms a thin film over the entire alveolar epithelium, reducing the surface tension of the air-liquid interface, which is an enormous force potentially capable of collapsing the lungs.

II. NASAL CAVITIES AND OLFACTORY EPITHELIUM

A. The nasal cavities modify incoming air, making it innocuous to the lower respiratory system.

1. The **external nares** have watery filtering secretions and large hairs.

2. Three plates of scalloped bones project into the nasal cavities, forming the superior, middle, and inferior nasal **conchae**. The nasal conchae create turbulent air flow, which traps particulate matter in the moist nasal mucosa.

3. The nasal mucosa is covered by a **pseudostratified ciliated columnar epithelium**.
   a. Glands secrete mucus and watery secretions to coat and moisten the epithelium.
   b. Cilia beat synchronously to move mucus and trapped particulate debris into the digestive tract.
   c. The mucosal lining is associated with an extensive vascular bed that warms, cools, and moistens inspired air.

B. The olfactory epithelium. The upper posterior portion of the nasal cavities contain two patches of olfactory epithelium just lateral to the nasal septum.

1. The olfactory epithelium is moistened by serous secretion products from **Bowman's glands**. Ducts from these glands empty onto the surface of the olfactory mucosa.

2. The **olfactory mucosa** also has a pseudostratified ciliated columnar epithelium, which makes the mucosa quite tall. This epithelium is modified for olfaction (Figure 16-1).

3. The olfactory epithelium contains four types of **cells**.
   a. **Olfactory cells** have very long, nonmotile cilia containing chemoreceptors, which recognize the structural differences of odoriferous substances (Figure 16-2).

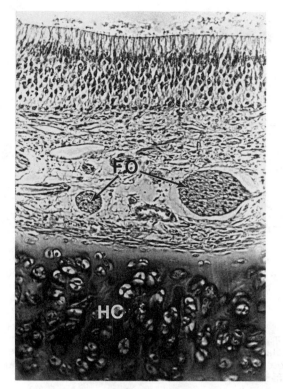

Figure 16-1. Light micrograph of the olfactory epithelium. At the top of the micrograph, the epithelium is tall and pseudostratified. Fila olfactoria (*FO*) are evident, and hyaline cartilage (*HC*) of the nasal septum is shown.

(1) Serous glands in the mucosa moisten these cilia to dissolve chemicals in inspired air. Receptor interaction with stimulant molecules causes depolarization of the olfactory cell membrane, with subsequent generation of an action potential.

(2) Olfactory cells are modified neurons. They have a dendrite and an axon, and they synapse with neurons in the olfactory bulb. Numerous epithelial cell axons are gathered into **fila olfactoria,** which carry the action potential through the cribriform plate of the ethmoid bone to the olfactory lobe of the central nervous system (CNS).

(3) Compared to other animals, the olfactory cells of humans have relatively short cilia, which results in a rather poorly developed sense of smell. For example, cats, which have a keen sense of smell, have cilia that are up to 80 μm long.

b. **Sustentacular cells** have apical microvilli and a well-developed apical Golgi complex and, therefore, look like secretory cells.

c. **Basal cells** are undifferentiated but have the ability to divide and differentiate into olfactory cells or sustentacular cells.

d. **Brush cells** are cells with apical microvilli (see Figure 16-2).

III. THE NASOPHARYNX AND LARYNX

A. **The nasopharynx** connects the nasal cavities to the larynx and the rest of the respiratory system. The mucosa of the nasopharynx is quite thin and has few glands.

1. The nasopharynx abuts on the oropharynx where the respiratory and digestive systems converge in the pharynx; however, the boundary between the nasopharynx and the oropharynx is histologically indistinct.

2. Most of the nasopharyngeal mucosa is covered by a pseudostratified ciliated columnar epithelium, although patches of stratified squamous nonkeratinized epithelium may spill over from the oropharynx.

3. A small **pharyngeal tonsil (adenoid)** is located in the dorsomedial portion of the nasopharynx.

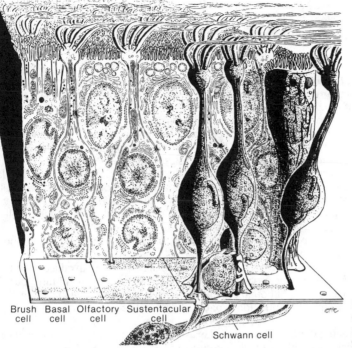

Brush   Basal   Olfactory   Sustentacular
cell   cell   cell   cell

Schwann cell

Figure 16-2. Diagram of the various cells of the olfactory epithelium. (Reprinted from Weiss L: *Histology*, 5th ed. New York, Elsevier, 1983, p 794.)

    4. Stratified squamous epithelium covers the lower pharynx and the **epiglottis**. The epiglottis is a flap of tissue containing elastic cartilage, which directs swallowed food into the esophagus and prevents it from entering the larynx.

B. The larynx is an expanded hollow portion of the respiratory system located between the nasopharynx and the trachea. It is composed of cartilages, ligaments, muscles, and a mucosal surface. The larynx prevents ingested solids and liquids from entering the respiratory system and also contains the vocal cords, which produce speech sounds.

    1. Gross anatomy
      a. The **glottis** is the laryngeal aperture; it is flanked by paired folds of the laryngeal mucosa known as the **vocal folds**.
      b. The laryngeal body is capped by the hyoid bone, contains a large broad **thyroid cartilage,** and rests on circular **cricoid cartilage** atop the trachea.
      c. The upper dorsal portion contains paired **arytenoid and corniculate cartilages**.
      d. The rim of the larynx contains **cuneiform cartilages** and dense bands of fibrous connective tissue and intrinsic laryngeal muscles. The ventral rim contains the epiglottis.
      e. The laryngeal mucosa has two conspicuous folds. The cranial fold is called the **ventricular fold** or **false vocal fold**. The caudal fold is called the **true vocal fold**. Deep recesses called the **laryngeal ventricles** exist between these two folds.

    2. Histology
      a. The mucosa is coated by a highly variable **epithelium**.
        (1) The epithelium is stratified squamous in the upper portion, over much of the epiglottis, and over the surface of the vocal folds, which is subjected to regular abrasion during speech.
        (2) Pseudostratified ciliated columnar epithelium covers most of the remaining laryngeal mucosa, although stratified columnar epithelium may exist in transitional zones.
      b. **Glands** that produce a protective mucus are scattered throughout the laryngeal mucosa and are particularly abundant in the ventricular folds.
      c. **Laryngeal cartilages** are hyaline (thyroid) or elastic (cuneiform, corniculate) and may ossify somewhat later in life. The arytenoid cartilage has elastic and hyaline portions.
      d. Laryngeal intrinsic and extrinsic **muscles** are ordinary skeletal muscle.

IV. THE TRACHEA carries air between the larynx and the bronchi. At its base, the trachea bifurcates to form the left and right main stem bronchi. The tracheal wall has four layers: an inner **mucosa,** a **submucosa,** a poorly differentiated **muscularis,** and an outer **adventitial layer** (Figure 16-3).

A. Tracheal mucosa

1. **Cells of the tracheal epithelium.** The trachea has a relatively large lumen lined by a pseudostratified columnar ciliated epithelium containing six or more cell types (Figure 16-4).

   a. **Goblet cells** synthesize and secrete mucous droplets. They have an apical dilation containing mucous droplets, a prominent Golgi apparatus, and a well-developed basal rough endoplasmic reticulum. Goblet cells have apical microvilli at their edges and centrioles among the mucous droplets. Under appropriate stimulation, goblet cells release mucous droplets and some strands of apical cytoplasm (**apocrine secretion**).

   b. **Ciliated cells** have numerous cilia that project into the mucus and move it toward the larynx. They also have a moderately large Golgi apparatus, a small amount of rough endoplasmic reticulum, and a few lysosomes and residual bodies (Figure 16-5).

      (1) Each cilium is anchored in a basal body, which is anchored in the ciliated cell apical cytoplasm.

      (2) Many mitochondria exist near the basal bodies, providing the adenosine triphosphate (ATP) required for ciliary beating.

      (3) Often, ciliated cells contain two basal centrioles not associated with cilia as basal bodies.

   c. Undifferentiated **short cells,** or **basal cells,** rest on the basement membrane but do not extend to the tracheal lumen. Although undifferentiated, these cells have the ability to divide and probably are a stem cell population capable of differentiating into other cell types in the epithelium.

   d. The tracheal epithelium contains two types of **brush cells**. (Brush cells are so-named because they have an apical microvillous brush border.)

      (1) The **brush₁ cell** has unusually long microvilli and is innervated by small afferent nerve fibers.

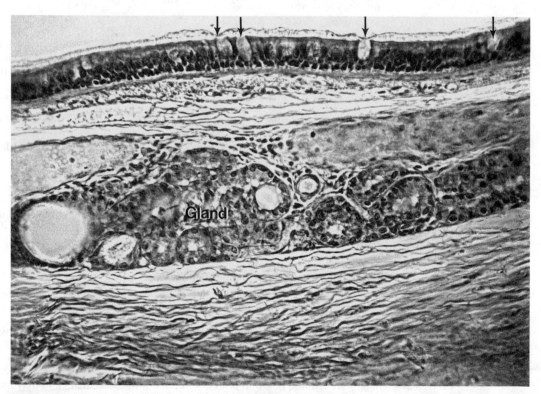

Figure 16-3. Light micrograph of the wall of the trachea. The mucosa has a pseudostratified ciliated columnar epithelium with goblet cells (*arrows*). The submucosa in the center of this micrograph contains a large gland.

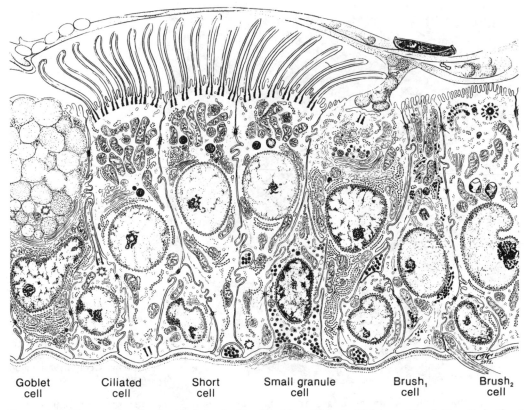

Figure 16-4. Diagram of tracheal epithelial cells. (Reprinted from Weiss L: *Histology*, 5th ed. New York, Elsevier, 1983, p 805.)

   (2) The **brush₂ cell** has a pair of apical centrioles and may be a short cell in the process of differentiating into a ciliated cell.
   e. Basal **small granule cells** are filled with cytoplasmic granules that are 100–300 nm in diameter. These granules are catecholamine-like materials that control the secretory activities of goblet cells and glands and that affect ciliary activity. Granules usually are located in the basal portion of the cells near mucosal blood vessels, which presumably transport their secretion products.
      (1) Small granule cells are part of the **amine precursor uptake and decarboxylation** system of cells **(APUD cells)**. Some tracheal APUD cells are short basal cells nestled among neighboring short cells; others are taller and have apical portions that reach the lumen.
      (2) Some tracheal APUD cells are innervated, while others are not. (This is also true for APUD cells elsewhere in the body, such as in the gastrointestinal tract or the islets of Langerhans in the pancreas.)
      (3) All tall tracheal epithelial cells are joined into a continuous epithelial sheet by apical junctional complexes.

 2. Other features of the tracheal epithelium
   a. Free **nerve endings** penetrate and enter the epithelium.
   b. The tracheal epithelium, like the laryngeal and bronchial epithelium, contains scattered **mucous glands** with serous demilunes.
   c. The tracheal epithelium rests on a thick **basement membrane**.

 3. Lamina propria. The tracheal mucosa also has a lamina propria containing a loose network of areolar connective tissue.
   a. The lamina propria contains an irregular weave of collagen and elastic fibers and an abundance of fibroblasts.
   b. Solitary lymphocytes, small nodular lymphocyte aggregates, plasma cells, macrophages, and granular leukocytes often are scattered throughout the lamina propria.

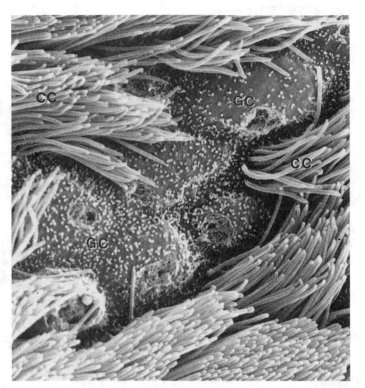

Figure 16-5. Scanning electron micrograph of the tracheal epithelium, showing ciliated cells (*CC*) and goblet cells (*GC*). (Micrograph courtesy of Dr. E.N. Albert, Department of Anatomy, George Washington University.)

4. **The elastic membrane,** a layer of elastic fibers, separates the mucosa and submucosa.

5. **The submucosa** extends from the elastic membrane to the perichondrium of the tracheal cartilages and consists principally of collagen fibers, fibroblasts, blood vessels, and lymphatic vessels.

B. Tracheal cartilage. The tracheal wall also contains many C-shaped **hyaline cartilages**.

1. The open portion of the **C** is directed posteriorly, toward the esophagus, and is bridged by connective tissue and bundles of smooth muscle fibers.

2. The tracheal cartilage perichondrium merges with the fat-laden connective tissue of the adventitia, which also contains blood vessels, nerves, and lymphatic vessels as well.

V. BRONCHI AND BRONCHIOLES

A. Bronchi. In most respects, the microscopic anatomy of larger bronchi is quite similar to the microscopic anatomy of the trachea.

1. Bronchi have a pseudostratified ciliated columnar **epithelium** containing numerous goblet cells, macrophages, and fibroblasts. However, the submucosa is thinner than the tracheal submucosa.

2. Bronchial **cartilages** have irregular shapes but still form a skeleton that helps maintain bronchial lumen patency.

3. The bulk of the bronchial wall is composed of **smooth muscle** fibers and **irregular cartilaginous plates**. As bronchi size decreases, the amount of cartilage present decreases, and the amount of smooth muscle in the mural portion of the bronchi increases.

4. Bronchi contain **glands** that are quite similar to tracheal glands. These glands sometimes contain myoepithelial cells.

5. In **distal smaller bronchi,** cartilages are smaller, glands are less conspicuous, and the pseudostratified ciliated columnar epithelium is lower columnar epithelium.

B. Bronchioles. The smallest bronchi communicate with a system of bronchioles.

1. **Large bronchioles** have a simple columnar **epithelium** of ciliated cells, bronchiolar cells, scattered brush cells, and a few small granule cells. They also contain interlacing smooth muscle cells and fibroblasts but no cartilage.
   a. **Ciliated cells** are more plentiful than bronchiolar cells in large bronchioles. However, as bronchioles become smaller (increasingly distal), the relative proportion of bronchiolar cells increases. Ciliated cells in the bronchioles are not as tall as ciliated cells in the trachea, but otherwise they have a similar ultrastructure.
   b. **Bronchiolar (Clara) cells** are characteristic of the bronchioles.
      (1) They have a peculiar domed apex that contains numerous apical granules of a complex and poorly characterized mixture of proteins, polysaccharides, and lipids.
      (2) They contain basal rough endoplasmic reticulum, apical smooth endoplasmic reticulum, a poorly developed Golgi apparatus, and apical granules of secretion product, and they secrete a surfactant-like material.
      (3) The lateral boundaries of bronchiolar cells are elaborately folded, and fenestrated capillaries are present just beneath the basement membrane, suggesting that bronchiolar cells help produce blood filtrates to moisten the bronchiolar epithelium.

2. **Terminal bronchioles** are the most distal simple bronchioles; they communicate with respiratory bronchioles.

3. **Respiratory bronchioles** have alveoli in their walls. These alveoli have the same microscopic anatomy as other alveoli (see VI). The remainder of the wall contains a cuboidal epithelium consisting principally of ciliated cells and bronchiolar cells; a basement membrane; and a thin layer of smooth muscle cells, fibroblasts, and collagen fibers.

VI. RESPIRATORY PORTION OF THE LUNGS

A. Distal airways

1. Respiratory bronchioles communicate with **alveolar ducts,** the most distal component of the conducting airways.

2. Alveolar ducts end in cul-de-sacs called **alveolar sacs** (Figure 16-6).

3. The walls of alveolar ducts and alveolar sacs are composed of **alveoli,** structures where gas exchange occurs. Each alveolus is surrounded by an anastomosing network of capillaries.

B. Alveolar epithelium and gas exchange. Alveoli are lined by **alveolar epithelium,** which consists of two distinct cell types joined in a continuous epithelial sheet by apical junctional complexes. Alveolar walls are specialized for gas transport.

1. The alveolar epithelium is exceedingly thin (usually much less than 1 μm thick) and rests on a typical basement membrane (Figure 16-7).

2. Capillary endothelial cells and connective tissue fibroblasts are intimately associated with the alveolar epithelium.

3. Gases in the alveolar space diffuse through the alveolar cell plasma membrane, the basement membrane, and the capillary endothelial cell plasma membrane into the blood. Gases in the blood diffuse back through these membranes into the alveolar space.

C. Cells of the alveolar epithelium (Figure 16-8)

1. **Squamous alveolar epithelial cells** (type I cells) are extremely flat cells primarily involved in lining the alveolar space. They are specialized for gas exchange.
   a. Type I alveolar cells have a nucleus that is hard to distinguish from a capillary endothelial cell nucleus or a connective tissue fibroblast cell nucleus.
   b. Their cytoplasm is unremarkable except that it contains a number of **pinocytotic vesicles**.

2. **Great alveolar cells** (type II cells) are joined by junctional complexes to the type I cells, forming an epithelium that is continuous with the bronchial and bronchiolar epithelium and with skin and gut. These rounded cells protrude into the alveolar spaces and have many secretory cell characteristics.

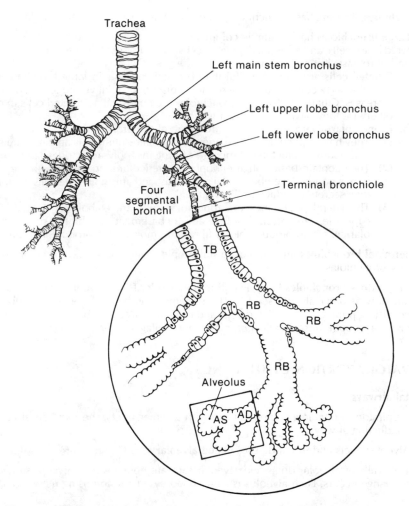

Figure 16-6. Diagram of the branching pattern of the airway in the respiratory system. The *circled area* is an acinus, which is sometimes defined as all units of the lung distal to the terminal bronchioles; an acinus is sometimes called a secondary lobule. The *boxed area* is a primary lobule, which is defined as all units distal to an alveolar duct. Labeled structures include respiratory bronchioles (*RB*), terminal bronchiole (*TB*), alveolar sac (*AS*), and alveolar duct (*AD*). (Reprinted from Johnson KE: *Histology: Microscopic Anatomy and Embryology.* New York, John Wiley, 1982, p 176.)

    a. Type II alveolar cells have a moderate amount of rough endoplasmic reticulum and an extensive Golgi apparatus.

    b. They contain **multivesicular bodies, multilamellar bodies** (or **cytosomes**), and intermediate structures (Figure 16-9). Cytosomes are a secretory product of type II cells, which spread over the surface of the alveoli. Cytosomes contain phospholipids, mucopolysaccharides, and proteins.

        (1) Dipalmatoyl lecithin, a predominant phospholipid in multilamellar bodies, is an active component of the **pulmonary surfactant**.

        (2) Polar phospholipid molecules in surfactant reduce the surface tension of the water layer on alveolar surfaces, preventing their collapse.

    c. Type II cells are conspicuously vacuolated. Vacuoles are the remnants of extracted cytosomes.

  D. Alveolar macrophages are part of the mononuclear phagocyte system. They continuously cleanse the epithelial surface and protect the alveolar epithelium from damage by microorganisms or other inhaled irritants by engulfing and digesting foreign material. They digest material by fusing the membranes of phagocytic vacuoles and lysosomes (sacs of hydrolytic enzymes).

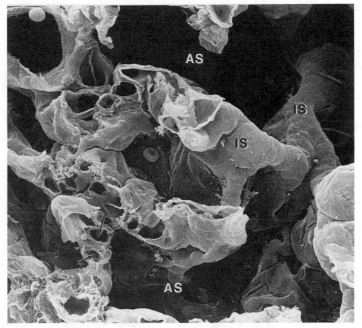

Figure 16-7. Scanning electron micrograph of the distal airways of the lung, showing alveolar sacs (*AS*), interalveolar septa (*IS*), and capillaries (*arrows*). (Micrograph courtesy of Dr. E.N. Albert, Department of Anatomy, George Washington University.)

E. Lung histology

1. Except at the hilus, the lungs are surrounded by a thin superficial connective tissue capsule. A simple squamous mesothelium rests on this capsule.

2. Pleural mesothelial cells have sparse apical microvilli, secrete watery lubricants into the pleural cavity, and rest on a thin basement membrane.

3. The mesothelium defines the two pleural cavity boundaries and allows the lungs to move freely next to the body wall and diaphragm.

4. Alveolar capillaries and alveoli are held together by thin strands of connective tissue. Where they are intimately associated, the connective tissue domain is little more than a collagen fibril or an attenuated fibroblast process.

5. Larger connective tissue domains surround large branches of the airway. For example, they hold bronchi and pulmonary arteries together.

6. Other connective tissue elements join veins, lymphatics, and nerves to other lung structures, and connective tissue septa divide the bronchopulmonary segments.

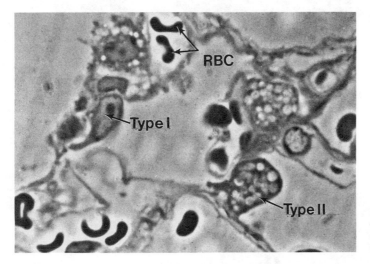

Figure 16-8. Light micrograph of alveoli, showing both types of alveolar cells (*Type I* and *Type II*) and erythrocytes (*RBC*) within capillaries.

Cytosomes

Multivesicular
body

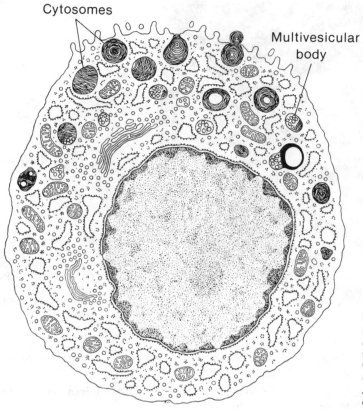

Figure 16-9. Diagram of the electron microscopic morphology of a type II alveolar cell, showing the multivesicular bodies and the numerous cytosomes, or multilamellar bodies. (Reprinted from Lentz TL: *Cell Fine Structure.* Philadelphia, WB Saunders, 1971, p 221.)

STUDY QUESTIONS

Directions: Each question below contains five suggested answers. Choose the **one best** response to each question.

1. Each of the following statements concerning the olfactory epithelium is true EXCEPT

(A) Bowman's glands secrete a serous product
(B) sustentacular cells contact olfactory cells
(C) olfactory cells have long motile cilia
(D) basal cells rest on a basement membrane
(E) axons project from the olfactory epithelium into the central nervous system

2. Each of the following statements concerning the alveolar epithelium is true EXCEPT

(A) type I and type II cells are joined by junctional complexes
(B) the epithelium rests on a well-developed basement membrane
(C) alveolar epithelial cells are closely associated with capillary endothelium
(D) ciliated cells are present in alveoli
(E) type II cells contain multilamellar bodies of surfactant

3. Each of the following statements concerning bronchi is true EXCEPT

(A) they contain numerous goblet cells
(B) they contain few smooth muscle cells
(C) they contain cartilage plates
(D) they contain numerous ciliated cells
(E) they contain small granule cells

4. Each of the following statements concerning goblet cells is true EXCEPT

(A) they contain a poorly developed Golgi apparatus
(B) they contain abundant rough endoplasmic reticulum
(C) they secrete mucous droplets
(D) they are abundant in the trachea
(E) they are not present in respiratory bronchioles

5. Each of the following statements concerning type I alveolar cells is true EXCEPT

(A) they are squamous cells
(B) they are specialized for gas exchange
(C) they are present in terminal bronchioles
(D) they rest on a basement membrane
(E) they are close to capillary endothelial cells

6. Each of the following statements concerning type II alveolar cells is true EXCEPT

(A) they secrete surfactant
(B) they contain multilamellar bodies
(C) they secrete substances rich in phospholipids
(D) they are tightly joined to type I cells
(E) they are squamous cells

Directions: The groups of questions below consist of lettered choices followed by several numbered items. For each numbered item select the **one** lettered choice with which it is **most** closely associated. Each lettered choice may be used once, more than once, or not at all.

Questions 7–10

Match each characteristic below with the most appropriate airways.

(A) Bronchi
(B) Large bronchioles
(C) Both
(D) Neither

7. Airways with smooth muscle in their walls

8. Airways with cartilage in their walls

9. Airways with ciliated cells in their mucosa

10. Airways with alveoli in their walls

Questions 11–15

Match each distal lung component described below with the appropriate lettered structure in the following micrograph.

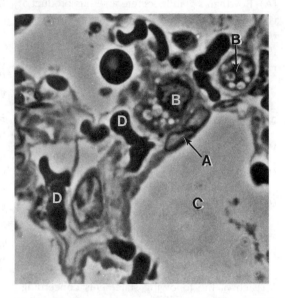

11. A cell that carries carbon dioxide and oxygen

12. A cell that secretes surfactant

13. A squamous type I cell

14. An area that frequently contains macrophages

15. A cell that contains phospholipid-rich multi-lamellar bodies

ANSWERS AND EXPLANATIONS

1. The answer is C. [*II B*] Olfactory cells have long cilia but they are not motile. Olfactory cells are neurons; each cell has an axon and a dendrite. Numerous axons of olfactory epithelial cells are gathered into fila olfactoria, which carry the action potential to the central nervous system. Serous secretions from Bowman's glands keep these cilia moist. The sustentacular cells of the olfactory epithelium are tightly bound to ciliated cells by tight junctional complexes. All cells in this pseudostratified epithelium, including basal cells, rest on the basement membrane.

2. The answer is D. [*VI C*] Alveolar epithelium does not contain ciliated cells. The alveolar epithelium is very thin and rests on a typical basement membrane. Capillary endothelial cells and connective tissue fibroblasts are intimately associated with the alveolar epithelium. This epithelium contains type I cells (squamous cells specialized for gas exchange) and type II cells (rounded cells that secrete surfactant), which are joined by junctional complexes to form a continuous epithelium.

3. The answer is B. [*V A 3*] Like the trachea, bronchi have an epithelium that contains goblet cells, ciliated cells, and granule cells. Bronchial walls contain an abundance of smooth muscle and irregularly shaped plates of cartilage that help maintain luminal patency. Bronchi differ from the trachea in that they have plates of cartilage rather then C-shaped cartilages and more smooth muscle.

4. The answer is A. [*IV A 1 a*] Goblet cells secrete large amounts of mucous glycoproteins and, therefore, have abundant endoplasmic reticulum and a well-developed Golgi apparatus. Goblet cells are present in the trachea, bronchi, and large bronchioles. Terminal bronchioles contain Clara cells instead of goblet cells. Respiratory bronchioles do not contain goblet cells or Clara cells; instead, their walls contain alveoli—cells involved with gas exchange.

5. The answer is C. [*VI C 1*] Type I alveolar cells are part of the alveolar epithelium. They are thin cells specialized for gas exchange, which rest on a basement membrane. Type I cells are present in alveoli and respiratory bronchioles, but they are not present in terminal bronchioles. The basement membrane of the alveolar epithelium is thin and closely associated with the thin basement membrane of capillaries.

6. The answer is E. [*VI C 2*] Type II alveolar cells are part of the alveolar epithelium; they are joined to type I alveolar cells by extensive junctional complexes. Unlike squamous type I cells, type II cells are rounded cells containing an abundance of multilamellar bodies. Type II cells create surfactant by synthesizing and secreting multilamellar bodies that are rich in phospholipid. The surfactant reduces the surface tension at the gas-liquid interface in alveoli.

7–10. The answers are: 7-C, 8-A, 9-C, 10-D. [*V A, B*] The chief difference between bronchi and large bronchioles is that the bronchioles lack mural cartilage. Both have smooth muscle in their walls and goblet and ciliated cells in their mucosa. Large bronchioles lead to terminal bronchioles, which lead to respiratory bronchioles. Respiratory bronchioles are the first portion of the conducting airway with mural alveoli for gas exchange.

11–15. The answers are: 11-D, 12-B, 13-A, 14-C, 15-B. [*VI B, C, D*] Alveoli (*C*) in the lungs are lined by an epithelium that contains type I cells and type II cells. Type I cells (*A*) are extremely attenuated squamous cells specialized for gas exchange. Type II cells (*B*) are rounded cells containing phospholipid-rich multilamellar bodies of secretion product (surfactant). Alveoli are adjacent to capillaries, which are filled with erythrocytes (*D*). Macrophages are abundant in alveoli, where they phagocytose and destroy inspired debris such as bacteria.

I. GENERAL FEATURES OF THE GASTROINTESTINAL TRACT

A. Components and functions

1. **Upper gastrointestinal tract**
   a. The upper gastrointestinal tract consists of the **oral cavity,** which includes the lips, tongue, oral mucosa, and salivary glands; the **pharynx;** and the **esophagus**.
   b. The upper portion of the digestive tract is involved in food intake, preparation of food for swallowing and digestion, partial digestion of complex carbohydrates, and conveyance of food from the oral cavity to the stomach.

2. **Lower gastrointestinal tract.** The components and functions of the lower gastrointestinal tract are discussed in Chapter 18.

B. Structure. The particular anatomic features of each gastrointestinal segment are dictated by the segment's function in the intake, digestion, assimilation, and elimination of food. The following are examples of upper gastrointestinal tract features.

1. The **oral cavity** has 32 teeth that crush and tear food, a tongue that forms food into a bolus, and glands that moisten food and initiate carbohydrate digestion.

2. The **esophagus** has a stratified squamous mucosal epithelium, which resists the abrasion associated with the passage of semi-solid masses of food. It also has abundant mural muscular tissue, which helps pass food to the stomach.

C. Tissue organization in hollow visceral organs. Many visceral organs have similar tissue organization. This section describes general aspects of tissue organization that apply to most visceral organs. Figure 17-1 illustrates the cardinal structural features of a gastrointestinal organ.

1. **Lumen and mucosa.** Many visceral organs are hollow structures with a central cavity called the **lumen**. In the gastrointestinal tract, food passes through the lumen. The **mucosa** is the layer of tissue adjacent to the lumen.
   a. The mucosa consists of a **mucosal epithelium** at the boundary of the lumen, an epithelial basement membrane, a **lamina propria** or loose areolar connective tissue domain, and a **muscularis mucosae,** which is a thin layer of smooth muscle fibers that forms a boundary between the mucosa and the submucosa. The mucosa may have intrinsic **glands** or glands that extend into the submucosa.
   b. The **mucosal epithelium** (and associated glands) **is highly variable** and may undergo abrupt transitions in type. For example, an abrupt transition from stratified squamous esophageal epithelium to simple columnar gastric epithelium occurs at the gastroesophageal junction. No intermediate zones exist at these transitions; however, boundaries may have irregular shapes.
   c. The mucosal lamina propria contains loose areolar connective tissue that is rich in fixed fibroblasts and wandering leukocytes. The lamina propria also has an abundance of capillaries and lymphatic vessels. Lamina propria lymphoid tissue can range from solitary lymphocytes to large lymphoid nodules.

2. **Submucosa.** The submucosa, which surrounds the mucosa, contains connective tissue with many collagenous fibers and, in some instances, elastic fibers. The submucosa connects the mucosa to the next outermost layer, the **muscularis externa**.

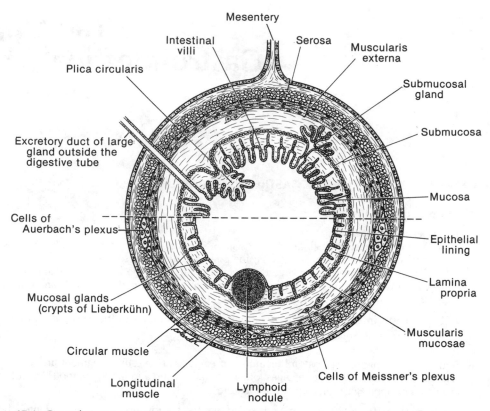

Figure 17-1. General tissue organization of moist visceral organs, such as those in the gastrointestinal tract. (Reprinted from Bloom W, Fawcett DW: *A Textbook of Histology*, 10th ed. Philadelphia, WB Saunders, 1975, p 599.)

3. **Muscularis externa.** In most sections of the gastrointestinal tract (and in other visceral organs), the muscularis contains several layers of smooth muscle cells arranged in diverse patterns that correspond to the shape and function of the organ.
   a. The muscles in this layer mix and propel food in the gastrointestinal tract. Autonomic innervation stemming from the **myenteric** or **Auerbach's plexus** controls smooth muscle motor activity.
   b. In the pharynx and upper esophagus, the muscularis externa contains skeletal muscle that is under voluntary control.

4. **Adventitia.** This is the outermost mural layer of visceral organs. In some organs, the adventitia is a **serosa** or coating of reflected peritoneal mesothelium. The adventitia consists of connective tissue and fat and often contains large blood vessels that penetrate the outer layers of the organ to the capillary beds in the lamina propria.

5. **Glands.** Digestive organs and tubular visceral organs have numerous associated glands. These glands can be contained in, or can empty onto, the mucosal epithelium.
   a. The simplest glands are unicellular, micro-secreting glands called **goblet cells** (see Chapter 16 IV A 1 a). Goblet cells and other closely related cells are prominent in the stomach and intestine.
   b. Small glands usually are contained within the mucosa. Moderately large glands may extend into the submucosa.
   c. Large extrinsic glands such as salivary glands, the liver, and the pancreas have ducts that empty into the gastrointestinal tract.
   d. Glandular secretions are as complex and varied as glandular structure, and they perform many essential functions in the digestive process. These secretions contain several kinds of mucous secretions, serous secretions with digestive enzymes, ions, bile salts, secretory immunoglobulins, and bacteriostatic substances.

II. THE ORAL CAVITY

A. The lips protect the opening of the oral cavity. They have a cutaneous (outer) layer, a labial (inner) layer, and a pigmented boundary called the vermillion.

    1. The **cutaneous layer** is covered by stratified squamous keratinized epithelium, like the rest of the skin, and has hairs and eccrine sweat glands.

    2. The **vermillion** is covered by stratified squamous keratinized epithelium that does not have hairs or glands.

    3. The **labial layer** is covered by nonkeratinized stratified squamous epithelium and contains numerous small mucous glands.

    4. The lips contain abundant skeletal muscle (**orbicularis oris**).

B. The lining of the oral cavity

    1. The oral cavity is lined by stratified squamous epithelium that is either keratinized (gingiva, hard palate) or nonkeratinized (soft palate, buccal mucosa).

    2. The lateral boundaries of the oral cavity are histologically similar to the lips; they contain skeletal muscle and have an outer layer of skin and an inner layer of nonkeratinized stratified squamous epithelium.

C. Teeth have a crown covered by **enamel** (a secretion product of **ameloblasts**); the crown rests on **dentin** (a secretion product of **odontoblasts**). Teeth also have a **pulp cavity,** which contains blood vessels and nerves, and **roots,** which anchor the teeth to the jaw bones.

    1. Ameloblasts (Figure 17-2) are tall cells that form an epithelial layer with a basement membrane within each developing tooth. They secrete an unusual extracellular matrix that calcifies to form enamel.

        a. As the cell prepares to secrete enamel, the rough endoplasmic reticulum and Golgi apparatus migrate toward the basal surface of the cell.

        b. Secretory granules containing the extracellular organic matrix of enamel accumulate in the basal portion of the cell.

            (1) The organic portion of enamel contains proteins and glycoproteins different than collagen and keratin.

            (2) The inorganic portion is predominantly hydroxyapatite but also contains trace amounts of other calcium salts. The crystalline structure of this hydroxyapatite differs from the crystalline structure of bone and dentin hydroxyapatite.

        c. The secretory granules are released into the forming enamel region, where the organic material calcifies almost immediately.

        d. Mature enamel differs from bone in that it lacks ameloblast processes. This accounts for the extreme hardness of enamel.

    2. Odontoblasts (see Figure 17-2) secrete dentin, a living, bone-like component of teeth. They are extremely elongated cells that form an epithelioid layer in the developing tooth.

        a. Odontoblasts recede centripetally as they secrete predentin, an uncalcified extracellular matrix that calcifies to form dentin. Consequently, the connective tissue pulp becomes progressively smaller.

        b. Odontoblast processes are left behind as these cells create progressively thicker layers of dentin.

D. The tongue

    1. Layers. The tongue is composed of three orthogonally arranged layers of skeletal muscle fibers, a superficial layer of connective tissue where the boundary between the submucosa and lamina propria is indistinct, and a dorsal and ventral covering of stratified squamous epithelium that is modestly keratinized on the dorsal (upper) surface (Figure 17-3).

    2. Papillae. The anterior two-thirds of the tongue is covered by numerous papillae (i.e., surface projections). In some cases, annular depressions surround the papillae. The posterior third of the tongue is closely associated with paired lingual and palatine tonsils and does not have papillae. There are four types of papillae on the tongue (Figure 17-4).

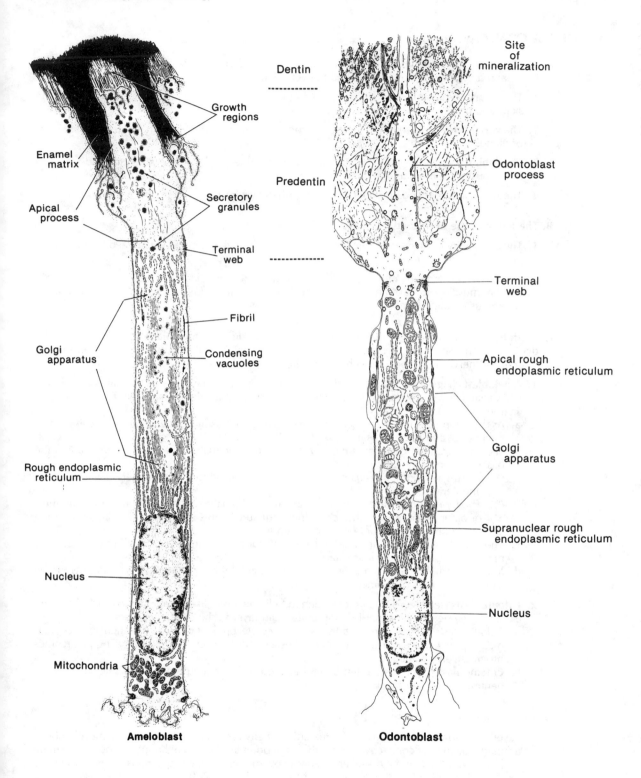

Ameloblast

Odontoblast

Figure 17-2. Diagram of the ultrastructure of the ameloblast and odontoblast. (Reprinted from Weinstock DL, Leblond CP: *Cell Biology* 60:92, 1974; *J Cell Biol* 51:26; 1971.)

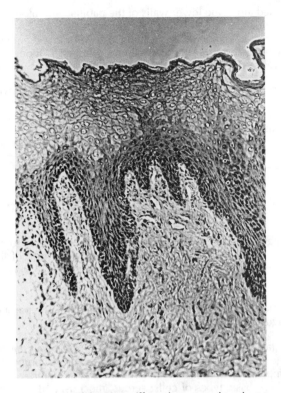

Figure 17-3. Light micrograph of the keratinized stratified squamous epithelium of the dorsal surface of the tongue.

a. **Filiform papillae,** the most abundant type, form parallel rows that diverge from the midline of the tongue. These flame-like structures are covered by an irregular epithelium and contain a core of lamina propria, which may be extensively branched. They do not contain taste buds.

b. **Fungiform papillae** are abundant and are scattered among the filiform papillae. They are mushroom-shaped structures with a constricted base and an expanded, flattened upper section. The lamina propria usually contains a large primary core and several secondary branches. These papillae contain a few taste buds.

c. **Circumvallate papillae** exist only in the V-shaped boundary between the anterior and posterior sections of the tongue.
   (1) The tongue usually has nine of these large round papillae, which are surrounded by a conspicuous collar and a deep furrow.

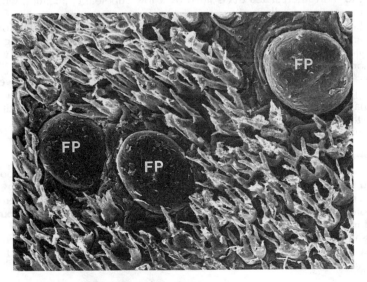

Figure 17-4. Scanning electron micrograph of the long, thin, bifurcated, filiform papillae (throughout micrograph) and three round, blunt, fungiform papillae (*FP*) of the tongue. (Micrograph courtesy of Dr. E.N. Albert, Department of Anatomy, George Washington University.)

      (2) Hundreds of taste buds are located along the lower walls of the papillae and along the walls of the circumferential recess.

      (3) Nearby **glands of von Ebner** secrete a watery, proteinaceous material into the circumferential crypt, which probably helps dissolve taste constituents of ingested food.

   d. Foliate papillae are very poorly developed in humans and are rarely observed.

  3. Taste buds. Each taste bud is a group of epithelial cells arranged around a small central cavity like slices in an orange.

   a. A **taste pore** is located on the apical surface of the taste bud. A nerve fiber from the basal surface synapses with some of the taste bud cells and enters the taste pore.

   b. Taste bud cells have apical microvilli that project into the taste pit and presumably contain chemoreceptors for the four primary taste sensations: sweet, salt, bitter, and acid.

III. SALIVARY GLANDS

A. Minor salivary glands. The oral cavity contains numerous minor salivary glands. In addition to the labial glands and the glands of von Ebner, other minor salivary glands exist in the root of the tongue. These lingual salivary glands produce a mixture of mucousal and serousal secretions.

B. Major salivary glands. The oral cavity contains three types of major salivary glands: the parotid, submandibular (submaxillary), and sublingual glands.

  1. General structure. The major salivary glands are well-circumscribed, surrounded by a connective tissue capsule, and drained by a large duct that empties into the oral cavity. All are compound tubuloacinar glands that contain a complex system of ducts that drain secretory acini.

   a. Cells. The acini are composed of three basic types of cells: serous, mucous, and myoepithelial cells. The relative number of serous and mucous cells varies widely among the types of glands.

      (1) Serous cells have a basal nucleus, a prominent basal rough endoplasmic reticulum, a prominent apical Golgi complex, numerous mitochondria, and numerous apical zymogen granules containing PAS-positive vesicles of glycoprotein secretion product.

      (2) Mucous cells are similar to the goblet cells located elsewhere in the gastrointestinal tract.

      (3) Myoepithelial cells are stellate cells situated between secretory cells of the acini and the acinar basement membrane. They are rich in cytoplasmic microfilaments that contain actin and probably are contractile elements that help express acinar secretion products.

   b. Duct system

      (1) Acini connect to short **intercalated ducts** lined by a low cuboidal epithelium, which may contain myoepithelial cells.

      (2) Intercalated ducts drain into **striated ducts (secretory ducts),** which modify the ionic composition and tonicity of saliva. They are lined by columnar cells containing an acidophilic cytoplasm and have prominent basal surface invaginations associated with many mitochondria.

      (3) Striated ducts empty into large **interlobular ducts** in the connective tissue septa between lobules of the gland. Distal sections of interlobular ducts have a columnar epithelium, larger ducts have a stratified columnar epithelium, and the main duct has a stratified squamous epithelium.

      (4) Interlobular ducts join the **primary duct,** which empties into the oral cavity.

  2. Type-specific structure

   a. Parotid glands contain only serous acini. They have long interlobular ducts and poorly developed striated ducts (Figure 17-5).

   b. Submandibular glands have many **mucous acini** associated with **serous demilunes**. Serous demilunes are more numerous in these glands than in sublingual glands. Submandibular glands have short intercalated ducts and extensive prominent striated ducts (Figure 17-6).

   c. Sublingual glands are composed almost exclusively of mucous acini associated with serous demilunes (far fewer than in submandibular glands). The intercalated and striated ducts of sublingual glands are poorly developed.

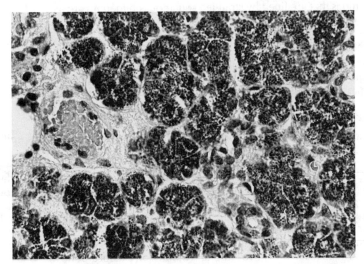

Figure 17-5. Light micrograph of the parotid salivary gland, showing numerous darkly-stained serous acini.

3. **Saliva** is a complex mixture of water, ions, enzymes, glycoproteins, proteoglycans, sialomucins, sulfomucins, immunoglobulins, and bacteriostatic substances.
   a. Saliva is a hypotonic solution that is secreted continuously under the control of the autonomic nervous system. Approximately 1 L of saliva is secreted each day.
   b. The ionic composition of saliva is considerably different from that of blood. Saliva contains some nuclease activity and one major digestive enzyme, **α-amylase,** which initiates carbohydrate digestion in the oral cavity.
   c. Saliva is rich in secretory immunoglobulin A and peroxidase substances, which limit bacterial flora in the oral cavity.

IV. THE PHARYNX AND ESOPHAGUS

A. The pharynx

1. The oropharynx is lined by a nonkeratinized stratified squamous epithelium. It has a well-developed lamina propria containing many lymphocytes and lymphoid nodules. The palatine, lingual, and pharyngeal tonsils (see Chapter 12 IV A) are close to the pharynx.

2. Instead of a muscularis mucosae, the pharynx has a thick prominent layer of elastic fibers.

3. The muscularis externa consists of several overlapping layers of skeletal muscle.

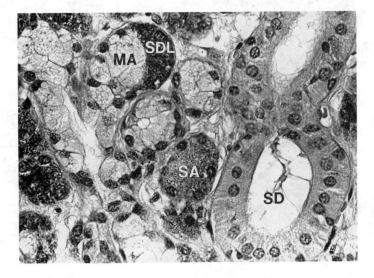

Figure 17-6. Light micrograph of the submandibular salivary gland, showing mucous acini (*MA*), serous acini (*SA*), a mucous acinus with a serous demilune (*SDL*), and a striated duct (*SD*).

4. The oropharyngeal portions that are lined by stratified squamous epithelium are associated with large mucous glands, which often penetrate deep into the muscularis externa.

B. The esophagus passes food from the oropharynx to the stomach.

1. Distensibility. The esophagus is quite distensible due to a series of longitudinal folds in the mucosa and part of the submucosa. These folds flatten as a food bolus approaches and then reform after the bolus passes. Distension of the esophagus stimulates muscular contraction, which helps propel food toward the stomach.

2. Mucosa

   a. The esophageal mucosa is covered by a thick **nonkeratinized stratified squamous epithelium** that rests on a well-developed basement membrane (Figure 17-7).

   (1) In some areas, especially near the top of the esophagus and at the junction between the esophagus and stomach, the epithelium has deep glandular invaginations. Most of these glands are located in the mucosa, although some extend into the submucosa. In both instances, the glands contain only mucous acini.

   (2) The acini in submucosal glands are connected to the surface epithelium by ducts, the largest of which are lined by a stratified squamous or cuboidal epithelium.

   b. The **lamina propria mucosae** in the esophagus is not as cellular as it is in more distal regions of the gastrointestinal tract. It does not have lymphoid infiltration except around mucous gland orifices.

   c. The **muscularis mucosae** is quite pronounced and begins at about the level of the cricoid cartilage, where it replaces the pharyngeal elastic fibers, which run longitudinally.

3. Submucosa. The submucosa is rather hypocellular, contains some mucous glands, and carries an abundant blood supply.

4. Muscularis externa. The esophageal muscularis externa has several anatomic variations from the pattern found elsewhere in the gastrointestinal tract.

   a. In the upper quarter of the esophagus, the muscularis externa contains a circular or obliquely spiraling inner layer of **skeletal muscle fibers** surrounded by a longitudinal outer layer of **skeletal muscle fibers**. Muscle movement in the upper quarter of the esophagus is under voluntary control.

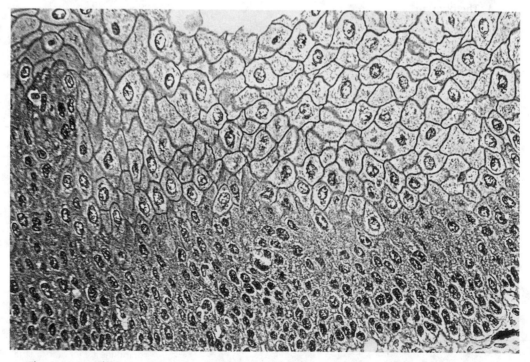

Figure 17-7. Light micrograph of the nonkeratinized stratified squamous epithelium of the esophagus.

    b. The next distal quarter of the esophagus has intermingled skeletal and smooth muscle fibers. Progressing distally, the number of smooth fibers gradually increases.

    c. In the lower half of the esophagus, muscle fibers are exclusively smooth, but they are still arranged in circular inner layers and longitudinal outer layers.

5. Adventitia. The esophageal adventitia is a thin layer of collagenous and elastic fibers, which passes through the mediastinum between the pleural cavities and pierces the diaphragm.

    a. The adventitia does not have a mesothelium until it pierces the diaphragm.

    b. The short pregastric segment of the adventitia below the diaphragm is covered by a thin reflection of the peritoneal mesothelium and, thus, is properly called a **serosa**.

STUDY QUESTIONS

Directions: Each question below contains four or five suggested answers. Choose the **one best** response to each question.

1. Each of the following statements concerning the teeth is true EXCEPT

(A) ameloblasts secrete dentin and enamel
(B) dentin is softer than enamel
(C) odontoblasts secrete a bone-like extracellular matrix
(D) enamel does not have cells or cellular processes
(E) ameloblasts secrete the extracellular matrix of enamel basally

2. Each of the following statements concerning tissue structure in the pharynx is true EXCEPT

(A) it is coated with stratified squamous epithelium
(B) its mucosal epithelium is nonkeratinized
(C) it contains numerous small glands
(D) it has a poorly defined muscularis mucosae
(E) its muscularis externa contains skeletal muscle

3. Each of the following statements concerning the esophagus is true EXCEPT

(A) its upper quarter contains skeletal muscle
(B) its lower half does not contain skeletal muscle
(C) it has numerous mucous glands
(D) its muscularis mucosae is poorly defined

4. Each of the following statements concerning the parotid salivary gland is true EXCEPT

(A) it is predominantly composed of serous acini
(B) it has poorly developed striated ducts
(C) it produces saliva that lacks digestive enzymes
(D) its secretion product is hypotonic
(E) its secretion product contains antibacterial substances

5. Each of the following statements concerning taste buds is true EXCEPT

(A) taste buds are scattered over much of the surface of the tongue
(B) filiform papillae contain taste buds that are sensitive to salt
(C) taste buds are abundant in circumvallate papillae
(D) taste buds have associated nerve endings
(E) the epithelial cells in taste buds have microvilli

Directions: The groups of questions below consist of lettered choices followed by several numbered items. For each numbered item select the **one** lettered choice with which it is **most** closely associated. Each lettered choice may be used once, more than once, or not at all.

Questions 6–9

Match each description of a structural component of a salivary gland with the appropriate lettered structure in the micrograph below.

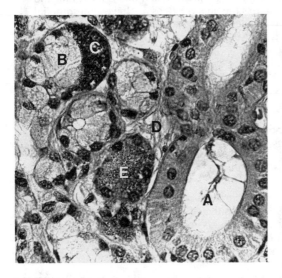

6. Conveys acinar secretions toward the oral cavity

7. Holds acini together

8. Secretes mucus

9. An acinus that secretes a watery, protein-rich material

Questions 10–14

Match each description of a component or region of the esophageal mucosa with the appropriate lettered area of the micrograph below.

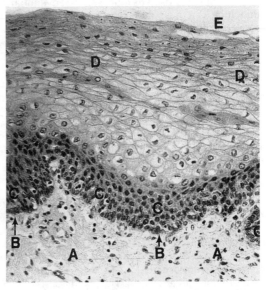

(Reprinted from Johnson KE: *Histology: Microscopic Anatomy and Embryology.* New York, John Wiley, 1982, p 207.)

10. Cells that are sloughed from the epithelium

11. Basement membrane between the epithelium and the lamina propria

12. Region where glands and lymphocytes may be abundant

13. Actively dividing layer of cells

14. Loose areolar connective tissue

ANSWERS AND EXPLANATIONS

1. The answer is A. [*II C 1*] The crown of each tooth contains enamel. The enamel layer rests on a layer of dentin, which surrounds the pulp cavity. Odontoblasts are the cells that secrete predentin, an extracellular matrix that calcifies to form dentin. Ameloblasts are the cells that basally secrete the extracellular matrix of enamel. Enamel is devoid of cells and cellular processes, which partly accounts for enamel being much harder than dentin. In contrast, dentin is more like bone; it has a bone-like calcified extracellular matrix and odontoblast processes throughout.

2. The answer is A. [*IV A*] The pharynx has a nonkeratinized stratified squamous epithelium and numerous small glands in its mucosa. The adventitial surface of the pharynx does not have a mesothelial coating. It lacks a muscularis mucosae but has numerous elastic fibers in the region where a muscularis mucosae would exist. Skeletal muscle is abundant in the muscularis externa of the pharynx.

3. The answer is D. [*IV B 2*] The esophagus has skeletal muscle in the upper quarter, intermingled skeletal and smooth muscle in the second quarter, and only smooth muscle in the lower half. The esophagus has numerous mucous glands in its mucosal and submucosal layers and a well-defined muscularis mucosae consisting of smooth muscle fibers.

4. The answer is C. [*III B 2 a, 3*] The parotid salivary gland consists primarily of serous acini and secretes saliva. The gland has poorly developed striated ducts. Saliva is hypotonic and contains secretory immunoglobulins and other antibacterial substances as well as the digestive enzyme α-amylase for carbohydrate digestion.

5. The answer is B. [*II D 2 a, 3*] Filiform papillae do not contain taste buds. Taste buds are groups of epithelial cells arranged like the segments of an orange. The epithelial cells in each taste bud have apical microvilli and are associated with a nerve fiber that conducts sensory impulses to the central nervous system. Taste buds are sparse, but present, in fungiform papillae and, therefore, are widely distributed on the tongue's surface. They are most abundant in circumvallate papillae.

6–9. The answers are: 6-A, 7-D, 8-B, 9-E. [*III B 2 b; Figure 17-6*] This is a light micrograph of a submandibular salivary gland. The gland consists of numerous mucous acini (*B*), which secrete mucus, and a few serous demilunes (*C*), which secrete a watery, protein-rich material that is associated with mucous acini. The gland also contains serous acini (*E*). Acinar secretions drain into a striated (secretory) duct (*A*) of the gland, which conveys the secretory products toward the oral cavity. Connective tissue (*D*) holds the acini together.

10–14. The answers are: 10-D, 11-B, 12-A, 13-C, 14-A. [*IV B 2*] This is a micrograph of the esophageal mucosa. The esophagus has a stratified squamous epithelium that rests on a basement membrane (*B*). The basal cells of the epithelium (*C*) proliferate, and the apical cells (*D*) are sloughed. The lamina propria (*A*) is a loose areolar connective tissue that may contain glands and lymphocytes. The esophageal lumen (*E*) is the cavity through which food passes on its way from the oral cavity to the stomach.

18
The Lower Gastrointestinal Tract

I. INTRODUCTION

A. Components. The lower gastrointestinal tract consists of the **stomach;** the **small intestines,** which include the duodenum, jejunum, and ileum; and the **large intestines,** which include the cecum, vermiform appendix, colon, rectum, and anal canal.

B. Functions

1. In the **stomach,** food entering from the esophagus is liquified and partially digested by gastric secretions to form **chyme.**
   a. Protein digestion begins and the digestion of carbohydrates, initiated in the oral cavity by salivary α-amylase, continues.
   b. Sphincters at both ends of the stomach isolate it from the rest of the gastrointestinal tract. Peristaltic action of gastric smooth muscle churns the chyme until it reaches the appropriate state of digestion.

2. Food passes from the stomach into the **small intestines,** where most of the digestion and absorption of food constituents occurs. In the duodenum, chyme is neutralized and mixed with pancreatic digestive enzymes and hepatic emulsifying agents, which promote lipid digestion.

3. In the **large intestines,** undigested food constituents are dehydrated and mixed with mucus. Feces pass out of the body through the rectum and anal canal.

C. Regulation. The digestive process is regulated by the **enteroendocrine system**. The lower gastrointestinal tract contains many types of endocrine cells. The secretions of these cells control the motility and secretion of gastrointestinal tract components. Many enteroendocrine cells are part of the **APUD** (amine precursor uptake and decarboxylation) **system.**

II. THE STOMACH

A. Basic anatomy

1. The stomach is a large sac-like organ in the peritoneal cavity between the esophagus and small intestines. It consists of a **cardiac antrum** (which receives the esophagus), a large dome-like **fundus,** a main body, or **corpus,** and a funnel-shaped **pylorus.**

2. The mucosal surface of the stomach is lined by a **simple columnar epithelium**. A sharp transition from the stratified squamous esophageal epithelium to the simple columnar gastric epithelium occurs at the gastroesophageal junction (Figure 18-1).

3. The stomach is covered by a peritoneal reflection and is suspended from the body wall by a **mesogastrium**.

B. Gastric secretions

1. The stomach secretes a large amount of acidified solution of **pepsin,** a proteolytic enzyme with an acidic pH optimum.

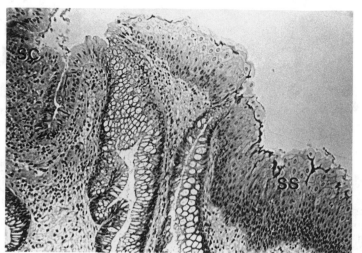

Figure 18-1. Light micrograph of the gastroesophageal junction, showing the stratified squamous (SS) esophageal epithelium adjacent to the simple columnar (SC) gastric epithelium.

2. The mucosal surface of the stomach has a thick protective coating of **mucus,** which is secreted by epithelial cells in the gastric mucosa.

3. Gastric secretions also include a glycoprotein called **gastric intrinsic factor,** which forms a complex with vitamin B_{12} that facilitates B_{12} absorption in the duodenum.

4. APUD cells in the gastric epithelium secrete **peptides** that control the motility and secretory activity of the gastrointestinal tract.

C. Mucosa

1. When the stomach is empty, the surface has a series of conspicuous mucosal and submucosal folds called **rugae**.
   a. Rugae disappear when the stomach is distended with food.
   b. Rugae form deep gutters along the curvature of the fundus, the lesser and greater curvatures of the stomach, and into the pyloric stomach.

2. The mucosal surface also contains numerous **gastric foveolae,** or **pits,** which are lined by deep surface epithelial glandular invaginations called **gastric glands** (Figures 18-2 and 18-3).
   a. The cardiac antrum has gastric pits that communicate with short, coiled cardiac glands consisting primarily of mucus-secreting cells.
   b. The pylorus contains extremely deep foveolae with coiled mucous glands.
   c. The fundus and corpic stomach have foveolae with long straight gastric glands.

3. The simple columnar **epithelium** that lines the gastric mucosa consists of several types of exocrine and endocrine cells and rests on a typical basement membrane.

4. The gastric **lamina propria** contains scattered lymphatic nodules, lymphocytes, and plasma cells.

5. The stomach has a well-developed **muscularis mucosae** within the rugae. Slips of muscularis mucosae may be closely associated with gastric glands, and the contraction of these smooth muscle cells may help gastric glands express their secretions.

D. Glands are present throughout the fundus and corpus. The main gastric glands contain endocrine cells and four types of exocrine cells: mucous surface cells, mucous neck cells, parietal cells, and chief cells.

1. **Mucous surface cells** are the most superficial cells in the glands (see Figure 18-2).
   a. They form a simple columnar epithelium that covers the gastric mucosa and extends a short distance into the gastric glands.
   b. They contain numerous apical mucous droplets, a well-developed apical Golgi apparatus, and a basal nucleus and rough endoplasmic reticulum. The mucus is a glycoprotein; consequently, these cells stain intensely with the periodic acid-Schiff (PAS) reaction.

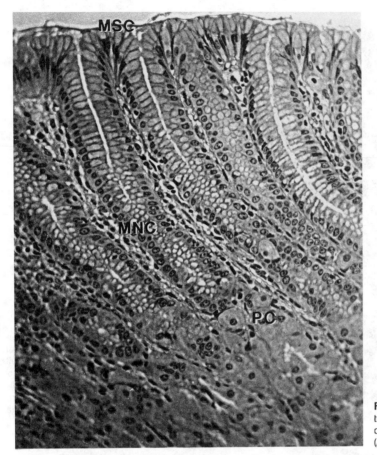

Figure 18-2. Light micrograph of gastric pits, showing mucous surface cells (*MSC*), mucous neck cells (*MNC*), and parietal cells (*PC*).

    c. Mucous surface cells secrete a continuous mucous film that prevents the acidified proteinase solutions in the stomach from ulcerating the gastric mucosa.

2. Mucous neck cells are slightly different cells that exist deeper in the gastric pits (see Figure 18-2).
    a. They are mixed with acid-secreting parietal cells in the upper portion of the gastric pits.
    b. Their mucus has different chemical properties from the mucus secreted by mucous surface cells.
    c. Mucous neck cells have more rough endoplasmic reticulum, a more prominent Golgi apparatus, and larger, less dense mucous granules than mucous surface cells.

3. Parietal cells (oxyntic cells; see Figure 18-2) are large, round, and intensely acidophilic cells that have a large central nucleus. They are most heavily concentrated in the upper and middle portions of gastric glands, where they are mixed with mucous neck cells and chief cells, respectively.
    a. Parietal cells secrete highly concentrated hydrochloric acid.
    b. Parietal cells have an elaborate system of apical surface invaginations called **intracellular canaliculi,** which communicate with the gastric lumen.
    c. The canaliculi are lined by interdigitating microvilli and are closely associated with numerous mitochondria and an abundance of smooth endoplasmic reticulum. These features facilitate the active transport of hydrogen ions across a large concentration gradient.
    d. Parietal cell acidophilia results from the presence of numerous large, round mitochondria and a plethora of smooth endoplasmic reticulum membranes.

4. Chief cells (peptic cells) are most prevalent in the lower portions of gastric glands (Figure 18-4).
    a. They secrete **pepsinogen,** a precursor of the proteolytic enzyme **pepsin**.
    b. They contain zymogen granules, a prominent apical Golgi complex, abundant basal rough endoplasmic reticulum, and a usual number of mitochondria.

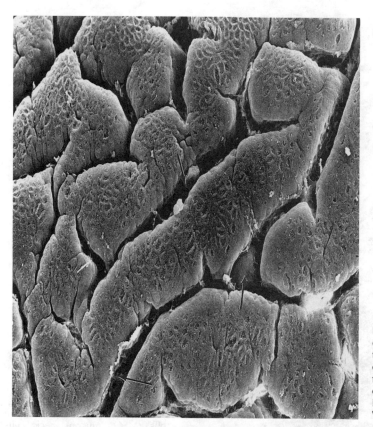

Figure 18-3. Scanning electron micrograph of the luminal surface of the stomach, showing gastric pit openings (*arrows*) and the apical surfaces of mucous surface cells. (Micrograph courtesy of Dr. E.N. Albert, Department of Anatomy, George Washington University.)

    c. The conversion of pepsinogen to pepsin occurs most actively at an acid pH. Pepsin has a strongly acidic pH optimum for this proteolytic conversion.

  5. Endocrine cells. Gastric glands also contain a dispersed group of endocrine cells that are important for the physiologic regulation of gut function.

E. Submucosa and muscularis externa

  1. The **submucosa** contains many collagenous and elastic fibers.

  2. The gastric **muscularis externa** is composed of three layers instead of the more common two layers that exist in the esophagus and intestines.
    a. General arrangement of layers. The following descriptions are only partially accurate due to the highly irregular shape of the stomach.
      (1) The inner layer is obliquely arranged.
      (2) The middle layer is roughly circular in the corpus.
      (3) The outer layer is roughly longitudinal to the corpus long axis.
    b. In the cardiac and pyloric sections, the circular layers of smooth muscle fibers are well-developed and form sphincters.
    c. The myenteric plexus, the nerve plexus that coordinates the churning motion of the stomach, is located between the circular and longitudinal layers.

F. Adventitia. The outermost layer of the stomach consists of a thin layer of connective tissue cells and fibers. It is classified as a serosa because it is coated by a mesothelium of the visceral peritoneum lining, which is continuous with the parietal peritoneum at the mesogastrium.

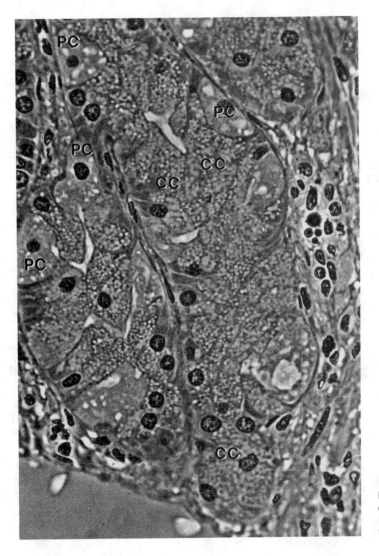

Figure 18-4. Light micrograph of the deep portion of the gastric pits, showing numerous chief cells (*CC*) and a few parietal cells (*PC*).

III. THE SMALL INTESTINES

A. Basic anatomy

1. The small intestines are approximately 6.25 m long and are composed of three main regions: the **duodenum** (0.25 m long), the **jejunum** (2.4 m long), and the **ileum** (3.6 m long).

2. Most of the digestive process is performed in the small intestines by pancreatic enzymes, and it is here that nutrient absorption is completed.

3. The submucosa of the small intestinal wall has circular folds called **plicae circulares**. Unlike gastric rugae, plicae circulares are permanent structures.

B. Mucosa

1. **Villi.** The luminal surface of the small intestine has numerous mucosal folds called villi (Figure 18-5).
   a. Villi are finger-like projections or flattened, leaf-like protrusions present throughout most of the small intestine.
   b. Villi and plicae circulares increase the absorptive surface area of the mucosal lining.

2. **Crypts of Lieberkühn** (small glandular invaginations between villi) extend to the muscularis mucosae but not beyond it.

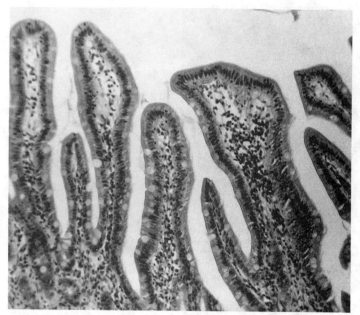

Figure 18-5. Light micrograph of jejunal villi.

C. Epithelial cells. Epithelium in the small intestine contains four types of cells. These cells, in descending order of importance, are: absorptive cells, goblet cells, Paneth cells, and undifferentiated columnar cells (Figures 18-6 and 18-7). This epithelium undergoes rapid turnover.

    1. Absorptive cells are tall columnar cells that perform several important digestive functions.
        a. Absorptive cells have an apical brush border of microvilli (Figure 18-8) that are covered by a **glycocalyx**. The glycocalyx is a thick, fuzzy glycoconjugate-rich external coat composed

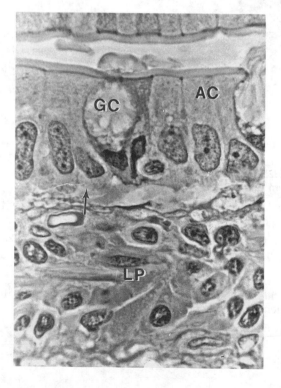

Figure 18-6. Light micrograph of the epithelium in the small intestine. It consists of columnar absorptive cells (*AC*) and goblet cells (*GC*). Beneath the basement membrane (*arrow*) is a loose areolar connective tissue domain called the lamina propria (*LP*).

of a meshwork of minute filaments that insert into the plasma membrane. Cell surface glycoproteins (and mucus from goblet cells) protect epithelial cells from digestion by luminal enzymes.

    (1) The glycocalyx contains **disaccharidases** and **dipeptidases** for the terminal digestion of carbohydrates and proteins, respectively.

    (2) Absorptive cell membranes also contain substantial amounts of ATPase and alkaline phosphatase as well as **enterokinase,** which absorbs and activates pancreatic **trypsinogen,** converting it to actively proteolytic **trypsin**.

b. Each microvillus contains many actin-containing microfilaments aligned parallel to its long axis. The microfilaments insert apically into the absorptive cell membrane, which covers the microvillus, and basally into the microfilament **terminal web** (aligned perpendicular to the long axis of the microvillus) in the cytoplasm beneath the microvilli.

c. Absorptive cells are tightly joined to each other and to adjacent goblet cells by apical **junctional complexes**.

    (1) Terminal web filaments are part of a continuous network of filaments associated with apical junctional complexes.

    (2) The terminal web probably contains actin and myosin throughout and α-actinin where it inserts into lateral surface membranes at the junctional complex **zonula adherens**.

d. Absorptive cells have an important **role in the transport of digested fats** from the intestinal lumen to the lacteal.

    (1) Pancreatic lipolytic enzymes secreted into the intestinal lumen break down triglycerides into fatty acids and monoglycerides, which are then transported into epithelial cells by a poorly understood mechanism that differs from simple pinocytosis.

    (2) Brush border **triglyceride synthetase** activity aids the uptake and transport of triglyceride breakdown products.

    (3) In epithelial cells, triglyceride breakdown products are resynthesized into triglycerides.

    (4) Small lipid-rich vesicles accumulate in the apical cytoplasm in the cisternae of the smooth endoplasmic reticulum and rough endoplasmic reticulum.

    (5) Triglycerides combine with cholesterol esters and proteins to make lipoprotein micelles, which are then expelled from the **lateral borders** of the cell.

    (6) Lipoprotein micelles in the extracellular compartment are called **chylomicrons** and are visible in the light microscope. Chylomicrons move through the epithelial basement membrane, through the connective tissue domain of the lamina propria, into lymphatic capillaries, then to the thoracic duct, and finally into the systemic circulation.

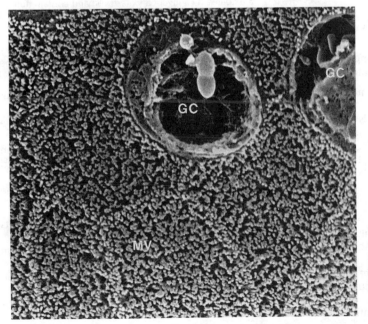

Figure 18-7. Scanning electron micrograph of the apical surface of the small intestinal epithelium, displaying absorptive cell microvilli (*MV*) and the orifices of two goblet cells (*GC*). (Micrograph courtesy of Dr. E.N. Albert, Department of Anatomy, George Washington University.)

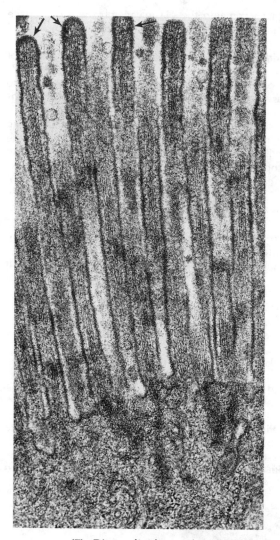

Figure 18-8. Electron micrograph of microvilli on the tall absorptive cells of the small intestinal epithelium. They are coated by a glycocalyx (*arrows*) and contain numerous microfilaments in their cores (visible in this micrograph).

(7) Dietary lipids travel to the liver where they combine with other proteins to form **very low density lipoproteins (VLDLs),** a major transport form of dietary triglycerides in the systemic circulation.

2. **Goblet cells** secrete a mucous layer that covers the small intestinal mucosa and protects it from cytolytic luminal contents.
   a. Goblet cell mucus is a glycoprotein composed predominantly of carbohydrate with some covalently bound sulfate.
   b. The protein component is synthesized in the basal rough endoplasmic reticulum and then transported apically to the Golgi apparatus, where it is glycosylated, sulfated, and packed into mucous droplets.
   c. The cell stores such large quantities of droplets that they distend the cell apex, creating the goblet shape for which the cell is named.

3. **Paneth cells** control intestinal microbial flora and are present in the base of intestinal crypts throughout the small intestine.
   a. They contain large acidophilic granules rich in **lysozyme,** a protein for hydrolysing bacterial cell walls.
   b. Because they secrete proteins, Paneth cells have a preponderance of rough endoplasmic reticulum, which causes intense cytoplasmic basophilia.
   c. Paneth cells also have a large apical Golgi apparatus that synthesizes and packages lysozyme-containing granules.

4. Undifferentiated columnar cells are a population of dividing stem cells that exist in the crypts of Lieberkühn. They have a moderately basophilic cytoplasm.
  a. Undifferentiated columnar cells divide and then differentiate into absorptive cells, goblet cells, enteroendocrine cells, and perhaps Paneth cells.
  b. Most differentiated daughter cells migrate up the villi, live for about 5 days, and then are sloughed from the epithelium into the luminal contents.

D. Lamina propria. The small intestinal lamina propria lies beneath the epithelium and basement membrane. The lamina propria contains the arterioles, a rich capillary plexus, venules, and blind-end lymphatic capillaries (called **lacteals**) that serve each villus. The lamina propria also has slips of smooth muscle tissue and a typical loose areolar connective tissue containing wandering leukocytes.

E. Submucosa and muscularis externa

  1. The **submucosa** is a broad connective tissue domain beneath the muscular mucosae.

  2. The **muscularis externa,** which surrounds the submucosa, is composed of two prominent layers of smooth muscle cells.
  a. The inner layer wraps around the lumen circumference.
  b. The outer layer parallels the lumen long axis.
  c. The myenteric plexus **(Auerbach's plexus)** is located between the layers of the muscularis externa.

  3. The muscularis externa is covered by another connective tissue domain. The outer boundary of most of the small intestines is delimited by a serosa covered by mesothelium.

F. Small intestinal regions. The three small intestinal regions have distinct histologic features that relate to their function.

  1. Duodenum
  a. The duodenum has short, leaf-like villi and small plicae circulares.
  b. The submucosa of the duodenum has numerous **Brunner's glands** connected to the crypts of Lieberkühn. These glands secrete large amounts of mucus and bicarbonate to protect the duodenal mucosa and to neutralize the acidic chyme arriving from the pylorus (Figure 18-9).
  c. The duodenal mucosa has relatively few goblet cells.

  2. Jejunum. The jejunum has long, finger-like villi and prominent plicae circulares. It does not contain Brunner's glands.

  3. Ileum
  a. The ileum has short, stout villi, except near the ileocecal valve, where there are no villi.

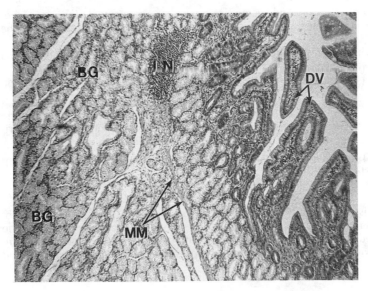

Figure 18-9. Light micrograph of the duodenum. Duodenal villi (*DV*) project into the lumen of the small intestine. The muscularis mucosae (*MM*) is the boundary between the mucosa and the submucosa. The submucosa contains Brunner's glands (*BG*). The micrograph also shows a small lymphoid nodule (*LN*) near the muscularis mucosae.

b. Most of the ileum has small plicae circulares; however, the section near the large intestine has no plicae circulares.

c. Goblet cells are most prominent in the ileum, and the ileum contains large lymph nodes called **Peyer's patches** (see Chapter 12 IV B).

4. Lymphoid tissue is present in the lamina propria and submucosa throughout the small intestine. The amount of lymphoid tissue increases from the duodenum to the ileum.

IV. THE LARGE INTESTINES

A. Basic anatomy

1. This terminal portion of the gastrointestinal tract begins at the end of the ileum (the ileocecal valve) and extends to the anus, a distance of approximately 1.5 m in adult humans. It includes the **cecum** (a sac-like pouch near the ileocecal valve); the **vermiform appendix** (a short diverticulum near the cecum; the ascending, transverse, descending, and sigmoid portions of the **colon;** the **rectum;** and the external orifice called the **anus.**

2. The large intestines are specialized for the processing of food after digestion. Undigested food residues are dehydrated by fluid resorption, mixed with mucus, and formed into feces for expulsion from the body.

B. Mucosa

1. The large intestines do not have villi. The mucosal epithelium is composed of columnar absorptive cells with microvilli and numerous goblet cells (Figure 18-10).

2. The mucosal surface is flat and pock-marked by numerous shallow (0.5 mm) **crypts** (Figure 18-11) that are lined by goblet cells and look like small straight glands that extend down to the muscularis mucosae.

3. The lamina propria is rich in lymphocytes, plasma cells, and macrophages. Plasma cells secrete large amounts of **secretory immunoglobulins,** which help control the hundreds of kinds of enteric flora that exist in the large intestine.

4. The lamina propria also contains lymphoid nodules of various sizes, which commonly extend into the submucosa.

C. Muscularis externa and serosa

1. The **muscularis externa** of the colon and cecum has a peculiar arrangement. The inner layer forms an interrupted circle, and the outer layer is three thick bands of smooth muscle called the **tenia coli.**

a. The bands are roughly equidistant from each other and have thin layers of longitudinal smooth muscle between them.

b. Tonic contractions of the teniae cause a conspicuous sacculation of the colonic walls called **haustra.**

2. The **adventitia** is a thin connective tissue layer. On the anterior surface of the large intestine, the adventitia is covered by a mesothelium. The posterior surface of the large intestine is retroperitoneal in some locations and, thus, not completely covered by a mesothelium.

3. The adventitia contains large globules of adipose tissue called the **appendices epiploicae,** which hang down from the colon.

D. The vermiform appendix is histologically similar to the rest of the colon; however, it does not contain tenia. The mucosa contains absorptive and goblet cells and is especially rich in the enteroendocrine cells called **EC cells.** The appendix contains numerous lymphoid nodules that are more or less confluent with one another and that occupy most of the lamina propria and submucosa.

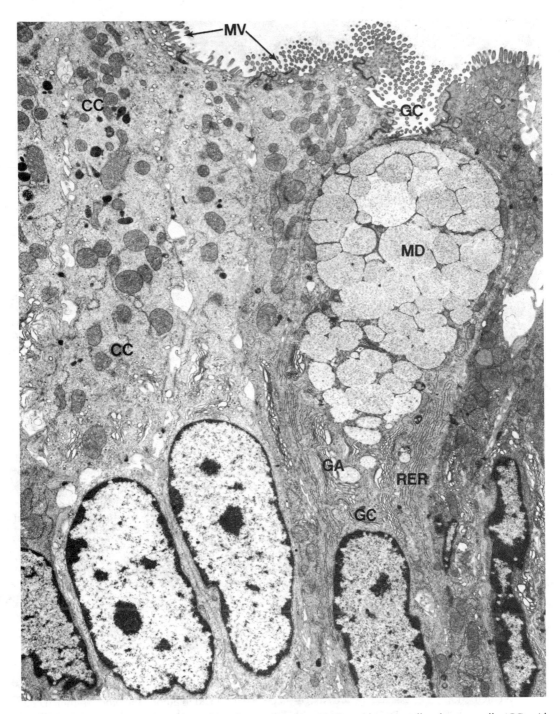

Figure 18-10. Electron micrograph of the colonic mucosal epithelium, showing tall columnar cells (*CC*) with numerous apical microvilli (*MV*) and a goblet cell (*GC*). The goblet cell contains abundant rough endoplasmic reticulum (*RER*), a well-developed Golgi apparatus (*GA*), and many apical mucous droplets (*MD*).

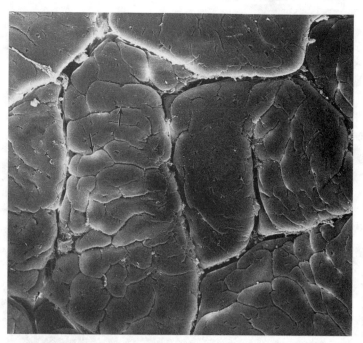

Figure 18-11. Scanning electron micrograph of the colonic luminal surface showing the entrance to crypts (*arrows*). (Micrograph courtesy of Dr. E.N. Albert, Department of Anatomy, George Washington University.)

E. The rectum and anus

1. The upper portion of the rectum has a series of transverse surface folds; the lower rectum, or **anal canal,** has longitudinal folds called **anal columns**.

2. The rectal **mucosa** is similar to the colonic mucosa.

3. The longitudinal smooth muscle of the **muscularis externa** is a continuous sheet: it does not contain tenia.

4. In the anal canal, the columnar epithelium suddenly becomes stratified squamous keratinized epithelium. The circular layers of the muscularis externa are thicker, forming the **internal anal sphincter,** and a circular band of skeletal muscle forms the **external anal sphincter.**

5. The skin near the anus has apocrine sweat glands.

6. Many blood vessels surround the anus. Veins of the anus form a prominent **hemorrhoidal plexus,** which can be easily afflicted with varicosities.

V. ENTEROENDOCRINE CELLS are widely distributed throughout the stomach and intestines and are part of the APUD system.

A. Morphology

1. Enteroendocrine cells usually are wedge-shaped cells tucked in among the other epithelial cells. Some enteroendocrine cells contact the lumen (using their surface microvilli); other cells do not. All cells rest on the basement membrane and contain prominent granules. There are many morphologic classes of granules.

2. Enteroendocrine cells synthesize polypeptide hormones such as **gastrin, cholecystokinin,** and **secretin**. The precise physiologic role of each hormone in digestion is not always well-delineated.

3. Some enteroendocrine cells secrete **candidate hormones,** polypeptides that have a well-defined set of pharmacologic properties but a poorly understood physiologic role in digestion. **Motilin** and **VIP** (vasoactive intestinal peptide) are examples of candidate hormones.

B. Gastric enteroendocrine cells

1. **G cells** secrete **gastrin,** a polypeptide that stimulates secretion of hydrochloric acid (HCl) by parietal cells. G cells are present throughout the lower portion of the stomach and are particularly prominent in the pyloric antrum.

2. **EC cells** secrete **serotonin,** which influences gut motility. EC cells are scattered throughout the gastric mucosa.

3. **A cells** secrete **glucagon** and are present only in the upper one-third of the gastric mucosa.

4. **D cells** secrete **somatostatin** and are present in the upper and lower portions of the stomach. D cells are sparse, if present at all, in the middle portion.

C. Intestinal enteroendocrine cells

1. **D cells** secrete **somatostatin** and are present in the duodenum. A subclass of D cells secretes VIP, which regulates water and ion secretion and gut motility.

2. **EC cells,** prevalent throughout the intestines, secrete **serotonin, motilin,** and **substance P**. These hormones are thought to regulate gut motility.

3. **G cells,** present in the duodenum and pylorus, secrete **gastrin,** which regulates parietal cell secretion of HCl.

4. **I cells,** present in the small intestines, secrete **cholecystokinin,** a peptide that influences pancreatic secretion and gallbladder emptying.

5. **K cells,** present in the small intestines, secrete **gastric inhibitory peptide (GIP),** a peptide that is antagonistic to gastrin.

6. **L cells,** present in the small intestines and colon, secrete **glucagon,** a polypeptide that alters hepatic glycogenolysis.

7. **S cells** secrete **secretin,** a peptide that modifies pancreatic and biliary water and ion secretion.

STUDY QUESTIONS

Directions: Each question below contains five suggested answers. Choose the **one best** response to each question.

1. Each of the following statements concerning lipid digestion and transport is true EXCEPT

(A) dietary triglycerides are degraded by pancreatic lipases in the lumen
(B) free fatty acids are transported into intestinal epithelial cells by pinocytosis
(C) triglycerides combine with proteins in the smooth endoplasmic reticulum
(D) chylomicrons contain triglycerides and proteins
(E) dietary lipids are combined with proteins in the liver

2. Each of the following statements concerning the stomach is true EXCEPT

(A) mucous neck cells are abundant in the upper portion of gastric pits
(B) parietal cells secrete hydrochloric acid
(C) the pylorus does not contain gastric pits
(D) chief cells secrete pepsinogen
(E) the adventitia is coated with a mesothelium

3. Each of the following statements concerning the small intestines is true EXCEPT

(A) jejunal villi are longer than other small intestinal villi
(B) the ileum does not contain Brunner's glands
(C) Brunner's glands are present in the jejunum
(D) Peyer's patches are most prominent in the ileum
(E) the muscularis externa of the duodenum has two layers

4. Each of the following statements concerning parietal cells is true EXCEPT

(A) they have a deeply invaginated apical surface
(B) they are acidophilic
(C) they have many mitochondria
(D) they secrete gastric intrinsic factor
(E) they secrete pancreozymin

Directions: The groups of questions below consist of lettered choices followed by several numbered items. For each numbered item select the **one** lettered choice with which it is **most** closely associated. Each lettered choice may be used once, more than once, or not at all.

Questions 5–9

Match each of the characteristics listed below with the appropriate enteroendocrine cell type.

(A) G cells
(B) EC cells
(C) I cells
(D) S cells
(E) None of the above

5. Secrete serotonin and are present in the stomach and small intestines

6. Secrete gastric inhibitory peptide and are present in the stomach and small intestines

7. Secrete gastrin, a peptide that regulates parietal cell function

8. Secrete cholecystokinin and are present only in the small intestines

9. Secrete a peptide that modifies the ionic composition of pancreatic and hepatic secretions and are present only in the small intestines

Questions 10–14

Match each description listed below with the appropriate lower gastrointestinal tract structure.

(A) Brunner's glands
(B) Appendices epiploicae
(C) Haustra
(D) Tenia coli
(E) Plicae circulares

10. Present throughout most of the small intestines

11. Sacculations in the wall of the colon

12. Submucosal structures that are homologous to rugae

13. Mucus- and bicarbonate-secreting glands that exist only in the duodenal submucosa

14. Longitudinal band of smooth muscle fibers in the muscularis externa of the colon

Questions 15 –17

Match each cell type described below with the appropriate lettered structure in the micrograph.

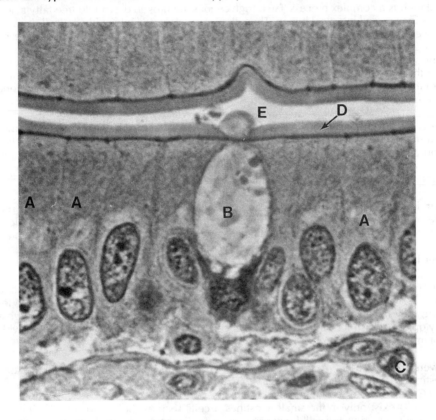

15. Secretes large amounts of mucus that protects the mucosal epithelium
16. Has a thick glycocalyx containing disaccharide digesting enzymes
17. Important in lipid absorption

ANSWERS AND EXPLANATIONS

1. The answer is B. [*III C 1 d*] Lipid digestion and transport from the gastrointestinal tract lumen into systemic circulation is a complex process. First, triglycerides are broken down into free fatty acids and monoglycerides in the small intestinal lumen. Then, they are transported into epithelial cells by a poorly understood, receptor-mediated transport process—not by pinocytosis. In the cell, triglyceride break-down products are resynthesized into triglycerides. Triglycerides are combined with proteins in the smooth endoplasmic reticulum and then expelled as lipoprotein micelles called chylomicrons. Chylo-microns move through the basement membrane and lamina propria into lymphatic capillaries and eventually into the systemic circulation, which carries them to the liver. One important liver function is modifying serum lipids.

2. The answer is C. [*II C 2*] Gastric pits are present throughout the stomach. Gastric pits have mucous surface cells at the luminal surface of glands. The neck of gastric glands contains mucous neck cells; the middle portion contains an abundance of parietal cells, which secrete hydrochloric acid; and the base contains an abundance of chief cells, which secrete pepsinogen. The stomach is a wholly peritoneal organ and is therefore coated with a mesothelium; no portion is retroperitoneal.

3. The answer is C. [*III F*] Brunner's glands are present only in the duodenum. Villi are present throughout the small intestines; however, they are longest in the jejunum. Villi are surface projections that increase the absorptive area. Peyer's patches are small lymphoid nodules in the gastrointestinal tract. They are most prominent in the ileum. Throughout the small intestines, the muscularis externa has a circular inner layer and a longitudinal outer layer.

4. The answer is E. [*II D 3*] Pancreozymin (cholecystokinin) is a secretion product of enteroendocrine cells. Parietal cells secrete hydrochloric acid into the stomach. To accomplish this impressive feat of active transport, they have surface invaginations and many mitochondria. The mitochondria give the parietal cell cytoplasm a distinct acidophilia. Parietal cells also secrete gastric intrinsic factor, which is involved in vitamin B_{12} absorption.

5–9. The answers are: 5-B, 6-E, 7-A, 8-C, 9-D. [*V B, C*] G cells and EC cells are present in the stomach and small intestines. G cells secrete gastrin, which regulates the secretion of hydrochloric acid by gastric parietal cells. EC cells secrete serotonin, which regulates peristalsis.

I cells and S cells exist only in the small intestines. I cells secrete cholecystokinin, a peptide that influences pancreatic secretion and gallbladder emptying. S cells secrete secretin, a peptide that modifies the ionic composition of pancreatic and hepatic secretions.

Gastric inhibitory peptide is secreted by K cells, which also exist only in the small intestines.

10–14. The answers are: 10-E, 11-C, 12-E, 13-A, 14-D. [*III A 3, F 1 b; IV C 1*] Plicae circulares exist throughout the length of the small intestines. They are permanent mucosal folds that increase the intestinal absorptive area. Brunner's glands are present only in the duodenum. They secrete mucus and bicarbonate. The large intestines have three longitudinal bands of smooth muscle, each called a tenia coli. When these bands of muscle contract, they produce sacculations in the wall of the colon, called haustra. The appendices epiploicae are strips of adipose tissue that hang down from the colonic serosa.

15–17. The answers are: 15-B, 16-A, 17-A. [*III C*] This is a micrograph of the epithelium present in both the small and large intestines. The small intestinal epithelium contains more columnar absorptive cells than goblet cells; the large intestinal epithelium contains more goblet cells than columnar absorptive cells.

Columnar absorptive cells (A) have an apical microvillous brush border projecting apically into the intestinal lumen (E). The brush border is coated with a thick glycocalyx (D), which contains enzymes that digest disaccharides and dipeptides. Columnar absorptive cells are important for lipid absorption.

Goblet cells (B) secrete a mucus that protects the mucosal epithelium from digestion by luminal enzymes. The epithelium in the fundus has parietal and chief cells in addition to columnar absorptive cells and goblet cells.

The lamina propria is a loose areolar connective tissue beneath the intestinal epithelium. It contains many fibroblasts (C).

19
The Liver, Gallbladder, and Pancreas

I. INTRODUCTION

A. Functions of the liver

1. The liver is the largest exocrine gland in the body; it secretes large amounts of bile. Bile salts emulsify fats in the small intestines, and the liver metabolizes lipids absorbed in the gastrointestinal tract.

2. The liver also performs endocrine functions. It synthesizes and secretes blood proteins such as serum albumin and transferrin (but not antibodies), and it regulates blood sugar level by storing massive amounts of glycogen.

3. The liver is the primary site for the detoxification and elimination of body wastes and poisons.

B. Functions of the gallbladder

1. The gallbladder, located behind the liver, stores and concentrates bile. Bile from the liver drains into the **right and left hepatic ducts,** then into the **common hepatic duct,** and finally into the gallbladder.

2. When food enters the duodenal lumen, bile drains from the gallbladder into the duodenum through a **cystic duct** and a **bile duct,** which enter at the duodenal papilla.

C. Functions of the pancreas. Like the liver, the pancreas performs exocrine and endocrine functions. Pancreatic **acinar cells** secrete most of the enzymes necessary for digestion, and endocrine cells in the **islets of Langerhans** secrete insulin and glucagon.

II. THE LIVER

A. Basic anatomy

1. The liver is covered by the peritoneal mesothelium, except at the **bare area** where the liver contacts the diaphragm.

2. The mesothelium covers a thin connective tissue **capsule** composed of collagen fibers, some elastic fibers, fibroblasts, and small blood vessels.

3. The inferior surface of the liver has a deep concave recess with a central hilus called the **porta hepatis,** where blood vessels, lymphatic vessels, and nerve fibers enter the liver.

4. A large mass of connective tissue, continuous with the liver capsule, supports vessels and nerve fibers in the liver. Projections from this connective tissue mass, called **trabeculae,** extend into the parenchyma and branch extensively, carrying the blood vessels, nerves, bile ducts, and lymphatics that serve the lobules of the liver.

B. Blood supply (Figure 19-1). The liver receives approximately 25% of the entire cardiac output, or about 1 ml of blood each minute for every gram of liver tissue.

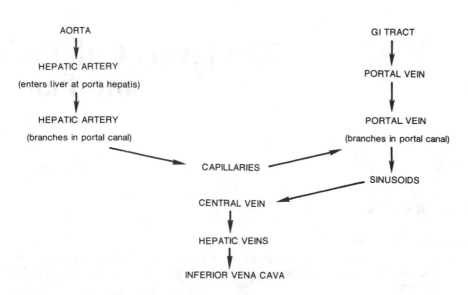

Figure 19-1. Diagram of the flow of blood through the liver. (Reprinted from Johnson KE: *Histology: Microscopic Anatomy and Embryology.* New York, John Wiley, 1982, p 237.)

1. **Afferent blood supply.** The blood supply to the liver enters at the **porta hepatis**.
   a. **Portal vein.** About 75% of the liver's blood is supplied by the portal vein, which drains the gastrointestinal tract and, therefore, carries nutrient-rich blood.
      (1) Branches of the portal vein follow connective tissue bundles into the lobes and then into lobules, where they branch further in the portal canals.
      (2) The smallest branches of the portal vein enter the sinusoids.
   b. **Hepatic artery.** The remaining 25% of the liver's blood is supplied by the hepatic artery.
      (1) Branches of the hepatic artery follow the trabeculae, supplying the connective tissue elements with blood. Small branches enter the **portal canals,** carrying blood to the capillary beds and to the cells of the portal canal.
      (2) Although a few small branches of the hepatic artery drain directly into sinusoids, most capillary beds derived from the hepatic artery drain into the venules carrying blood to portal veins, which carry the drainage to sinusoids.

2. **Efferent blood supply.** Most of the blood that flows through the liver passes through the sinusoids.
   a. **Sinusoids** are vascular spaces radiating from the portal vein, which carry blood from the portal canal to the **central vein** (Figure 19-2).
      (1) Sinusoids are lined by **discontinuous capillaries**. The large intercellular gaps in the walls of discontinuous capillaries allow the serum albumin synthesized within liver parenchymal cells to pass into the systemic circulation.
      (2) The lining of sinusoids also contains phagocytic cells called **Kupffer cells**. These cells commonly contain ingested debris, whole cells, or iron deposits from destroyed red cells.
      (3) Plates of parenchymal cells usually exist close to sinusoids. The gap between sinusoidal epithelial cells and the parenchymal cells is called the **space of Disse** (Figure 19-3).
         (a) Numerous microvilli on the surface of parenchymal cells project into the space of Disse, and occasional collagen and reticular fibers can be found in the space.
         (b) The basement membrane underlying the epithelial cells is thin or absent.
   b. **Venous circulation.** From the sinusoids, blood flows into central veins and then into hepatic veins, which merge with the inferior vena cava.

C. **Lobules.** The liver has several grossly visible lobes demarcated by deep fissures and prominent connective tissue septa. Each lobe contains many lobules **(acini),** which are the basic functional units of the liver. Since the liver parenchyma in humans appears continuous, with no distinct boundaries between lobules, histologists have imposed imaginary boundaries to define liver lobules in three ways.

1. **Classical lobules** are roughly hexagonal. They are centered on a central vein and have a **portal canal** in each corner. Each portal canal contains a **hepatic artery,** a **portal vein,** and a **hepatic bile duct** joined into a unit by connective tissue.

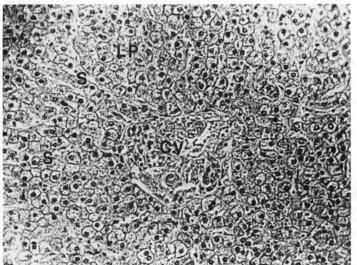

Figure 19-2. Light micrograph of a central vein (*CV*) receiving blood from sinusoids (*S*) between cords of liver parenchymal (*LP*) cells.

 a. This description is based on blood-flow in the lobule. Blood enters the lobule through the portal canals and drains through the central vein.

 b. Since most glands are centered around their draining ducts and vascular supply, some histologists object to this definition because it is inconsistent with lobular organization in other exocrine glands.

2. Portal lobules are triangular. Each triangle has a portal canal in the center and a central vein at each apex. Although this definition is consistent with the lobular organization of other glands (i.e., the vascular supply and draining duct are situated at the center), it does not describe the smallest unit of functional organization in the liver.

3. Liver acini are rhomboidal and have a central vein at each end and a portal canal approximately in the middle of each side. This definition describes tissue that is supplied by a terminal branch of the portal vein and a terminal branch of the hepatic artery, and that is drained by a terminal branch of the bile duct (see III C).

D. Liver parenchymal cells are cuboidal epithelial cells arranged in anastomosing plates and cords. In classical lobules, the plates radiate from the central vein and the cords alternate with sinusoids. Liver parenchymal cells are complex structures that have important roles in the varied functions of the liver.

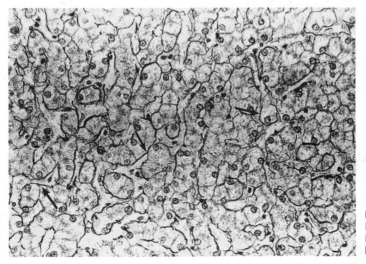

Figure 19-3. Light micrograph of the liver stained to reveal reticular fibers. Reticular fibers are visible at the boundary of liver sinusoids (*arrows*).

1. **Rough endoplasmic reticulum.** Parenchymal cells contain abundant cytoplasmic rough endoplasmic reticulum, which enables them to perform their primary function of synthesizing and secreting all serum proteins (e.g., serum albumin, microglobulins, transferrin, ceruloplasmin) except immunoglobulins. Parenchymal cells also synthesize protein components for serum lipoproteins.

2. **Smooth endoplasmic reticulum.** Liver parenchymal cells also contain abundant smooth endoplasmic reticulum—an organelle that is intermediate between the rough endoplasmic reticulum and the Golgi apparatus. Newly synthesized proteins pass through the smooth endoplasmic reticulum before entering the Golgi apparatus.
   a. The smooth endoplasmic reticulum contains enzymes involved in **cholesterol biosynthesis**.
      (1) Cholesterol is a precursor for other steroids and a component of cell membranes.
      (2) Cholesterol is also a component of **very low density lipoprotein (VLDL),** a serum lipid-protein complex synthesized in the liver.
         (a) VLDL is composed of a glycoprotein apoprotein capsule mixed with cholesterol and phospholipid, surrounding a triglyceride core.
         (b) Rough endoplasmic reticulum synthesizes most of the apoprotein, and smooth endoplasmic reticulum synthesizes the cholesterol, triglycerides, and phospholipid. The final stages of VLDL synthesis and packaging occur in the Golgi apparatus.
   b. The smooth endoplasmic reticulum also contains other enzymes, including a **mixed-function oxidase system,** which is involved in drug and poison detoxification; enzymes that conjugate glucuronides to bilirubins for bile salt formation; enzymes that break down glycogen; and enzymes that deiodinate T_4 to form T_3 (the active thyroid hormones; see Chapter 21 I A 1).

3. **Golgi apparatus.** Several areas in liver parenchymal cells contain well-formed stacks of three to five Golgi membranes.
   a. The Golgi apparatus glycosylates most of the serum proteins before they are secreted. The abundant glycogen granules in liver parenchymal cells may represent stored protein glycosylation precursors.
   b. The final stages of VLDL synthesis and packaging may occur on Golgi membranes.
   c. The Golgi apparatus is involved in the turnover of the parenchymal cell surface and in the formation of lysosomes and microbodies within the cell.

4. **Lysosomes** usually are in close proximity to the Golgi apparatus and the bile canaliculi. Lysosome morphology is highly variable because they can contain a variety of partially destroyed materials. The primary function of lysosomes is to destroy worn cellular components, although some evidence indicates that lysosomes also are crucial for iron recovery and the turnover of glycogen stores.

5. **Microbodies** are smaller than lysosomes but still are bounded by a unit membrane. They contain enzymes that remove hydrogen peroxidase and that metabolize alcohol and lipids.

6. **Mitochondria.** Each liver parenchymal cell has hundreds of mitochondria evenly distributed throughout its cytoplasm. These mitochondria produce the adenosine triphosphate (ATP) necessary for mitochondrial synthetic activities.

III. BILE SECRETION.
Bile is a secretion product that aids digestion by emulsifying dietary lipids. Bile is secreted by liver parenchymal cells into bile canaliculi, the small extracellular spaces between parenchymal cells at the termination of the biliary space.

A. Bile composition

1. Bile is a complex mixture of water, salts, detergents, and glucuronides of bilirubin. These components are waste products that are eliminated from the body only after they aid digestion.

2. Bile constituents are cytolytic and must be kept out of the bloodstream.

B. Bile canaliculi
run between parenchymal cells. In classical lobules, they radiate away from the central vein toward the portal canal (Figure 19-4).

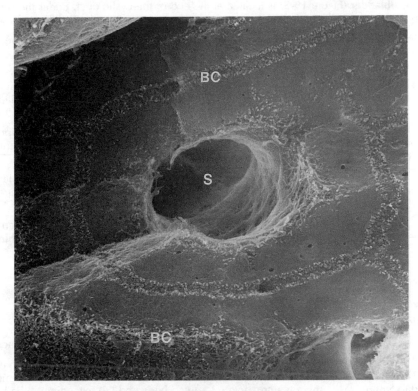

Figure 19-4. Scanning electron micrograph of fractured liver parenchymal cells, showing a sinusoid (*S*) and bile canaliculi (*BC*). (Micrograph courtesy of Dr. E.N. Albert, Department of Anatomy, George Washington University.)

1. Parenchymal cells have surface protrusions that project into the canaliculi and **occluding junctions** all around the canaliculi. These occluding junctions block cytolytic bile constituents from passing into the bloodstream.

2. Adjacent to occluding junctions, liver parenchymal cells form desmosomes and large **gap junctions**.

3. The cytoplasm of parenchymal cells near the canaliculi is rich in actin-containing **microfilaments**.

4. Canaliculi empty into short ductules, which also drain hepatocytes. These ductules are lined by squamous ductular cells that are continuous with the cuboidal epithelial cells of hepatic ducts in the portal canals.

C. **Bile ducts.** Bile drains from the ductules into interlobular bile ducts in the portal canals. Smaller ducts drain into larger intrahepatic ducts as bile flows toward the porta hepatis.

1. **Intrahepatic ducts**
   a. Bile duct **epithelium** is a continuous layer of tightly coupled cells resting on a basement membrane.
      (1) The smallest ducts have a cuboidal epithelium. As the lumen of the intrahepatic ducts increases, the epithelium becomes increasingly columnar. Near the porta hepatis, the larger branches of the intrahepatic duct are lined by tall columnar cells with many apical microvilli.
      (2) Areas of epithelium in larger ducts may contain goblet cells and slips of smooth muscle that contract to restrict the diameter of the lumen.
   b. The **walls** of intrahepatic ducts contain collagen and elastic fibers that merge into the connective tissue of trabeculae and portal canals.

2. **Extrahepatic bile ducts** have a mucosa with tall columnar epithelial cells and a thin lamina propria, a submucosa that contains regularly spaced mucous glands, a well-developed muscularis externa containing smooth muscle cells, and a typical adventitia covered by mesothelium.

D. The gallbladder (Figure 19-5) is located at the end of the cystic duct, under the right lobe of the liver. It is a distensible cul-de-sac with a capacity of approximately 50 ml when distended.

1. When distended, the gallbladder wall is relatively smooth; however, when the gallbladder is empty, the wall has conspicuous folds called **Rokitansky-Aschoff crypts,** which can extend into the gallbladder muscularis externa.

2. The **mucosa** is a simple columnar epithelium containing cells with many apical microvilli and conspicuous lateral compartments between adjacent cells. Bile in the gallbladder is highly concentrated, probably due to water absorption at microvilli and along lateral cell boundaries.

3. The **lamina propria mucosae** is thin and unremarkable. It contains numerous fibroblasts, small blood vessels, and leukocytes and may have glands containing goblet cells and enteroendocrine cells.

4. The **muscularis externa** contains several layers of smooth muscle fibers arranged in longitudinal bundles. In some locations, they form sphincters; in other locations they form spiral valves (e.g., at the union of the gallbladder and cystic duct).

5. The gallbladder discharges bile into the duodenum when **cholecystokinin** (a secretion of **I cells** in the small intestine) stimulates gallbladder contraction and relaxation of the **sphincter of Oddi**.

IV. THE PANCREAS

A. Basic anatomy

1. The pancreas is a conspicuous retroperitoneal organ that lies behind the stomach and weighs approximately 100 g.

2. The pancreas has a **head** and **uncinate process** on its right side—where it empties into the duodenum by the **main pancreatic duct**—and a **body** and **tail** extending to the left toward the hilus of the left kidney.

3. In about 10% of humans, the pancreas also is drained by an **accessory pancreatic duct,** which enters the duodenum caudal to the main pancreatic duct.

B. The exocrine pancreas

1. **Basic histologic features** (Figure 19-6)
   a. The exocrine pancreas is divided into **lobes** by connective tissue septa. Each lobe is divided into several indistinct **lobules** by gossamer slips of connective tissue.
   b. Blood vessels and nerves ramify distally along the branches of the **pancreatic duct system** from the large ducts toward the pancreatic acini.
      (1) The **acinus** is the main functional unit of the exocrine pancreas. Each acinus is composed of many pyramidal epithelial cells tightly joined to each other by junctional complexes and surrounded by a basement membrane. Acini are round.

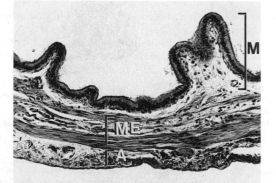

Figure 19-5. Light micrograph of the wall of the gallbladder, showing the mucosa (*M*), the muscularis externa (*ME*), and the adventitia (*A*).

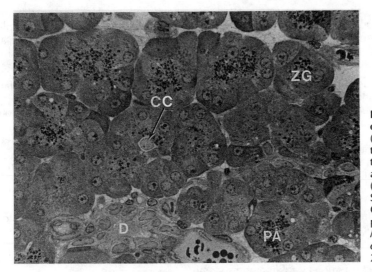

Figure 19-6. Light micrograph of the exocrine pancreas. Pancreatic acini (*PA*) are composed of cells that contain zymogen granules (*ZG*). Centroacinar cells (*CC*) connect pancreatic acini to larger ducts (*D*). (Micrograph courtesy of Dr. F.J. Slaby, Department of Anatomy, George Washington University. Reprinted from Johnson KE: *Histology: Microscopic Anatomy and Embryology.* New York, John Wiley, 1982, p 239.)

    (2) One pole of each acinus connects to **centroacinar cells,** the most distal cells of the duct system. Centroacinar cells drain into **intralobular ducts,** which in turn drain several different acini of a lobule.

    (3) Intralobular ducts empty into larger **interlobular ducts** in the connective tissue septa between lobes.

    (4) Pancreatic ducts are lined by **simple cuboidal epithelium,** which becomes taller and may contain goblet cells near the duodenum.

2. Pancreatic acinar cells comprise more than 80% of the pancreas. These cells are specialized to secrete the proteins used in the digestive process.

    a. Secretion products

        (1) Pancreatic acinar cells secrete enzymes in **active and inactive forms**. Inactive enzymes are converted to active forms by **enterokinase** (a secretion product of pancreatic acinar cells) in the duodenum.

            (a) Trypsinogen and **prophospholipase** are inactive enzymes secreted by the acinar cells.

            (b) Amylase, triacylglycerol lipase, DNAase, and **RNAase** are active enzymes secreted by the acinar cells.

            (c) Pancreatic acinar cells also secrete **trypsin inhibitor**.

        (2) Three factors provide the pancreas and pancreatic ducts with **protection from autodigestion**.

            (a) Pancreatic acinar cells are joined by **junctional complexes,** which prevent digestive enzymes from escaping across the secretory or ductal epithelium.

            (b) Some enzymes are secreted in inactive forms, becoming active only after they enter the duodenum.

            (c) The pancreas secretes trypsin inhibitor and may secrete other protective substances.

    b. Cell biology

        (1) Pancreatic acinar cells contain **one prominent nucleus** with abundant euchromatin and a conspicuous nucleolus. These features support mRNA and rRNA synthesis. The mRNA for digestive enzymes moves from the nuclear sap into the cytoplasm and binds to ribosomes.

        (2) The basal portion of the cell is broader than the apical portion and is packed with parallel cisternae of rough endoplasmic reticulum. Mitochondria and free ribosomes occupy the space between cisternae.

            (a) Polypeptide chain synthesis begins on rough endoplasmic reticulum membranes and nascent chains protrude into the lumen of the rough endoplasmic reticulum.

            (b) Many pancreatic enzymes are glycoproteins. Glycosylation begins in the rough endoplasmic reticulum and is completed in the Golgi apparatus.

        (3) Ribosome-free, membrane-delimited **transitional vesicles** bud from the rough endoplasmic reticulum and transport proteins to the convex forming face of the Golgi complex, where the vesicles fuse with Golgi membranes.

    (a) As secretory proteins are glycosylated and passed through the Golgi membranes, small condensing vacuoles form on the apical concave surface of the Golgi complex.

    (b) Condensing vacuoles fuse with one another, concentrating the secretion product, and form mature, stable **zymogen granules,** which are stored in the apical cytoplasm of acinar cells (Figure 19-7).

3. Regulation of exocrine secretion. Pancreatic exocrine secretion is a complex process regulated by five or more well-known hormones.

    a. Acidic chyme entering the duodenum stimulates enteroendocrine cells to secrete hormones into the bloodstream. **I cells** secrete **cholecystokinin,** and **S cells** secrete **secretin**.

        (1) In acinar cells, cholecystokinin stimulates zymogen granule exocytosis and the release of bile into the duodenum.

        (2) Secretin stimulates centroacinar cells and ductal cells to secrete alkaline, bicarbonate-rich, pancreatic fluid.

        (3) As acidic chyme is neutralized, pancreatic trypsinogen is activated to trypsin by enterokinase. Trypsin, in turn, activates other proenzymes and starts digestion.

        (4) Bicarbonate from duodenal Brunner's glands and pancreatic secretions neutralize stomach acid, establishing the pH optimum for pancreatic enzymes, and digestion continues. The rise in pH also causes cholecystokinin and secretin secretion to subside.

    b. Nervous reflex stimulates the islets of Langerhans to release **pancreatic polypeptide,** which counters the effects of cholecystokinin and secretin.

    c. **K cells** in the duodenum release **gastric inhibitory polypeptide (GIP),** which stops the secretion of gastric acid and stimulates **B cells** in the islets of Langerhans to release **insulin**.

    d. **D cells** release **vasoactive intestinal peptide (VIP),** which stimulates epithelial cells to secrete electrolytes and water.

C. The endocrine pancreas. Endocrine cells comprise only about 2% of the volume of the pancreas but are an important population of cells. Pancreatic endocrine cells form the **islets of Langerhans,** which are separate from the pancreatic exocrine lobules.

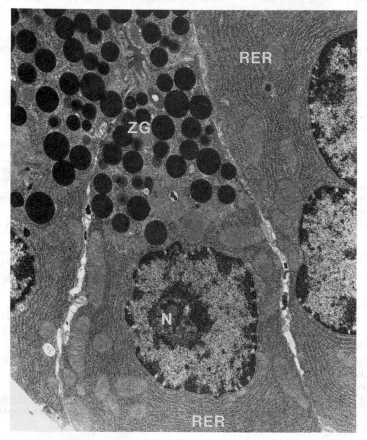

Figure 19-7. Transmission electron micrograph of a pancreatic acinar cell. It contains abundant basal rough endoplasmic reticulum (*RER*), a prominent nucleolus (*N*), and apical zymogen granules (*ZG*). (Micrograph courtesy of Dr. F.J. Slaby, Department of Anatomy, George Washington University.)

1. **Basic histologic features** (Figure 19-8)
   a. The islets of Langerhans are surrounded by delicate slips of connective tissue, lack ducts, and are supplied with a rich capillary plexus to drain the hormones from the islet tissues.
   b. Each islet is 0.1–0.2 mm in diameter and contains thousands of cells.
   c. Blood flows to islet capillaries through arterioles that branch directly from intralobular arteries.
      (1) Islets have fenestrated capillaries (typical of all endocrine tissues), which facilitate hormone transport.
      (2) The capillaries supply the peripheral A cells and D cells first, then the centrally located B cells, and then the acini.
      (3) Blood supplied by the intralobular arteries contains high concentrations of hormones. Some of these hormones (e.g., VIP) directly stimulate secretion by islet cells.

2. **Cell types.** The islets of Langerhans contain six types of cells: A cells, B cells, and D cells are the major endocrine cells; C cells, EC cells, and PP cells are minor endocrine cells.
   a. **A cells** secrete **glucagon**. They comprise about 15% of the islet endocrine cell population and usually are located along the periphery of islets. A cells have an irregularly shaped nucleus and secretory granules that contain glucagon.
   b. **B cells** secrete **insulin**. They comprise about 70% of the islet endocrine cell population and are centrally located in islets. B cells have large, round nuclei.
   c. **D cells** comprise about 10% of the islet endocrine cell population and are located at the islet periphery, close to A cells. Two types of D cells exist: one secretes **somatostatin,** and the other secretes **VIP**.
   d. **C cells, EC cells, and PP cells** comprise about 5% of the islet endocrine cell population. The function of C cells is not known. EC cells secrete serotonin, and PP cells secrete **pancreatic polypeptide**.

3. **Secretion products.** Insulin and glucagon are the chief hormones produced in islet tissue.
   a. **Insulin** has two polypeptide chains (A and B) and a molecular weight of 6000. Insulin stimulates glucose transport and metabolism in many target tissues, most notably liver and adipose tissues, and stimulates glycogen synthesis, thereby lowering the blood glucose concentration.
   b. **Glucagon** is a single polypeptide chain with a molecular weight of 3500. Its effects are antagonistic to the effects of insulin. Glucagon stimulates glycogen breakdown and glucose synthesis in the liver, thereby increasing the blood glucose concentration.

4. **Regulation of endocrine secretion. Somatostatin** inhibits the secretion of insulin and glucagon.
   a. Rising blood glucose levels stimulate insulin secretion, which reduces the amount of glucose in the blood by stimulating glucose metabolism in target organs.
   b. In contrast, falling blood glucose levels stimulate glucagon secretion, which increases the amount of glucose in the blood by stimulating glucose synthesis.

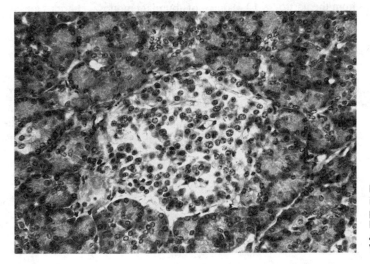

Figure 19-8. Light micrograph of an islet of Langerhans in the human pancreas. (Reprinted from Johnson KE: *Histology: Microscopic Anatomy and Embryology.* New York, John Wiley, 1982, p 241.)

STUDY QUESTIONS

Directions: The groups of questions below consist of lettered choices followed by several numbered items. For each numbered item select the **one** lettered choice with which it is **most** closely associated. Each lettered choice may be used once, more than once, or not at all.

Questions 1–5

For each structure described below, select the appropriate lettered structure in the diagram.

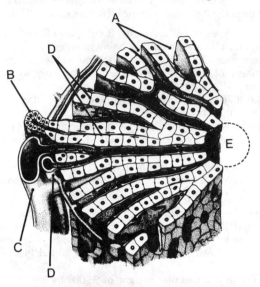

Reprinted from Bloom W, Fawcett D: *A Textbook of Histology.* Philadelphia, WB Saunders, 1975, p 694.

1. Its lining contains Kupffer cells

2. Empties into the hepatic vein

3. Its branches communicate with the hepatic sinusoids

4. Lined by cuboidal epithelial cells

5. Carries blood rich in nutrients from the gastrointestinal tract

Questions 6–10

Match each description or function listed below with the most appropriate polypeptide.

(A) Enterokinase
(B) Trypsinogen
(C) Insulin
(D) Glucagon
(E) Somatostatin

6. Secreted by D cells in islet tissue

7. Secreted by pancreatic acinar cells in an inactive form

8. Secreted by B cells in islet tissue

9. Secreted by pancreatic acinar cells; activates proenzymes

10. Secreted by A cells in islet tissue

Questions 11–15

Match each description or function listed below with the most appropriate organelle.

(A) Glycogen granules
(B) Rough endoplasmic reticulum
(C) Smooth endoplasmic reticulum
(D) Golgi apparatus
(E) Zymogen granules

11. Causes the cytoplasmic basophilia of pancreatic acinar cells

12. Source of abundant monosaccharides for glycoprotein synthesis in liver parenchymal cells

13. Site of cholesterol biosynthesis in liver parenchymal cells

14. Receives partially glycosylated secretion products; completes glycosylation and transfers polypeptides to condensing vacuoles

15. Site where proximal polysaccharide units couple with glycoproteins for secretion

ANSWERS AND EXPLANATIONS

1–5. The answers are: 1-A, 2-E, 3-C, 4-B, 5-C. [*II B, C*] This is a diagram of the portal canal, sinusoids, and central vein in a portion of a liver lobule. Sinusoids (*A*) are lined by phagocytic Kupffer cells. The central vein (*E*) empties into the hepatic vein. Branches of the hepatic artery nourish hepatic connective tissue and empty into sinusoids. The bile canaliculi (*D*) drain into the bile ducts (*B*), which are lined by a cuboidal epithelium. The portal vein (*C*) drains nutrient-rich blood from the gastrointestinal tract into the liver.

6–10. The answers are: 6-E, 7-B, 8-C, 9-A, 10-D. [*IV B 2 a, C 2–4*] Enterokinase and trypsinogen are secreted by pancreatic acinar cells. Enterokinase is active when secreted. It cleaves polypeptides from proenzymes such as trypsinogen to produce catalytically active proenzymes such as trypsin. Trypsinogen is inactive when secreted.

The islets of Langerhans contain A cells, B cells, D cells, and several minor cell types. A cells secrete glucagon, B cells secrete insulin, and D cells secrete somatostatin.

11–15. The answers are: 11-B, 12-A, 13-C, 14-D, 15-B. [*II D 1–3; IV B 2 b (3) (b)*] Pancreatic acinar cells secrete proteins and, thus, have an abundant rough endoplasmic reticulum (RER). This abundant RER causes the cytoplasmic basophilia.

Glycogen granules are abundant in liver parenchymal cells. They are a form of stored glucose that can be used for energy production or glycoprotein synthesis.

In liver parenchymal cells, polypeptide synthesis and the addition of proximal sugar residues to glycoproteins occurs on the rough endoplasmic reticulum. Glycoprotein synthesis is completed in the Golgi apparatus with the addition of distal sugar residues. Cholesterol biosynthesis and drug detoxification occurs on the smooth endoplasmic reticulum.

Zymogen granules are membrane-delimited packets of concentrated enzymes or proenzymes in the apical region of pancreatic acinar cells.

20
The Pituitary and Pineal Glands

I. INTRODUCTION

A. The pituitary gland

1. **Basic anatomy.** The pituitary gland (**hypophysis cerebri**) is a small organ located at the base of the brain, adjacent to the third ventricle.
   a. The pituitary gland is shaped somewhat like a mushroom. It is about 1.5 cm wide and weighs slightly more than 1 g. Multiparous women have larger pituitary glands.
   b. The pituitary lies at the base of the third ventricle of the brain in a small depression called the **sella turcica,** deep within a recess in the sphenoid bone.
   c. The dura mater is discontinuous at the pituitary gland; however, it surrounds the stalk of the pituitary and closes the sella with a membranous covering called the **diaphragma sellae**.

2. **Subdivisions.** By gross inspection, the pituitary has a pinkish (vascularized) anterior lobe and a whitish (neuronal) posterior lobe containing nerve fibers. Microscopic examination reveals striking regional variations within the pituitary gland that subdivide it into the **adenohypophysis** (anterior lobe) and **neurohypophysis** (posterior lobe).

3. **Functions**
   a. Despite its small size, the pituitary gland is the master control gland of the endocrine system. It regulates the basal metabolic rate, other endocrine organs (e.g., adrenal cortex), the reproductive system, mammary glands, and the overall growth of the body.
   b. In most instances, target organs produce hormones that alter the function of the pituitary gland. This feedback mechanism regulates the production of pituitary hormones.

B. The pineal gland

1. **Basic anatomy.** The pineal gland (**epiphysis cerebri**) is a small endocrine organ located above the third ventricle of the brain. In adult humans, the pineal gland is smaller than the pituitary gland.

2. **Functions.** The pineal gland receives both hormonal and neuronal inputs that regulate the secretory activities of its main parenchymal cell: the **pinealocyte.** The role of the human pineal gland is obscure; however, it is clear that its secretory activity is important for regulating certain biologic rhythms.

II. THE ADENOHYPOPHYSIS

A. Hormones. The adenohypophysis produces the following hormones:

1. Prolactin

2. Follicle-stimulating hormone (FSH)

3. Luteinizing hormone (LH)

4. Thyrotropin (thyroid-stimulating hormone; TSH)

5. Growth hormone (GH)

6. Adrenocorticotropin (adrenocorticotropic hormone; ACTH)

7. Lipotropins (lipotropic hormones; LPH)

8. Melanocyte-stimulating hormone (MSH)

9. Endorphins

B. **Adenohypophysial cell types.** The adenohypophysis consists of three principal types of hormone-secreting parenchymal cells: **acidophils, basophils,** and **chromophobes.** Each type of cell has unique staining characteristics related to its function, which can be used to identify the type of cell. The unique staining characteristics are caused by differences in the chemical composition of the hormones contained in granules in each type of cell (Figures 20-1 and 20-2).

1. **Acidophils** are subdivided into two classes: **lactotropes** (mammotropes) and **somatotropes.** Lactotropes produce prolactin; somatotropes produce GH.

2. **Basophils** are subdivided into three classes: **corticotropes, thyrotropes,** and **gonadotropes.** Corticotropes produce ACTH and LPH; thyrotropes produce TSH; and gonadotropes produce LH and FSH.

3. **Chromophobes** probably are acidophils or basophils that have lost their specific staining properties after releasing their hormone-containing granules.

C. **Adenohypophysial cell histophysiology.** Adenohypophysial secretory cells are epithelial cells arranged in cords or follicles that are surrounded by a dense anastomosing network of fenestrated capillaries. These capillaries convey secreted hormones into the systemic circulation, which carries them to target organs such as the ovaries and testes.

1. **Basophils**
   a. **Corticotropes** are large ovoid cells with prominent granules that stain immunohistochemically for their hormones. Their cytoplasm contains distinctive 6-nm to 8-nm filaments. They are the most common type of basophil and are most plentiful in the anterior central part of the pars distalis.
      (1) Corticotropes in the pars distalis produce the polypeptide hormone ACTH, which stimulates the adrenal cortex to secrete glucocorticoids.
      (2) ACTH is produced by the type of basophil that also produces a precursor to MSH called LPH.
      (3) LPH and ACTH share a large number of amino acids and have extensive sequence homologies.
   b. **Gonadotropes** secrete LH and FSH. They have moderately large granules, which are intermediate in size to the large granules of somatotropes and the small granules of thyrotropes. Gonadotropes are small and sparse during childhood when gonadal function is slight.
      (1) FSH and LH are glycoprotein hormones that stimulate the gonads. FSH stimulates follicular growth in females and stimulates the spermatogenic epithelium in males. LH is required for ovulation in females and stimulates androgen secretion by testicular Leydig cells in males. LH is sometimes called **interstitial cell stimulating hormone,** or **ICSH,** for this reason.
      (2) Gonadal steroid levels exert negative feedback on gonadotropes. Thus, during pregnancy, when placental steroid secretion is high, production of FSH and LH is low and gonadotropes exhibit functional regression.
   c. **Thyrotropes** secrete the glycoprotein hormone TSH. Granules in thyrotropes are smaller than the large granules of somatotropes and the intermediate granules of gonadotropes.
      (1) **Normal function.** Thyroxine (produced by the thyroid gland), thyroxine-releasing hormone (TRH; produced by the hypothalamus), and TSH comprise a simple negative feedback loop. High levels of thyroxine inhibit TRH production by the hypothalamus. The resulting low levels of TRH reduce the production of TSH. Low levels of thyroxine stimulate TRH production in the hypothalamus, which stimulates TSH secretion. TSH then stimulates the thyroid to secrete thyroxine into the circulation.
      (2) **Primary thyroid failure.** In patients with primary thyroid failure, thyrotropes commonly are enlarged and filled with drops of colloid. Chronic stimulation of thyrotropes by TRH occurs because, without normal thyroid function, the feedback inhibition of TRH production by thyroxine is lost. Consequently, releasing hormones are constantly produced and thyrotropes cannot accumulate large numbers of granules due to the chronic stimulation.

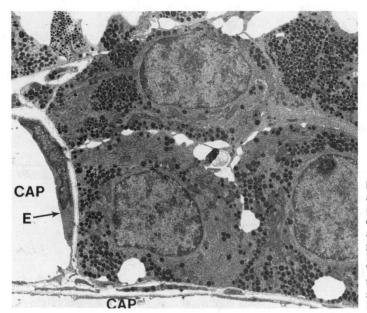

Figure 20-1. Electron micrograph of a cluster of hormone-secreting cells in the adenohypophysis. This group of cells is close to capillaries (*CAP*) and an endothelial cell (*E*). (Micrograph courtesy of Dr. R.J. Walsh, Department of Anatomy, George Washington University. Reprinted from Johnson KE: *Histology: Microscopic Anatomy and Embryology.* New York, John Wiley, 1982, p 291.)

(3) **Hyperthyroidism.** When thyroid function is chronically elevated, high levels of thyroxine cause decreased production of releasing factor, which results in a lack of thyrotrope stimulation. Thyrotropes undergo regression during hyperthyroidism that is unrelated to excessive thyrotropin secretion.

2. Acidophils
 a. **Lactotropes** (mammotropes) produce prolactin, a low molecular weight polypeptide hormone that stimulates glandular development and milk secretion in the mammary glands.
  (1) The lactotropes of pregnant and lactating women contain scattered 40-nm to 60-nm granules and abundant rough endoplasmic reticulum.
  (2) Women with prolactinoma have an enormous titer of circulating prolactin and numerous acidophilic lactotropes in the adenohypophysis.

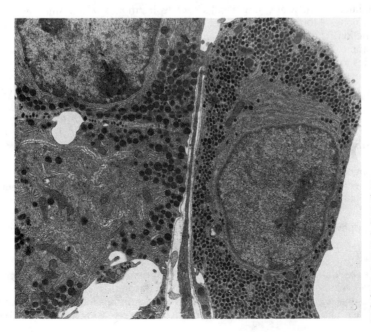

Figure 20-2. Electron micrograph of two kinds of cells in the adenohypophysis. Notice that the granules are different sizes and that the cells contain a different number of granules. (Micrograph courtesy of Dr. R.J. Walsh, Department of Anatomy, George Washington University. Reprinted from Johnson KE: *Histology: Microscopic Anatomy and Embryology.* New York, John Wiley, 1982, p 291.)

    b. Somatotropes in the pars distalis secrete the polypeptide hormone GH. They contain many 35-nm granules and a well-formed rough endoplasmic reticulum.
        (1) GH causes the body to grow. Patients lacking GH are small in stature and designated **pituitary dwarfs**.
        (2) Excessive secretion of GH during childhood results in growth to an abnormally large size. In adults, somatotrope hyperactivity causes **acromegaly,** which results in enlargement of bones in the face and limbs.

D. Adenohypophysial subdivisions. The adenohypophysis is subdivided into a **pars distalis, pars intermedia,** and **pars tuberalis**.

  1. Pars distalis
    a. The pars distalis is the largest adenohypophysial subdivision. It consists of many acidophils, basophils, and chromophobes arranged in anastomosing cords, small clusters, and follicles.
    b. Clusters of hormone-secreting cells are surrounded by fenestrated capillaries, which are supported by a delicate mesh of collagenous and reticular fibers.
    c. The pars distalis contains little or no connective tissue.

  2. Pars tuberalis and pars intermedia
    a. The pars tuberalis is a collar of tissue around the infundibular stalk. The infundibular stalk connects the neurohypophysis to the rest of the brain.
    b. The pars intermedia lies between the pars distalis and the pars nervosa (see III A).
    c. The pars tuberalis and pars intermedia are not well-developed in humans. Both parts are primarily basophilic cells mixed with a few chromophobes. Most of the basophils in the pars tuberalis are gonadotropes.

E. Regulation of adenohypophysial function

  1. Releasing hormones and the hypothalamohypophysial portal system. Releasing hormones secreted by the neurohypophysis control the secretory activity of the adenohypophysis.
    a. Releasing hormones are secreted by neurons that have their cell bodies in the **supraoptic** and **paraventricular nuclei** of the brain.
    b. The axons of these neurons project away from the nuclei and end on capillaries in the median eminence of the neurohypophysis. Releasing hormones are secreted into capillaries from the axonal terminations near the capillaries.
    c. The hypothalamohypophysial portal system conveys releasing hormones from the capillaries near these axonal terminations to the capillaries surrounding the cords and follicles of epithelial cells in the pars distalis.

  2. Hormone-target cell interaction. Many target cells secrete hormones when stimulated by adenohypophysial hormones. Hormones from target cells provide feedback that helps regulate adenohypophysial function.
    a. In females, for example, FSH and LH secreted by gonadotropic basophils in the pars distalis stimulate the ovaries to produce steroids from developing follicles and theca interna cells.
    b. Ovarian steroids provide the hypothalamus with feedback that regulates the level of **gonadotropin releasing hormone (GnRH)** and, thus, the amount of LH and FSH secreted by the pars distalis.

III. THE NEUROHYPOPHYSIS

A. Subdivisions. The neurohypophysis is subdivided into the **pars nervosa** (infundibular process), **infundibular stem,** and **median eminence**. The median eminence is part of the **tuber cinereum,** which makes up part of the third ventricle of the brain and is poorly developed in humans.

B. Hormones. The neurohypophysis produces **antidiuretic hormone (ADH), oxytocin,** and **releasing hormones**.

  1. Releasing hormones
    a. Releasing hormones regulate the secretion of hormones from specific adenohypophysial cells (e.g., TRH stimulates release of TSH from anterior pituitary basophils).
    b. Releasing hormones that are important in the control of the pars distalis include **TRH, GnRH** (also called **LH releasing hormone,** or **LH-RH**), and **corticotropin releasing hormone (CRH)**.

   (1) TRH has been isolated and synthesized; it consists of pyroglutaminyl-histidinylproline-amide.

   (2) GnRH/LH-RH is a decapeptide, and CRH consists of 41 amino acids.

2. ADH and oxytocin

   a. ADH (sometimes called **vasopressin**) alters renal distal convoluted tubules and collecting ducts, making them permeable to water. This reduces urine volume and increases urine tonicity.

   b. Oxytocin causes the release of milk from the lactating mammary gland. It also stimulates contraction of uterine smooth muscle after birth.

C. Dominant features of the neurohypophysis

   1. Hypothalamohypophysial tract. The upper central portion of the pars nervosa contains a large bundle of unmyelinated nerve fibers called the hypothalamohypophysial tract. These nerve fibers are axons of neurons in the supraoptic nucleus of the hypothalamus, the paraventricular nucleus, and the tuber cinereum.

   a. Some of these axons end on capillary loops in the infundibular stem. Other nerve fibers pass through the infundibular stem into the pars nervosa, where they branch to form the cores of indistinct lobules. These lobules are surrounded by thin septa of connective tissue and many capillaries.

   b. Nerve terminals end near the periphery of the lobules, close to the capillaries.

   2. Neurosecretory neurons

   a. Axon terminals in the pars nervosa often are distended with an abundance of neurosecretory material consisting of aggregates of membrane-bound granules that are 100–300 nm in diameter.

   (1) These membrane-bound granules contain oxytocin and ADH and associated carrier proteins called **neurophysins** (Figure 20-3).

   (2) Oxytocin and ADH are stored in separate granules.

   b. Neurosecretory material is synthesized in the cell bodies of the supraoptic and paraventricular nuclei. Cells synthesizing ADH and oxytocin are distributed in different parts of these nuclei.

   c. ADH and oxytocin are synthesized as large macromolecules that encompass the hormone and its respective neurophysin.

   (1) As the macromolecules are transported to nerve terminals, the covalent bond between hormone and neurophysin is cleaved, although the two components remain closely associated.

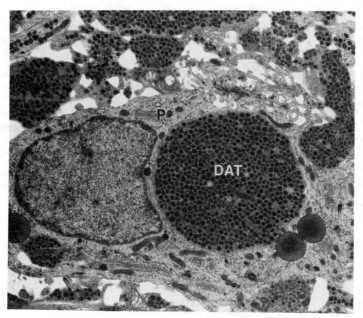

Figure 20-3. Electron micrograph of a pituicyte (*P*) and distended axon terminal (*DAT*) of an axon in the pars nervosa of the pituitary gland. The distended axon terminal is filled with electron-dense granules, which are secretion products of the neurohypophysis coupled to neurophysins. (Micrograph courtesy of Dr. R.J. Walsh, Department of Anatomy, George Washington University. Reprinted from Johnson KE: *Histology: Microscopic Anatomy and Embryology.* New York, John Wiley, 1982, p 295.)

(2) The hormone and its carrier neurophysin are discharged into the bloodstream at nerve terminals.

3. **Herring bodies.** The pars nervosa also contains enormously distended axon terminals called Herring bodies.

4. **Pituicytes,** the predominant intrinsic cells of the pars nervosa, are neuroglial cells that are interspersed between the nerve terminals (see Figure 20-3).

IV. THE PINEAL GLAND

A. General characteristics

1. The pineal gland is located above the third ventricle of the brain. It measures about 4 mm wide and 7 mm long. At maximum development, it weighs about 100–150 mg.

2. The pineal gland is connected to the brain by the **pineal stalk,** which contains an evagination of the roof of the third ventricle called the **pineal recess**.
   a. The pineal stalk contains nerve processes from adjacent regions of the brain; however, it is not clear if a functional link exists between these nerve fibers and pineal endocrine function.
   b. The pineal stalk is surrounded by a connective tissue capsule that is continuous with the pia-arachnoid meninges.

3. Pineal gland function is regulated by hormonal and neuronal stimulation.

B. Pinealocytes are the main parenchymal cells of the pineal gland. Pinealocytes secrete **melatonin** and peptides that are similar to the peptides formed in the hypothalamohypophysial axis.

1. Pinealocytes have large polymorphic nuclei with conspicuous nuclear indentations and one or more prominent nucleoli.

2. Pinealocytes are joined to each other by complex intercellular junctions with desmosomes and gap junctions, which allow small molecules to pass between adjacent cells for intercellular communication.

3. Pinealocytes have two processes that extend from the cell body.
   a. One is a short, thin process that ends on adjacent pinealocytes. The other is longer and thinner and ends near the space around blood vessels. The longer processes often contain vesicles and dense bodies, which may be secretory structures.
   b. Both cytoplasmic projections contain microtubules and have many large mitochondria around their nuclei.

4. Melatonin (5-methoxy-*N*-acetyl tryptamine) helps regulate LH levels. The level of melatonin varies with the time of day, appears to vary with the seasons, and is sensitive to norepinephrine and cyclic adenosine 3′,5′-monophosphate (cAMP).

5. The pineal gland produces low levels of peptides similar to ADH and oxytocin and their associated neurophysins. It also contains substantial levels of GnRH and has a role in the onset of puberty (pineal tumors in males are associated with precocious puberty).

C. Corpora arenacea. The pineal gland contains calcified concretions (called **corpora arenacea** or **brain sand**), which contain hydroxyapatite. In humans, these concretions usually become more prominent with age. They are a medial intracranial landmark visible in x-rays, computed tomography scans, and magnetic resonance imaging.

STUDY QUESTIONS

Directions: The groups of questions below consist of lettered choices followed by several numbered items. For each numbered item select the **one** lettered choice with which it is **most** closely associated. Each lettered choice may be used once, more than once, or not at all.

Questions 1–5

Match each description below with the most appropriate structure.

(A) Pars nervosa
(B) Median eminence
(C) Pars tuberalis
(D) Pars intermedia
(E) Pars distalis

1. The largest portion of the adenohypophysis

2. The portion of the adenohypophysis between the pars distalis and the pars nervosa

3. The portion of the neurohypophysis containing distended axon terminals rich in neurophysins and oxytocin

4. A portion of the adenohypophysis; contains most of the capillaries that convey releasing hormones

5. A portion of the adenohypophysis; forms a collar around the infundibular stalk and process

Questions 6–11

Match each description below with the most appropriate adenohypophysial cell type.

(A) Corticotrope
(B) Gonadotrope
(C) Somatotrope
(D) Thyrotrope
(E) Lactotrope

6. A basophil that secretes a hormone that regulates spermatogenesis and ovulation

7. The acidophil that secretes prolactin

8. A basophil that secretes a hormone that regulates basal metabolic rate

9. A basophil that secretes a hormone that regulates the ionic composition of blood and urine

10. An acidophil; the volume of its secretion product is altered in pituitary dwarfs

11. The basophil with the smallest granules

Questions 12–14

Match each description below with the most appropriate structure or structures.

(A) Hypothalamohypophysial portal vessels
(B) Adenohypophysial capillaries
(C) Both
(D) Neither

12. Contain releasing hormones

13. Are fenestrated vascular channels

14. Surround follicles and cords of acidophils and basophils

ANSWERS AND EXPLANATIONS

1–5. The answers are: 1-E, 2-D, 3-A, 4-E, 5-C. [*II D; III A*] The pars nervosa and median eminence are part of the neurohypophysis. Distended axon terminals rich in neurophysin-oxytocin complexes terminate in the pars nervosa. The pars tuberalis is part of the adenohypophysis. It forms a collar around the infundibular stalk. The pars intermedia is part of the adenohypophysis and lies between the pars nervosa and the pars distalis. The pars distalis is the largest portion of the adenohypophysis. Most of the adenohypophysial capillaries of the hypothalamohypophysial portal system end here.

6–11. The answers are: 6-B, 7-E, 8-D, 9-A, 10-C, 11-D. [*II C*] Corticotropes are basophils that secrete adrenocorticotropic hormone (ACTH). ACTH regulates adrenal cortical steroid production and, thus, regulates blood and urine ionic composition. Corticotropes have large granules and abundant cytoplasmic filaments.

Gonadotropes are basophils with medium-sized granules. Their granules are larger than the small granules of thyrotropes and smaller than the large granules of corticotropes. Gonadotropes secrete luteinizing hormone (LH) and follicle-stimulating hormone (FSH), which are hormones that help control spermatogenesis and ovulation.

Somatotropes are acidophils that secrete growth hormone (GH). GH secretion is decreased in pituitary dwarfs.

Thyrotropes are basophils with small granules. Thyrotropes secrete thyroid-stimulating hormone (TSH), a hormone that regulates basal metabolic rate.

Lactotropes are acidophils that secrete prolactin, a hormone that stimulates mammary gland development and lactation.

12–14. The answers are: 12-C, 13-B, 14-B. [*II D 1, E 1*] Hypothalamohypophysial portal vessels join capillary beds in the neurohypophysis and the adenohypophysis. They convey releasing hormones, which are synthesized in the neurohypophysis, to the adenohypophysis. Adenohypophysial capillaries are fenestrated capillaries surrounding the follicles and cords of acidophils and basophils in the adenohypophysis. Both hypothalamohypophysial portal vessels and adenohypophysial capillaries contain releasing hormones.

21
The Thyroid, Parathyroid, and Adrenal Glands

I. INTRODUCTION

A. Functions of the thyroid and parathyroid glands. The thyroid and parathyroid glands have a central role in maintaining the metabolic state and internal milieu of the human body; their secretions regulate the basal metabolic rate and serum calcium concentration.

1. **Thyroid follicular epithelial cells** produce **thyroxine,** which regulates basal metabolism. These cells produce a 660-kD glycoprotein called **thyroglobulin,** an apical secretion product that is a stored form of the active thyroid hormone thyroxine.
   a. When a physiologic need for thyroxine exists, follicular epithelial cells degrade stored thyroglobulin into active thyroid hormones—the iodinated amino acids L-thyroxine (3,5,3′,5′-tetraiodothyronine), which is called T_4, and triiodothyronine (3,5,3′-triiodothyronine), which is called T_3.
   b. T_4 and T_3 increase oxidative metabolism and have important functions in the development of the organism.

2. **Parafollicular cells (C cells)** in the thyroid gland produce the polypeptide hormone **calcitonin,** which lowers serum calcium levels and inhibits bone formation.

3. The principal parenchymal cells of the **parathyroid glands** release a polypeptide hormone called **parathyroid hormone (PTH),** which is an antagonist of calcitonin. PTH increases the level of serum calcium by stimulating bone resorption. PTH and calcitonin also affect the kidneys and the gastrointestinal tract to regulate calcium excretion and uptake.

4. **Control of thyroid and parathyroid function.** The secretory activity of these endocrine glands is partially controlled by interactions between target organs, the hypothalamus, and the adenohypophysis.
   a. For example, T_4 regulation of blood thyroxine levels involves two other hormones. Thyroxine levels provide feedback to the hypothalamus, which is the source of **thyrotropin releasing hormone (TRH).**
   b. TRH stimulates thyrotropes in the pars distalis to release **thyroid-stimulating hormone (TSH).**
   c. TSH stimulates thyroid follicular epithelial cells to phagocytose stored thyroglobulin, degrading it intracellularly and then releasing it into the bloodstream as T_3 and T_4.
   d. When T_3 and T_4 levels in the blood fall, the hypothalamus releases TRH, thus stimulating TSH secretion and a subsequent rise in T_3 and T_4 levels.

B. Functions of the adrenal glands. The adrenal glands are paired glands that lie above each kidney. Their functions are very important. Each adrenal gland consists of a cortex and a medulla.

1. **Adrenal cortex**
   a. The adrenal cortex synthesizes a complex array of **steroids** that regulate carbohydrate metabolism, ion composition of the internal milieu, and sexual function.
   b. The cortex receives endocrine control from the adenohypophysis in the form of **adrenocorticotropic hormone (ACTH),** and from the kidneys through the **renin-angiotensin system.**

2. **Adrenal medulla**
   a. The adrenal medulla synthesizes the catecholamines **epinephrine** and **norepinephrine.** These two hormones have widespread effects on heart rate, smooth muscle contraction, and carbohydrate and lipid metabolism.
   b. The medulla is controlled by nerve impulses.

II. THE THYROID GLAND

A. Microscopic anatomy (Figure 21-1)

1. **Connective tissue.** The thyroid gland is surrounded by a thin connective tissue capsule. Trabecular projections from the capsule form delicate septa that divide the gland into poorly defined lobules. Blood vessels from the capsule pass through the septa to the lobules and follicles.

2. **Thyroid follicles** are the predominant structures in the thyroid gland. They are closed cavities lined by a continuous simple layer of thyroid epithelial cells.
   a. Thyroid follicles contain **thyroglobulin,** a 660-kD glycoprotein in colloidal suspension. Thyroglobulin is a stored form of thyroxine and an exocrine secretion product of follicular epithelial cells.
      (1) Thyroid colloid stains bright magenta with the periodic acid-Schiff (PAS) reaction due to the abundant conjugated neutral sugars attached to the thyroglobulin protein backbone.
      (2) Thyroglobulin is **amphoteric:** it can take either a negative or a positive charge, depending on pH. Consequently, this glycoprotein stains with either acidic or basic dyes under appropriate conditions.
   b. **Thyroid follicular epithelium** lies on a basement membrane and varies in height. In empty undistended follicles it is columnar, in partially filled follicles it is cuboidal, and in completely filled and distended follicles it is squamous.
      (1) When thyroid follicular epithelial cells are at rest (i.e., not actively engulfing thyroglobulin), they form a simple squamous layer around a large reservoir of colloid.
      (2) When stimulated by TSH, these cells engulf the colloid, reducing the amount of colloid in the follicle, and the follicle shrinks. The cells become cuboidal and then columnar as the follicle empties.
      (3) Under normal functional conditions, some follicles are resting and others are active; therefore, follicles of many different diameters are present at all times.
   c. Numerous **fenestrated continuous capillaries** form an anastomosing, lace-like network around each follicle. These capillaries pick up thyroxine and transport it into the systemic circulation.

B. Ultrastructure of epithelial cells.
Thyroid follicular epithelium contains two types of cells: thyroid follicular epithelial cells and C cells.

1. **Thyroid follicular epithelial cells** are the predominant type of cell in the thyroid epithelium. They synthesize and metabolize thyroglobulin, and their ultrastructure closely reflects their secretory function.

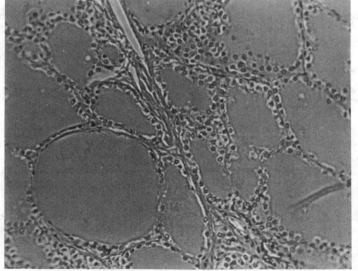

Figure 21-1. Light micrograph of a human thyroid gland. The large round structures are thyroid follicles filled with thyroglobulin. Each of the follicles is lined by a layer of follicular epithelial cells. (Reprinted from Johnson KE: *Histology: Microscopic Anatomy and Embryology.* New York, John Wiley, 1982, p 266.)

    a. Thyroid follicular cells have a **marked polarity,** which is especially evident when the cells are columnar.

        (1) Polarized cells have a rounded basal nucleus, an apical Golgi apparatus, and numerous mitochondria scattered throughout the cytoplasm.

        (2) Their apical surface is covered by numerous microvilli that vary in size and number, depending on the functional state of the thyroid gland.

            (a) The microvilli and broader lamellipodium-like surface projections engulf stored thyroglobulin by pinocytosis and phagocytosis.

            (b) Stimulation by TSH results in a sudden increase in the number of microvilli and other cell surface projections. Columnar cells exhibit many apical surface microvilli and folds, and apical cytoplasmic droplets of engulfed thyroglobulin.

    b. Each follicular epithelial cell is joined to neighboring cells by a well-developed apical **junctional complex**. The junctional complex isolates and seals the thyroglobulin-containing follicular compartment and prevents colloid from leaking into the adjacent glandular stroma.

    c. Follicular epithelial cells have several organelle features that are characteristic of cells involved in apical exocrine secretion.

        (1) Follicular epithelial cells have round basal nuclei with prominent nucleoli and abundant basal endoplasmic reticulum, which is used during thyroglobulin synthesis.

        (2) They have a large apical Golgi apparatus above the nucleus, which glycosylates and packages thyroglobulin and then transports it into the follicle.

    d. The cytoplasm contains numerous mitochondria, which provide ATP for protein synthesis and for engulfing thyroglobulin. Follicular epithelial cells also contain numerous lysosomes and phagolysosomes to degrade the engulfed thyroglobulin.

        (1) During normal function, the follicular epithelium can secrete thyroglobulin, or engulf it from the follicular reservoir and then degrade it into thyroxine in phagolysosomes.

        (2) Some engulfed thyroglobulin makes its way into the cytoplasm in coated pits and coated vesicles, which form at the base of microvilli.

    e. Under some circumstances, the basal portion of follicular epithelial cells is deeply infolded and lies on a delicate basement membrane. A thin connective tissue domain with fibroblasts and collagen fibers exists immediately adjacent to the basement membrane.

2. Parafollicular cells (C cells). Although encountered less frequently, C cells may exist apical to the basement membrane. They are larger than follicular cells and stain palely with most stains but show a strong affinity for silver stains, which react with their granules of calcitonin.

    a. C cells usually exist alone or in small clusters.

        (1) They are most often wedged between the follicular epithelial cells and the basement membrane; however, they also can exist as small groups just outside the basement membrane in the interfollicular stroma.

        (2) C cells in both locations have a highly irregular shape and a central nucleus.

    b. C cells have abundant rough (with attached ribosomes) endoplasmic reticulum, which may be flat cisternae or spirals.

    c. C cells have elongated mitochondria scattered throughout the cytoplasm, and contain a prominent Golgi apparatus (common in cells synthesizing protein for export).

    d. They also have membrane-delimited granules of secretion product that are 100–200 nm in diameter.

        (1) The secretory material is faintly granular and a clear halo exists between the edge of the secretory material and the membrane. The Golgi apparatus buds off similar, but smaller, vesicles containing secretory material.

        (2) The granules of C cells contain catecholamines and calcitonin and are rich in **monoamine oxidase,** an enzyme that metabolizes catecholamines.

    e. C cells originate in the neural crest and are part of the **APUD** (amine precursor uptake and decarboxylation) **system** of cells. (Other APUD cells include cells in the adrenal medulla and catecholamine-containing cells in the respiratory and gastrointestinal systems.)

C. Synthesis and secretion of thyroid hormones

1. Thyroglobulin synthesis

    a. Iodine and amino acids for thyroglobulin synthesis are transported across the basal plasma membrane of thyroid follicular epithelial cells. Presumably, the basal cell membrane infoldings increase the surface area available for transport into and out of these cells.

b. The polypeptide chains of thyroglobulin and the mannose residues of this glycoprotein are added in the rough endoplasmic reticulum.

(1) Mannose residues are present in the inner core of the glycosylated side chains, linked to the polypeptide backbone by hydroxylated amino acids such as threonine and serine.

(2) The remaining neutral sugars and sialic acid are added by glycosyltransferases present in Golgi apparatus membranes.

c. The Golgi apparatus is the production site for membrane-bound secretion vesicles of thyroglobulin. The vesicles fuse into large granules, move apically, and then are exocytosed into the follicular lumen.

2. Thyroperoxidase synthesis. Thyroperoxidase is another enzyme produced by the follicular epithelium. It is synthesized along the same route as thyroglobulin; however, it performs a much different function.

a. Thyroperoxidase near the follicular lumen is responsible for iodine oxidation and for coupling iodine to thyroglobulin. This process occurs near or within the follicular lumen.

b. Thyroperoxidase has been demonstrated at the luminal surface of the brush border of microvilli.

3. Thyroxine secretion. Under TSH stimulation, surface folds and microvilli of the follicular epithelium engulf and phagocytose stored thyroglobulin.

a. Phagosomes containing colloid fuse with the numerous lysosomes present in the cytoplasm. The action of actin-containing microfilaments and tubulin-containing microtubules is necessary for the phagocytosis and transport of colloid droplets.

b. Proteolytic enzymes stored in lysosomes are released into colloid-containing phagolysosomes to create phagolysosomes.

(1) Proteolytic enzymes degrade thyroglobulin, releasing the two iodinated amino acids T_4 and T_3.

(2) These two thyroid hormones are known generically as **thyroxine**.

c. T_4 and T_3 leave follicular epithelial cells and diffuse quickly into the dense networks of fenestrated continuous capillaries surrounding thyroid follicles and into the lymphatic vessels in the connective tissue stroma. From the capillaries and small lymphatics, thyroid hormones move into the systemic circulation.

d. Thyroxine has a striking effect on the basal metabolic rate. When thyroid hormone levels increase, the rate of metabolism increases; when hormone levels decrease, the rate of metabolism decreases.

D. Thyroid disorders. A normal level of thyroid hormone secretion is required for normal growth and development of the musculoskeletal system and for normal mentation.

1. Infantile hypothyroidism results in short stature, mental retardation, and abnormal reproductive function.

2. Colloid goiter results from a diet deficient in iodine, due to follicular hypertrophy and the excessive accumulation of colloid in follicles.

3. Graves' disease is a hyperthyroid condition characterized by follicular hyperplasia without colloid accumulation and high levels of circulating thyroid hormone, which may be caused by a peculiar circulating immunoglobulin causing chronic thyroid stimulation by a poorly understood mechanism.

III. THE PARATHYROID GLANDS

A. General characteristics

1. The parathyroid glands are embedded in the thyroid gland but are separated from it by a thin connective tissue capsule.

2. Connective tissue trabeculae from the capsule penetrate into the parenchyma of the glands, carrying blood vessels, lymphatics, and nerve fibers.

3. Deep within the glands, delicate reticular fibers connect the glandular cells to the surrounding capillaries.

    4. Adipocytes are scattered among the glandular elements and are especially numerous in older people.

B. Parathyroid parenchyma. The rest of each parathyroid gland consists of chief cells (principal cells), which constitute most of the parenchyma, and scattered oxyphilic cells (Figure 21-2).

    1. Chief cells are 7–10 μm in diameter, are roughly polygonal, and are packed together in nests or cords of cells. They have a palely stained, central nucleus and a cytoplasm dedicated to synthesizing protein for export.

      a. The cytoplasm of chief cells contains abundant rough endoplasmic reticulum arranged in plates, a juxtanuclear Golgi apparatus, scattered mitochondria, and granules of secretion product bounded by a unit membrane and containing a granular, electron-dense material.

      b. Some chief cells have one poorly developed cilium projecting away from the cell surface: a vestige of the epithelial origin of these cells.

    2. Oxyphilic cells are greatly outnumbered by chief cells. They are larger than chief cells, but they have smaller, more densely stained nuclei than chief cells and a distinctly acidophilic cytoplasm. Oxyphilic cells occur alone or in small clusters.

      a. In the electron microscope, oxyphilic cell cytoplasm exhibits many elongated mitochondria; however, it lacks the protein synthesis components so prominent in chief cells.

      b. With increasing age, the amount of adipose tissue and the number of oxyphilic cells increase.

    3. Other cells. Parathyroid glands also contain intermediate cell types of unknown significance.

IV. HORMONAL REGULATION OF SERUM CALCIUM

A. General features. The regulation of the calcium level in the blood and extracellular fluid is a crucial homeostatic process essential to life.

    1. Nerve conduction and muscle contraction cannot occur without normal levels of calcium. Cell adhesion and blood clotting also require normal levels of calcium.

    2. The hydroxyapatite crystals in bones are the main reservoir of stored calcium in the body.

      a. Bone is constantly broken down and reformed to regulate serum calcium and to replace worn-out bone.

      b. The serum calcium lost in excreted urine is replaced by dietary calcium.

    3. The uptake, excretion, and turnover of bone calcium are regulated by the antagonistic actions of calcitonin and PTH.

B. Calcitonin actions. Calcitonin is secreted by C cells in the thyroid gland.

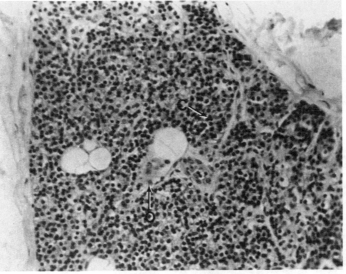

Figure 21-2. Light micrograph of a human parathyroid gland. Chief cells (*C*) and oxyphil cells (*O*) are visible. (Reprinted from Johnson KE: *Histology: Microscopic Anatomy and Embryology.* New York, John Wiley, 1982, p 269.)

1. Calcitonin is a polypeptide containing 32 amino acids. Its principal effect is to reduce the amount of calcium released from bones by inhibiting **osteocytic osteolysis**.

2. When serum calcium levels rise, calcitonin secretion occurs, bone resorption decreases, and serum calcium levels fall again.

C. PTH actions

1. PTH is synthesized in chief cells initially as a 115-amino acid polypeptide called **pre-pro-parathyroid hormone**.

2. A 25-amino acid polypeptide is cleaved from this polypeptide chain after it is completely synthesized and is passed into the endoplasmic reticular lumen. This **proparathyroid hormone** passes through the Golgi apparatus, where a 6-amino acid peptide is cleaved from it, releasing active PTH.

3. PTH has an effect that opposes the effect of calcitonin (i.e., PTH stimulates osteocytic osteolysis). In chronic hyperstimulation by PTH, bones become degraded by osteoclastic osteolysis.

4. PTH secretion not only elevates serum calcium by increasing bone resorption, it also reduces renal excretion of calcium.

5. PTH affects the synthesis of 1,25-dihydroxycholecalciferol (a metabolite of vitamin D), which indirectly affects serum calcium by increasing the efficiency with which dietary calcium is absorbed in the gastrointestinal tract.

V. THE ADRENAL GLAND (Figure 21-3)

A. Microscopic anatomy of the adrenal cortex

1. **General features**
   a. The adrenal gland is encapsulated by a layer of fibroblasts, collagen fibers, and elastic fibers. Trabecular elements from the capsule pass into the parenchyma of the gland, carrying blood vessels and lymphatics.
   b. A delicate network of reticular fibers surrounds parenchymal cells, binding them together and supporting the delicate vascular channels that pass around and between the cortical cells.

2. **Morphologic zonation.** The adrenal cortex has conspicuous morphologic and functional zonation involving three distinct layers (see V B 2 for a discussion of functional zonation).
   a. **Zona glomerulosa.** The outer cortical layer is the zona glomerulosa. It comprises roughly 15% of the thickness of the cortex.

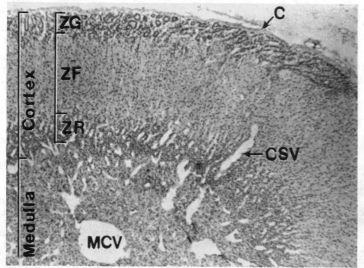

Figure 21-3. Light micrograph of a human adrenal gland. This gland has a thin connective tissue capsule (*C*), a cortex, and a medulla. The cortex is divided into a zona glomerulosa (*ZG*), zona fasciculata (*ZF*), and zona reticularis (*ZR*). A cortical shunt vessel (*CSV*) and a medullary central vein (*MCV*) are visible. (Reprinted from Johnson KE: *Histology: Microscopic Anatomy and Embryology.* New York, John Wiley, 1982, p 276.)

(1) Cells in the zona glomerulosa are arranged in rounded clusters 12–15 μm in diameter. Each cell has a small round nucleus and a few lipid droplets scattered in its cytoplasm.

(2) The zona glomerulosa sometimes is absent in the human adrenal gland. In such instances, it is replaced by cells of the zona fasciculata that extend out to the capsule.

b. Zona fasciculata. The middle cortical layer is the zona fasciculata, which comprises roughly 75% of the thickness of the cortex.

(1) Cells in the zona fasciculata are larger than those in the other two layers of the cortex. They have a diameter of 20 μm and are arranged in long straight cords of several cells surrounded by vascular sinusoids (Figure 21-4).

(2) Rows of cells are arranged in radial spokes projecting away from the adrenal medulla. The rows sometimes branch to form anastomoses with adjacent cords.

(3) The outstanding histologic feature of these cells is the large number of lipid droplets in their cytoplasm.

    (a) These lipids are poorly fixed by routine histologic fixatives. Furthermore, they are extracted by the solvents used to dehydrate and embed specimens in paraffin.

    (b) This artifactual extraction causes the cytoplasm to appear foamy. Some histologists call these cells **spongiocytes** because of this conspicuously vacuolated appearance.

c. Zona reticularis. The inner cortical layer, lying next to the medulla, is the zona reticularis. This layer comprises roughly 10% of the thickness of the cortex.

(1) Cells in the zona reticularis are smaller than cells in the zona fasciculata. They do not have abundant lipid droplets; however, they may have prominent cytoplasmic masses of brownish **lipofuscin** granules.

    (a) Lipofuscin granules are occasionally observed in the zona glomerulosa and the zona fasciculata, but they are most abundant in the zona reticularis.

    (b) Lipofuscin granules have acid phosphatase activity and, thus, are believed to be lysosome-related residual bodies of lipid metabolites.

(2) Cells in the zona reticularis are arranged in a reticulum rather than in rows.

3. Cortical cell ultrastructure

a. Cells in the adrenal cortex synthesize and secrete steroids; thus, their ultrastructure is similar to the ultrastructure of cells in the ovarian corpus luteum and testicular Leydig cells.

(1) For example, adrenal cortical cells often contain numerous lipid droplets, which probably are a stored form of cholesterol esters and other steroid precursors.

(2) They also have an unusual abundance of smooth endoplasmic reticulum and many large round mitochondria with **tubular cristae** rather than the lamellar cristae typical of many other mitochondria. The significance of this unusual mitochondrial architecture is obscure, but it is frequently encountered in steroid-secreting cells.

b. Cortical cells in different zones have striking differences in their ultrastructure.

(1) For example, cells in the zona glomerulosa have few lipid droplets; long, thin mitochondria; abundant smooth endoplasmic reticulum; and little rough endoplasmic reticulum. In contrast, cells in the zona fasciculata have many large lipid droplets.

(2) Cells in the zona reticularis are much like cells in the zona fasciculata except that they

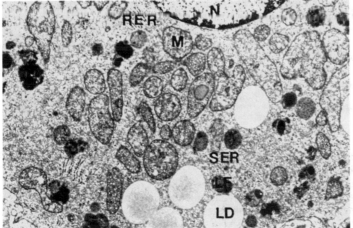

Figure 21-4. Electron micrograph of a cell in a human adrenal zona fasciculata. This cell has a prominent nucleus (*N*), abundant rough endoplasmic reticulum (*RER*) and smooth endoplasmic reticulum (*SER*), mitochondria (*M*), lipid droplets (*LD*), and lipofuscin droplets (*LF*). (Micrograph courtesy of Dr. J.A. Long, Department of Anatomy, University of California, San Francisco.)

have fewer lipid droplets, smaller mitochondria, and substantial deposits of **lipofuscin pigments,** which may be accumulated by-products of lipid metabolism.

   c. The ultrastructure of cortical cells changes according to their functional state.

   (1) For example, an injection of ACTH decreases the number of lipid droplets and increases the amount of smooth endoplasmic reticulum. This probably is a response to increased steroid biosynthesis and secretion.

   (2) The precise intracellular pathways for steroid synthesis and secretion are the subject of experimental investigations; however, most cell biologists believe that steroid secretion is continuous and does not occur through quantal granules such as those in the adenohypophysis or the parathyroid glands.

4. **Cortical sinusoids**

   a. A rich network of capillaries surrounds cortical cells. The shape of these capillary networks tends to reflect the arrangement of cortical parenchymal cells. Thus, in the zona glomerulosa the capillaries follow tortuous paths, whereas capillaries in the zona fasciculata are long, straight sinusoidal channels.

   b. In all locations, the capillaries are continuous **fenestrated capillaries.** Each of the fenestrations is closed by a thin diaphragm, which has a composition that differs from the composition of the plasma membrane.

   c. The capillaries are intimately associated with a thin, but continuous, basement membrane and with numerous delicate reticular fibers. The reticular fibers are part of a connective tissue domain that separates groups of cortical cells from each other and holds these cells together.

   d. Steroid-secreting cells in the adrenal cortex are surrounded by their own distinct basement membrane. This feature is not so unusual, because these cells are derived from coelomic epithelium and retain some features typical of epithelial cells after developing into the definitive adrenal cortex.

B. Histophysiology of steroid secretion

1. **Steroid classes.** Cells in the adrenal cortex secrete a wide variety of steroid hormones. These steroids are classified as:

   a. **Glucocorticoids** (e.g., cortisol; corticosterone), which have profound effects on carbohydrate metabolism and, thus, alter blood sugar levels

   b. **Mineralocorticoids** (e.g., aldosterone), which affect renal tubular function and, thus, regulate the ionic composition of blood and extracellular fluids

   c. **Weak androgens** (e.g., dehydroepiandrosterone), which are sex steroids

2. **Functional zonation.** The adrenal cortex has functional as well as morphologic zonation. For example, the zona glomerulosa is primarily responsible for the synthesis of **aldosterone,** one of the mineralocorticoids, and the zona fasciculata and zona reticularis are primarily responsible for the synthesis of **cortisol** and **dehydroepiandrosterone.** Also, some evidence indicates that the zona reticularis is more active in androgen secretion than the zona fasciculata.

   a. Steroids are a closely related family of molecules that are synthesized from cholesterol precursors.

   b. The enzymes required for steroid biosynthesis are localized in mitochondria or in smooth endoplasmic reticular membranes and may have zonal distributions.

   (1) For example, the enzymes used in the initial stages of steroid biosynthesis (e.g., **pregnenolone synthetase,** an enzyme that converts cholesterol to pregnenolone) are found in all cortical zones.

   (2) The enzymes used for the final stages of aldosterone synthesis exist only in the zona glomerulosa, which is the only layer that synthesizes aldosterone.

   (3) A key enzyme in the cortisol pathway, **17-α-hydroxylase,** exists only in the zona fasciculata and zona reticularis.

3. **Endocrine regulation of adrenocortical function.** Two endocrine organs regulate cortical function.

   a. **Renal regulation.** The zona glomerulosa, which secretes mineralocorticoids, is regulated by the **renin-angiotensin system.**

   (1) The renal juxtaglomerular apparatus releases **renin** in response to diminished blood pressure. Renin is an enzyme that converts **angiotensinogen** to **angiotensin I.**

   (2) A pulmonary enzyme then converts angiotensin I into **angiotensin II,** which stimulates the zona glomerulosa to secrete aldosterone.

      (3) Aldosterone, in turn, promotes sodium and water retention in the kidneys and, thus, increases blood pressure.
  b. Adenohypophysial regulation. The remainder of the adrenal cortex is controlled by ACTH, which is secreted by adenohypophysial basophils.
      (1) In hypophysectomized animals, the zona glomerulosa remains unchanged, but the zona fasciculata and zona reticularis show marked involution. ACTH administered in physiologic doses temporarily reverses this effect.
      (2) Excess amounts of ACTH cause a striking loss of lipid droplets and an increase in smooth endoplasmic reticulum as well as a dramatic rise in steroid levels in the blood.
      (3) The concentration of glucocorticoids in the blood has a direct effect on hypothalamic secretion of **corticotropin releasing hormone (CRH)** and hypophysial secretion of ACTH.
        (a) When glucocorticoid concentration is high, the secretion of CRH and ACTH is suppressed and the concentration of glucocorticoid falls.
        (b) When glucocorticoid levels fall, the feedback inhibition is removed and CRH is secreted, which in turn stimulates ACTH secretion.
      (4) ACTH directly stimulates the adrenal cortex to secrete glucocorticoids. Glucocorticoids have complex systemic effects, with a net result of raising the blood concentration of glucose.

C. The adrenal medulla

  1. Microanatomy
    a. The adrenal medulla is composed of masses of **catecholamine-secreting cells** surrounded by complex networks of blood vessels.
      (1) Catecholamine-secreting cells are sometimes called **chromaffin cells** because their granules stain a brownish color when exposed to solutions containing dichromate ions.
      (2) Interaction between the dichromate and the catecholamines produces deposits in cells. After aldehyde fixation, the catecholamine granules have specific autofluorescence.
    b. The medulla also contains the cell bodies of **preganglionic sympathetic neurons;** however, these are not always distinguishable from the catecholamine-secreting cells.
    c. In the electron microscope, adrenal medullary cells exhibit prominent (200 nm in diameter) membrane-bound granules.
      (1) These granules contain a moderately electron-dense material (a stored form of catecholamines) and enzymes responsible for converting dopamine into catecholamines.
      (2) Transmission electron microscopy reveals two morphologically distinct granules with different electron densities. The variant with greater electron density stores **norepinephrine;** the variant with less electron density stores **epinephrine**.
      (3) These granules are formed in a prominent perinuclear Golgi apparatus.
    d. Medullary cells also have moderate amounts of rough endoplasmic reticulum and unremarkable mitochondria.
    e. Synaptic contacts exist between the sympathetic preganglionic neurons and the parenchymal cells of the adrenal medulla. These synaptic contacts probably stimulate the secretory activity of medullary cells.
    f. The blood vessels that connect to the adrenocortical sinusoids contain large concentrations of corticosteroids and are closely associated with epinephrine-secreting medullary cells. In contrast, the cortical shunt vessels drain directly into the medullary central veins, where they are loosely associated with norepinephrine-secreting medullary cells.

  2. Histophysiology. Catecholamine secretion by medullary cells often has striking systemic effects.
    a. Epinephrine
      (1) Epinephrine secretion causes a dramatic increase in the rate of the heartbeat, cardiac output, and blood flow to skeletal muscles and a decrease in splanchnic perfusion. It also increases basal metabolic rate and stimulates glycogenolysis in the liver.
      (2) During periods of stress or fright, a sudden secretion of epinephrine prepares the body to deal with the stress in a response called the **fight or flight reaction**. This response causes increased cardiac output and increased skeletal muscle perfusion, which clearly relates to the body's ability to ward off a dangerous situation.
      (3) Also, epinephrine secretion can stimulate the release of ACTH, which promotes glucocorticoid secretion and, thus, indirectly stimulates gluconeogenesis.
    b. Norepinephrine causes vasoconstriction, thereby increasing blood pressure.
    c. The release of epinephrine and norepinephrine is controlled by the central nervous system through the preganglionic sympathetic fibers in the medulla.

STUDY QUESTIONS

Directions: Each question below contains five suggested answers. Choose the **one best** response to each question.

1. Which of the statements below best describes the histologic organization of the thyroid gland?

(A) Follicular epithelial cells are surrounded by an anastomosing network of continuous capillaries without fenestrations

(B) Follicular epithelial cells secrete thyroglobulin into capillaries

(C) Follicular epithelial cells exocytose thyroglobulin as a secretion product

(D) The thyroid gland is conspicuously lobulated, with dense connective tissue trabeculae

(E) Thyroid follicles connect to ducts

2. All of the following statements concerning the zona fasciculata are true EXCEPT

(A) cells have little smooth endoplasmic reticulum

(B) mitochondria are ovoid

(C) cells are sensitive to ACTH stimulation

(D) cells have numerous large lipid droplets

(E) it is the thickest cortical zone

Directions: The groups of questions below consist of lettered choices followed by several numbered items. For each numbered item select the **one** lettered choice with which it is **most** closely associated. Each lettered choice may be used once, more than once, or not at all.

Questions 3–5

Match each description below with the most appropriate hormone or hormones.

(A) Calcitonin
(B) Parathyroid hormone
(C) Both
(D) Neither

3. A polypeptide hormone secreted by parafollicular cells in the thyroid gland

4. A steroid hormone secreted by chief cells in the parathyroid glands

5. A hormone that stimulates bone degradation and retention of calcium in the renal system

Questions 6–9

Match each description below with the appropriate lettered site or structure in the following illustration of a thyroid cell.

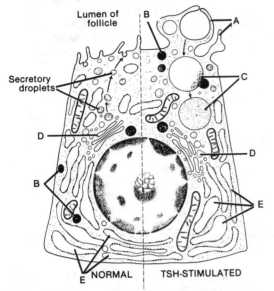

Reprinted from Long DW, Jones AL: *Recent Prog Horm Res* 25:315-380, 1969.

6. Results from fusion of a phagosome and a lysosome

7. Surface elaboration on the follicular apex, which increases in size in response to TSH

8. Lysosomes are synthesized here; thyroglobulin synthesis is completed here

9. Membrane-delimited packet of hydrolytic enzymes

Questions 10–14

Match each description below with the appropriate lettered site or structure in the following illustration of a cell in the zona fasciculata.

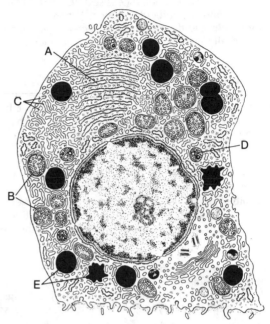

Reprinted from Lentz TL: *Cell Fine Structure*. Philadelphia, WB Saunders, 1971.

10. Stored form of cholesterol and cholesterol esters

11. Has tubular cristae and is prominent in the cells of steroid-secreting tissue

12. Lipofuscin

13. Rough endoplasmic reticulum

14. Many enzymes for steroid biosynthesis are located here

ANSWERS AND EXPLANATIONS

1. The answer is C. [*II A 2 a, C 1*] The thyroid gland is an endocrine organ and, therefore, has no ducts. The thyroid gland has no conspicuous lobulation and contains only sparse connective tissue. Thick connective tissue septa around the follicles would retard the passage of hormones from the follicles into the blood. Thyroid epithelial cells secrete a 660-kD glycoprotein called thyroglobulin (a stored form of thyroxine) into the follicular lumen. When necessary, follicular epithelial cells phagocytose stored thyroglobulin and then cleave and secrete thyroxine into the anastomosing network of capillaries that surrounds all follicles. Like capillaries in all endocrine tissues, these capillaries are fenestrated and they carry thyroxine into the systemic circulation.

2. The answer is A. [*V A 2 b; Figure 21-4*] The zona fasciculata is the intermediate zone in the adrenal cortex, lying between the superficial zona glomerulosa and the deep zona reticularis. It is the thickest cortical zone and is sensitive to adrenocorticotropic hormone (ACTH) stimulation. It is active in steroid synthesis and secretion and has an ultrastructure that reflects this function. The ultrastructure includes numerous lipid droplets, ovoid mitochondria with tubular cristae, and an abundance of smooth endoplasmic reticulum.

3–5. The answers are: 3-A, 4-D, 5-B. [*I A 2, 3; IV B, C*] Calcitonin and parathyroid hormone (PTH) have antagonistic effects in the regulation of serum calcium levels. Calcitonin is a polypeptide secreted by C cells in the thyroid gland. When serum calcium levels fall, calcitonin secretion is stimulated. Calcitonin causes resorption of bone and renal calcium retention and directly causes an increase in serum calcium.

PTH is a polypeptide (not a steroid) secreted by the chief cells of the parathyroid glands. PTH suppresses bone resorption and leads directly to a decrease in serum calcium.

6–9. The answers are: 6-C, 7-A, 8-D, 9-B. [*II B 1*] Thyroid-stimulating hormone (TSH) stimulation causes an elaboration of apical microvilli (A) and production of phagosomes. Phagolysosomes (C) are formed when phagosomes fuse with lysosomes (B). The Golgi apparatus (D) is involved in lysosome synthesis. Thyroglobulin synthesis begins in the rough endoplasmic reticulum (E) and is completed in the Golgi apparatus.

10–14. The answers are: 10-E, 11-B, 12-D, 13-A, 14-C. [*V A 2 b, 3*] This steroid-secreting cell contains both rough (A) and smooth (C) endoplasmic reticulum. The latter contains many enzymes involved in steroid synthesis. This cell has many lipid droplets (E), which are cholesterol and cholesterol esters used in steroid biosynthesis, as well as prominent mitochondria (B) with tubular cristae. Also, lipofuscin (D) is present in these cells but is much less abundant than in cells of the zona reticularis.

22
The Placenta and
Mammary Glands

I. INTRODUCTION. The placenta and mammary glands are not part of the female reproductive system; however, they are closely related to the reproductive function of the female reproductive system.

    A. Functions of the placenta. The fetus develops in a closed system; the placenta is the interface between mother and fetus. All material entering or leaving the fetal system passes through the placenta. The important functions of the placenta are:

        1. Respiration

        2. Nutrient assimilation

        3. Waste elimination

        4. Water, electrolyte, and immunoglobulin transport

        5. Hormone synthesis

    B. Functions of the mammary glands. Mammary glands are modified sweat glands. Their most important function is the production of **milk** for the infant. Milk has two important functions.

        1. Milk is a complete food for the infant; it provides all the essential nutrients and minerals.

        2. Milk also provides the infant with essential immunologic protection. It is rich in immunoglobulins and other bacteriostatic substances, which protect the infant against enteric infections.

II. THE PLACENTA

    A. The forming placenta. After ovum fertilization and cleavage, the compact mass of cells called the **morula** cavitates to form a **blastocoel.** At this time, the embryo is called a **blastocyst.**

        1. Trophoblast formation
            a. The blastocyst has a thin epithelial wall called the **trophoblast** and an **inner cell mass** (embryoblast), which projects into the blastocoel.
            b. The trophoblast forms the **chorion,** a shell of tissue around the developing embryo, and subdivides into a cellular layer called the **cytotrophoblast,** which begins to proliferate rapidly.
            c. Cytotrophoblastic cellular mitoses form new cells, which fuse to form a syncytial layer called the **syncytiotrophoblast** (Figure 22-1).
                (1) The syncytiotrophoblast, which forms the outermost shell of the developing embryo, is an invasive syncytium that degrades maternal endometrial epithelial cells and blood vessels.
                (2) As maternal blood vessels break down, maternal blood bathes the surface of the syncytiotrophoblast.

        2. Chorionic villi formation
            a. Cytotrophoblastic cells continue to divide and fuse to produce syncytiotrophoblast. They also differentiate into the mesenchymal cells that form blood vessels and connective tissue fibroblasts.
            b. Finger-like projections from the chorion and surface invaginations called **primary villi** begin to form. Primary villi lack connective tissue and fetal blood vessels (Figure 22-2).
            c. Primary villi become **secondary villi** as they are invaded by connective tissue cores.

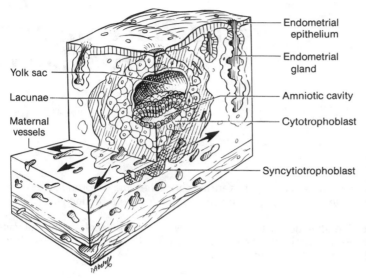

Yolk sac

Lacunae

Maternal vessels

Endometrial epithelium

Endometrial gland

Amniotic cavity

Cytotrophoblast

Syncytiotrophoblast

Figure 22-1. Human conceptus during implantation. Cytotrophoblast cells divide and fuse to form the syncytiotrophoblast, an invasive structure that facilitates conceptus implantation. The endometrial epithelium has almost completely healed over the implantation site. (Reprinted from Johnson KE: *Human Developmental Anatomy.* Baltimore, Williams & Wilkins, 1988, p 94.)

   d. Secondary villi become **tertiary villi** as they are invaded by developing fetal blood vessels (Figure 22-3).
   e. Many long villi develop on one side of the chorion, forming the **chorion frondosum,** which is the precursor of the definitive placenta. The other side of the chorion has few, very short villi, which comprise the **chorion laeve** or smooth part of the chorion.

B. The definitive placenta

1. The chorionic plate and decidual plate
   a. As villi develop, the placenta develops two tissue layers. The **chorionic plate** develops on the fetal side of the placenta and the **decidual plate** (basal plate) develops on the maternal side of the placenta (Figure 22-4).
   b. **Cytotrophoblastic cell columns** connect the two plates. Villi project from these columns into the intervillous space.

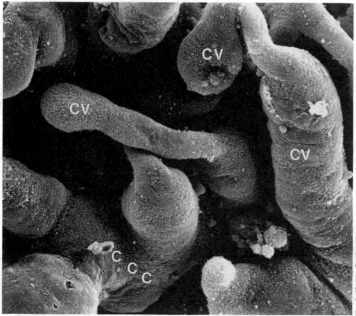

Figure 22-2. Scanning electron micrograph of human placental chorionic villi (*CV*) attached to cytotrophoblastic cell columns (*CCC*). This specimen was gathered by chorionic villus biopsy from a 10-week human pregnancy.

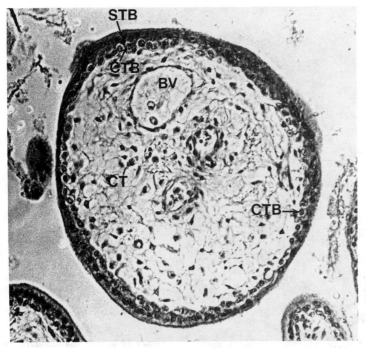

Figure 22-3. Light micrograph of a tertiary villus from a first-trimester human placenta. The syncytiotrophoblast (*STB*) is a syncytial layer surrounding a layer of cytotrophoblast (*CTB*). The core of the villus consists of fetal connective tissue (*CT*) and fetal blood vessels (*BV*).

    c. Maternal blood vessels (branches of uterine artery) penetrate the decidual plate and enter the intervillous spaces.

    d. The human placenta is called a **hemochorial placenta** because maternal blood bathes the syncytiotrophoblast.

2. The decidua

    a. Decidual cells. Certain endometrial stromal fibroblastic cells undergo dramatic transformation from stellate or fusiform fibroblasts to glycogen-rich polygonal **decidual cells** (Figure 22-5). Decidual cells form a barrier around the embryo and may help limit the invasiveness of the syncytiotrophoblast.

    b. Decidual regions. The decidua has three regions.

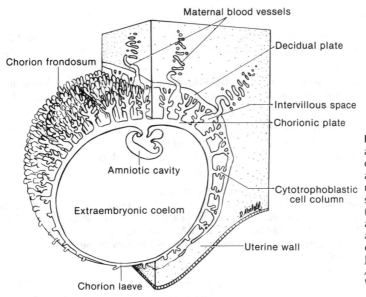

Figure 22-4. Chorionic development as it nears the formation of a placenta. The chorionic surface has an area of extensive villous development (*chorion frondosum*) and a smoother area with fewer villi (*chorion laeve*). The decidual plate and chorionic plate have formed and are connected by cytotrophoblastic cell columns. (Reprinted from Johnson KE: *Human Developmental Anatomy.* Baltimore, Williams & Wilkins, 1988, p 97.)

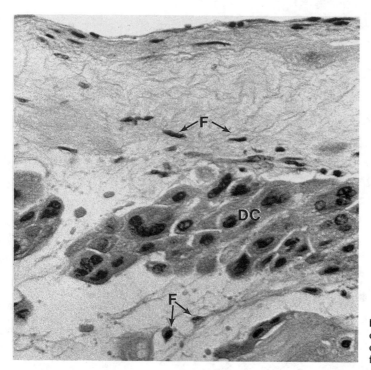

Figure 22-5. Light micrograph of decidual cells (*DC*). These large, polygonal cells differentiate from stromal fibroblasts (*F*).

(1) The **decidua basalis** is the portion that becomes incorporated into the definitive placenta. It is an integral part of the placenta and is delivered at term. Thus, the placenta contains maternal and fetal components.

(2) The **decidua capsularis** is the portion that encapsulates the implantation site. It expands as the embryo grows and then fuses with the decidua parietalis, obliterating the uterine lumen and forming the **decidua vera**.

(3) The **decidua parietalis** is the portion that forms everywhere in the endometrial wall except over the implantation site.

C. Functions of the placenta

1. Nutrient transport. All fetal nutrients including glucose, fatty acids, amino acids, nucleotides, vitamins, and minerals are derived from the maternal blood supply and are transported across the placenta.

2. Electrolyte transport. Water crosses the placenta by diffusion and pinocytosis. Salts lost in fetal urine are replaced by placental transport. Small ions enter the placenta by diffusion and pinocytosis mediated by small membrane-bound vesicles.

3. Gas transport. Oxygen bound to maternal hemoglobin diffuses across the placenta, enters fetal blood, and binds to fetal hemoglobin. Carbon dioxide bound to fetal hemoglobin diffuses across the placenta, enters the maternal blood, and binds to maternal hemoglobin.

4. Hormone synthesis

a. The placenta produces large amounts of **progesterone** and supplants the corpus luteum as a source of progesterone after the first trimester. Due to this steroid biosynthesis function, the syncytiotrophoblast contains abundant smooth endoplasmic reticulum and many mitochondria, which are essential for steroid biosynthesis and ATP production.

b. The placenta synthesizes large amounts of **human chorionic gonadotropin (hCG)**.

(1) hCG is a glycoprotein hormone that maintains the female reproductive system in a state suitable for pregnancy.

(2) hCG is synthesized on rough endoplasmic reticulum; glycosylation is completed in the Golgi apparatus. These organelles are well developed in the syncytiotrophoblast.

    c. The placenta produces another polypeptide hormone called **placental lactogen,** which stimulates the development of the mammary glands in preparation for lactation.

  5. Immunoglobulin transport
    a. Maternal immunoglobulin G (IgG) is transported across the placenta and enters fetal circulation functionally intact. Some of the vesicles in the syncytiotrophoblast are involved in IgG transport.
    b. Maternal IgG passively immunizes the fetus against the bacterial antigens that the fetus is exposed to at birth.
    c. After birth, maternal IgG is catabolized and replaced by fetal IgG, which is synthesized de novo as the infant is exposed to bacterial antigens.

D. Circulation in the placenta

  1. Maternal blood enters the intervillous space from branches of the uterine arteries. The oxygen in maternal blood diffuses through the walls of fetal capillaries in the placenta into fetal blood.

  2. Fetal capillaries drain into venules, which empty into the **umbilical vein**. The umbilical vein carries oxygen-rich blood through the umbilical cord to the **ductus venosus,** which bypasses the liver and empties into the inferior vena cava of the fetus.

  3. Oxygen-depleted blood returning from the fetal circulatory system flows through arterioles into the **umbilical arteries** (branches of the internal iliac arteries) and then into fetal capillary beds in the placenta.

III. THE MAMMARY GLANDS

A. Structure. Mammary glands are **compound tubuloalveolar glands**.

  1. Mammary glands have a complex duct system and numerous glandular acini.
    a. Approximately 20 large **lactiferous ducts** empty into each nipple. Lactiferous ducts branch several times and then connect to **lactiferous sinuses**.
    b. Mammary gland lobules consist of many acini. Lobules are drained by ducts that branch from the lactiferous sinuses.

  2. Lactiferous ducts are lined by stratified squamous epithelium near the nipple and by simple cuboidal epithelium in the more distal portions of ducts. Acini are comprised of simple cuboidal epithelial cells.

  3. In the pre-pubertal and non-lactating mammary gland, glandular cells and ductal cells are quite similar. During lactation, acinar cells differentiate into secretory cells that secrete milk, which distends the acini, and cells lining many of the ducts become secretory (Figure 22-6).

  4. During lactation, secretory cells become columnar. They have numerous apical microvilli, abundant rough endoplasmic reticulum, a well-developed Golgi apparatus, fat globules, and lysosomes.

  5. Myoepithelial cells are intraepithelial cells (i.e., they reside on the apical side of the basement membrane) in acini and ducts that contain numerous contractile microfilaments. These microfilaments contract to help express secreted milk from acini and ducts.

  6. The secretion of milk occurs by **apocrine secretion** (i.e., a small part of the apical portion of each secretory cell is released with the secretion product).

B. Function. Milk production is the most essential function of mammary glands. Milk has nutritive and immunologic functions.

  1. Nutritive function
    a. Casein is the predominant protein constituent of milk.
    b. Lactose is the predominant carbohydrate nutrient and osmotic element in milk.
    c. Milk is rich in a variety of **lipids**. The lipids are secreted from lipid droplets and form micelles in milk. Lipid micelles provide nutrition and help bind calcium and phosphate ions.
    d. Milk also provides an infant with **water, electrolytes, and vitamins**. It is particularly rich in calcium and phosphates, which are bound in lipid micelles and have an important role in the skeletal development of the infant.

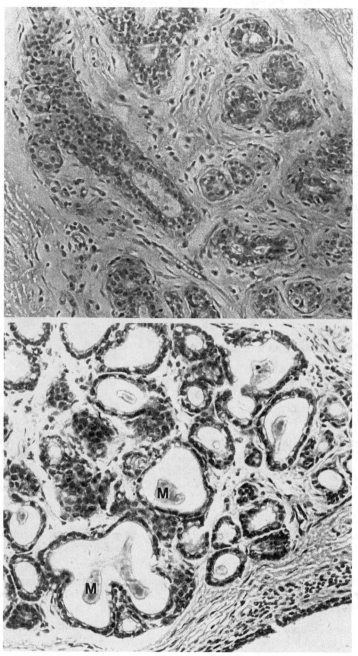

Figure 22-6. Light micrographs of non-lactating (*top*) and lactating (*bottom*) mammary glands. The distended acini of the lactating mammary gland contain milk (*M*).

2. Immunologic function

a. Milk is rich in **IgG** and **secretory immunoglobulin A (sIgA)**. sIgA is a dimeric form of IgA that is protected from enteric proteolysis by a secretory component polypeptide. These immunoglobulins protect the infant from the enteric infections that can cause diarrhea, dehydration, or malnutrition.

b. Milk contains nonspecific antibacterial substances such as **lactoperoxidase** and **lysozyme**.

c. Milk contains maternal lymphocytes, which presumably have an immunologic function; however, this function is poorly understood.

STUDY QUESTIONS

Directions: The groups of questions below consist of lettered choices followed by several numbered items. For each numbered item select the **one** lettered choice with which it is **most** closely associated. Each lettered choice may be used once, more than once, or not at all.

Questions 1–5

Match each description below with the most appropriate component of milk.

(A) Immunoglobulin G
(B) Secretory immunoglobulin A
(C) Casein
(D) Lactose
(E) Lipid micelles

1. A dimeric immunoglobulin that prevents diarrhea and enteric infections in the newborn

2. A multimolecular aggregate that binds calcium and phosphate in milk

3. The predominant protein constituent of milk

4. Manufactured by maternal plasma cells; enters fetal blood, where it provides passive bacterial Immunity for the infant

5. The major osmotic constituent of milk

Questions 6–9

Match each function below with the most appropriate syncytiotrophoblastic organelle.

(A) Rough endoplasmic reticulum
(B) Smooth endoplasmic reticulum
(C) Golgi apparatus
(D) Membrane bound vesicles
(E) Mitochondria

6. Involved in both steroid biosynthesis and ATP production

7. Involved in glycosylation of human chorionic gonadotropin (hCG) and production of secretory vesicles

8. Site where polypeptide chains for hCG and human placental lactogen are assembled

9. Involved in pinocytosis and immunoglobulin transport

Questions 10–14

Match each description below with the appropriate lettered structure in the micrograph.

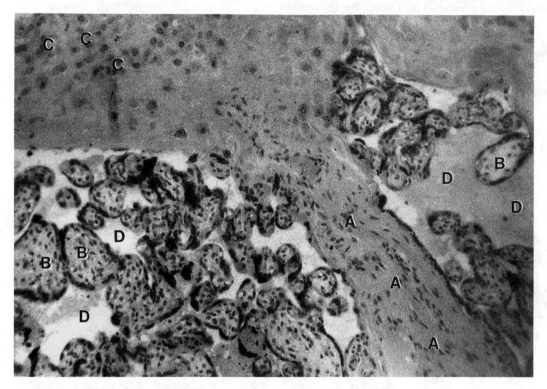

10. Contains fetal capillaries and is the site of gas exchange in the placenta

11. A cytotrophoblastic cell column

12. Receives blood from a branch of the uterine artery and contains many maternal erythrocytes

13. Carries maternal oxygen to the fetus

14. Decidual cells

ANSWERS AND EXPLANATIONS

1–5. The answers are: 1-B, 2-E, 3-C, 4-A, 5-D. [*III B*] Immunoglobulin G (IgG) is a component of breast milk. It is also manufactured by maternal plasma cells and transferred intact across the placenta into fetal circulation. IgG provides the fetus and newborn with passive immunity until fetal plasma cells become functional.

Secretory immunoglobulin A (sIgA) is a dimeric form of immunoglobulin A (IgA) associated with a secretory component that prevents proteolysis in the gastrointestinal tract of the infant. It is manufactured by maternal plasma cells in mammary gland connective tissues. sIgA passes into milk and coats enteric bacteria, preventing their attachment to epithelial cells in the gastrointestinal tract of the newborn. Babies lacking sIgA are prone to enteric infections and diarrhea.

Casein is the predominant protein constituent of milk and is manufactured on the endoplasmic reticulum of mammary gland secretory cells.

Lactose, an important nutrient, is a disaccharide and the major osmotic component of milk.

Lipid micelles are an important nutrient and perform a transport function by binding calcium and phosphate.

6–9. The answers are: 6-E, 7-C, 8-A, 9-D. [*II C 4*] The syncytiotrophoblast contains a complex collection of organelles, which perform the same general functions that they perform in other cells in the body.

The rough endoplasmic reticulum is involved in the synthesis of the polypeptide chains of secretory proteins, such as human chorionic gonadotropin (hCG) and placental lactogen.

In conjunction with mitochondria, the smooth endoplasmic reticulum has an important role in steroid biosynthesis.

The Golgi apparatus completes the glycosylation of hCG, a glycoprotein hormone.

Membrane-bound vesicles help transport substances across the syncytiotrophoblast in both directions. For example, pinocytosis and endocytosis transport water and IgG in, and exocytosis transports polypeptide hormones out.

Mitochondria are the source of ATP for many syncytiotrophoblast synthesis and transport functions and are involved in steroid biosynthesis.

10–14. The answers are: 10-B, 11-A, 12-D, 13-D, 14-C. [*II B*] The definitive placenta has two plates of tissue. The decidual plate is on the maternal side of the placenta, and the chorionic plate is on the fetal side of the placenta. Cytotrophoblastic cell columns (A) connect the decidual and chorionic plates.

Gas exchange occurs at the surface of the villi (B) that project from cytotrophoblastic cells into the intervillous space. Placental villi contain fetal capillaries surrounded by connective tissues and, therefore, contain fetal erythrocytes.

Oxygenated maternal blood from branches of the uterine arteries flows through the decidual plate into the intervillous space (D), where it bathes the villi. Fetal capillaries carry fetal erythrocytes, which carry maternal oxygen from the placenta to the fetus and carbon dioxide from the fetus to the placenta.

A layer of decidual cells (C) surrounds the fetus and fetal membranes. This layer limits the invasiveness of the syncytiotrophoblast.

23
The Female
Reproductive System

I. INTRODUCTION

A. Components. The female reproductive system is composed of internal and external genital organs. Organs of the female reproductive system function exclusively in sexual and reproductive capacities.

1. **Internal genitalia** include the:
   a. Ovaries
   b. Uterine (fallopian) tubes
   c. Uterus (including the cervix)
   d. Vagina

2. **External genitalia** include the:
   a. Labia majora and mons pubis
   b. Labia minora
   c. Clitoris
   d. Vestibular bulbs

3. The greater vestibular glands (Bartholin's glands) are **accessory organs** of the female reproductive system.

B. Functions. Internal components are exclusively involved in gamete elaboration and transportation. When fertilization occurs, they also maintain the conceptus.

1. The two ovaries regularly produce ova and briefly release them into the peritoneal cavity. Ovulation is a cyclical event, occurring once a month from menarche to menopause.

2. During the fertilization process, an ovum released from an ovary is swept into the uterine tubes and fertilized. Then, the developing conceptus is transported to the uterus.

3. The conceptus implants in the uterine cavity and establishes an intimate relationship with the maternal tissues.

4. If an ovum is not fertilized, menstruation occurs about 14 days after ovulation.

5. During menstruation, the uterus sheds most of the uterine mucosa; however, remnants of the mucosa persist after menstruation has ceased. These residual cells proliferate to produce new mucosal cells.

C. Hormonal regulation

1. Each month the female reproductive tract undergoes changes that prepare it for pregnancy. The **follicular phase** encompasses changes that occur prior to ovulation, and the **luteal phase** includes changes that occur after ovulation.

2. Four important hormones regulate these cyclic changes. These hormones are **follicle-stimulating hormone (FSH), luteinizing hormone (LH), estrogen,** and **progesterone**.

II. THE OVARIES

A. Histology

1. **Germinal epithelium.** The surface of each ovary is covered by a simple cuboidal epithelium called germinal epithelium, which is continuous with the simple squamous mesothelial covering of the mesovarium and uterus.

2. The **ovarian medulla** is located in the center of the ovary.
   a. The medulla is composed of loosely packed connective tissue, coiled blood vessels, nerves, and lymphatics.
   b. The medulla may also contain the vestigial **rete ovarii,** an anastomosing network of closed ducts lined by low cuboidal epithelium. The rete ovarii is homologous to the **rete testis** in males.

3. The **ovarian cortex** surrounds the medulla (Figure 23-1).
   a. The cortex contains **ovarian follicles** in different stages of development depending on a woman's age, number of pregnancies, health, and whether or not she is currently pregnant.
      (1) Ovarian follicles consist of an **oocyte** and a layer of **follicular epithelial cells** that varies in thickness.
      (2) In certain more mature follicles, an acellular, glycoconjugate-rich, periodic acid-Schiff (PAS)-positive layer called the **zona pellucida** surrounds the oocyte.
   b. **Stromal cells** and small blood vessels are present between the follicles.
   c. The stroma of the cortex forms a dense capsule called the **tunica albuginea,** which lies beneath the basement membrane of germinal epithelium.

B. The ovarian cycle

1. **General characteristics.** The 40-year period between menarche (approximately age 13) and menopause (approximately age 53) constitutes a woman's reproductive life. In a monthly cycle during a woman's reproductive life, one ovarian follicle develops full functional maturity and is ovulated.
   a. Follicular development to ovulation usually occurs only once during each ovarian cycle. Many follicles begin the final stages of development that culminate in ovulation.
      (1) Usually, only one follicle reaches full functional maturity and is ovulated; however, multiple ovulatory events can occur and can lead to multiple pregnancies.
      (2) Follicles that do not reach maturity undergo **atresia** and eventually die.
   b. Ovulation from either the right or left ovary is a random event; the ovaries do not necessarily alternate months.
   c. The ovarian cortex of postmenopausal women contains few follicles. Instead, it consists mainly of stromal elements and scar tissue from degenerate follicles.

2. **Follicular development** toward maturity and ovulation or senescent atresia depends on changes in the number of follicular cells, their arrangement around an antrum, and their relation to each other and to the oocyte.

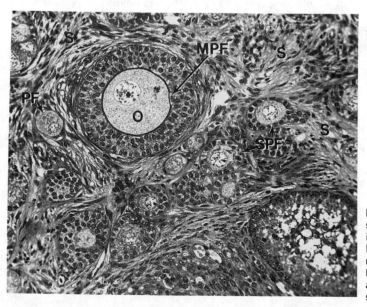

Figure 23-1. Light micrograph of a section of the ovarian cortex, showing primordial follicles (*PF*), single-layered primary follicles (*SPF*), and multilayered primary follicles (*MPF*). Each follicle contains an oocyte (*O*) and is embedded in connective tissue stroma (*S*).

a. Primordial and primary follicles
  (1) Follicular epithelial cells in **primordial follicles** form a thin unilaminar squamous shell around the oocyte.
  (2) The primordial follicle becomes a multilaminar **primary follicle** as follicular epithelial cells proliferate to form several layers.
  (3) As follicular development continues, a **zona pellucida** becomes interposed between the oocyte and the follicular epithelial cells **(granulosa cells)**.
    (a) The zona pellucida is formed by secretions from granulosa cells and the oocyte.
    (b) Microvilli on the surface of both the oocyte and granulosa cells project deeply into the zona pellucida. Granulosa cell microvilli may completely cross the zona pellucida and contact the surface of the oocyte.
  (4) As granulosa cells proliferate to form multiple layers, follicular diameter increases rapidly and the follicle sinks deep into the ovarian cortex. Small lakes of PAS-positive material called **Call-Exner bodies** appear between granulosa cells after many layers of these cells have formed.
    (a) Call-Exner bodies probably are precursors of the hyaluronate-rich fluid called **liquor folliculi** that begins to accumulate in a single large cavity called the **antrum**.
    (b) The antrum continues to grow as it accumulates liquor folliculi.
  (5) Stromal cells adjacent to each growing follicle form concentric layers called the **theca interna**.
    (a) Initially, theca interna cells are similar to other stromal cells; however, as the follicle matures, they differentiate into steroidogenic cells.
    (b) The theca interna merges with the surrounding stromal elements without a distinct boundary.
b. Secondary (antral) follicles (Figure 23-2). Multilaminar primary follicles become secondary follicles when a complete antrum forms.
  (1) In larger follicles, the antrum typically exists near the center of the follicle.
  (2) The ovum is eccentrically located on a small tuft called the **cumulus oophorus** and is surrounded by granulosa cells.

3. Follicular maturation. Until the maturing follicle becomes a secondary follicle, the oocyte is arrested in the first meiotic prophase.
  a. The oocyte completes meiosis I by forming a first polar body and then expelling it into the space between the zona pellucida and the oocyte.
  b. Granulosa cells undergo a final burst of proliferation, liquor folliculi production increases suddenly, and granulosa cells surrounding the oocyte lose their attachment to neighboring granulosa cells.
  c. The follicle bulges from the surface of the ovary. As the preovulatory surge of LH occurs, the follicle bursts, rupturing the germinal epithelium, and the secondary oocyte enters the peritoneal cavity.

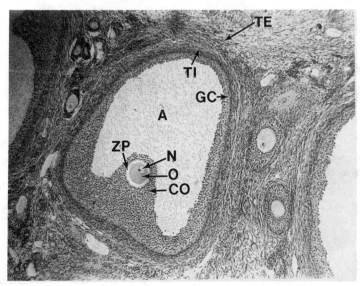

Figure 23-2. Light micrograph of an antral follicle. The antrum (*A*) is filled with liquor folliculi. Granulosa cells (*GC*) form a heap called the cumulus oophorus (*CO*) around the oocyte (*O*). The ovum is surrounded by a zona pellucida (*ZP*). The theca interna (*TI*) and theca externa (*TE*) are visible. *N* = nucleus of the oocyte. (Reprinted from Johnson KE: *Histology: Microscopic Anatomy and Embryology.* New York, John Wiley, 1982, p 305.)

d. Because the ovary and uterine tube are intimately related, the ovum, zona pellucida, associated follicular epithelial cells (now called the **corona radiata**), and attached liquor folliculi all are swept into the uterine tube.

C. Hormonal regulation of the ovarian cycle

1. Regulatory hormones
  a. The monthly cyclical changes in the female reproductive tract are controlled by FSH, LH, estrogen, and progesterone.
  b. FSH and LH are secreted by adenohypophysial gonadotropes. Estrogen and progesterone are synthesized in the ovary—principally in and around follicles.

2. Phases. The female reproductive cycle is divided into a preovulatory phase called the follicular phase, ovulation, and a postovulatory phase called the luteal phase.
  a. The **follicular phase** involves FSH stimulating the growing follicle to produce **estrogen** and **progesterone**. Estrogen then stimulates the growth of endometrial glands.
  b. The **luteal phase** encompasses the surge of LH, ovulation, conversion of the follicular remnants into a **corpus luteum,** and the secretion of **progesterone** from the corpus luteum.
    (1) Luteinization
      (a) During follicular luteinization, the basement membrane of the follicular epithelium disintegrates, granulosa cells collapse, and numerous blood vessels and theca interna cells migrate into the developing corpus luteum (Figure 23-3).
      (b) The corpus luteum is composed of **theca lutein cells** (derived from theca interna cells) and **granulosa lutein cells** (derived from granulosa cells) and many blood vessels. Granulosa lutein cells, like other steroid-secreting cells, have abundant smooth endoplasmic reticulum, ovoid mitochondria with tubular cristae, and numerous lipid droplets (Figure 23-4).
      (c) The corpus luteum secretes large amounts of steroids for 14 days after the midcycle LH surge. The principal hormone is progesterone, but androgens and estrogens are produced as well.
    (2) Progesterone stimulates endometrial glands to secrete the trophic substances that support the early development of an embryo, provided fertilization occurs.
    (3) The developing embryonic trophoblast produces hormones, principally **human chorionic gonadotropin (hCG)**. These hormones are present only when fertilization occurs.
    (4) If fertilization does not occur, the corpus luteum of the cycle regresses, progesterone support of the endometrium is lost, and most of the endometrium is sloughed as the menses. At menses, a new follicle begins to develop as the ovarian cycle begins again.
    (5) Follicular development is a continuous process in reproductively competent females and its initial phases are independent of gonadotrophin stimulation.

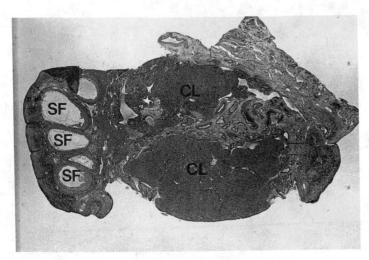

Figure 23-3. Light micrograph of a section of an ovary, showing numerous cortical secondary follicles (*SF*) and two corpora lutea (*CL*). The *upper* corpus luteum originated in a previous cycle; the *lower* corpus luteum originated in the most recent cycle. (Micrograph courtesy of Dr. B. Gulyas, National Institute of Child Health and Human Development.)

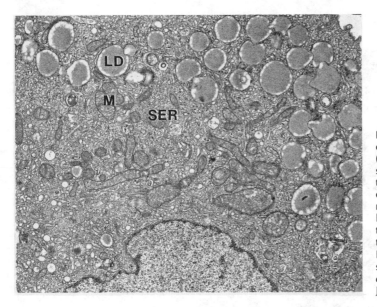

Figure 23-4. Electron micrograph of corpus luteum granulosa lutein cells (steroid-secreting cells). These cells synthesize progesterone. They contain mitochondria (*M*) with tubular cristae, abundant smooth endoplasmic reticulum (*SER*), and numerous lipid droplets (*LD*). (Micrograph courtesy of Dr. B. Gulyas, National Institute of Child Health and Human Development. Reprinted from Johnson KE: *Histology: Microscopic Anatomy and Embryology.* New York, John Wiley, 1982, p 306.)

III. THE UTERINE TUBES

A. Functions. The uterine tubes convey the ova from the ovary toward the uterus, transport spermatozoa from the uterus toward the ovary, and produce trophic substances that ensure the development of the fertilized conceptus.

B. Segmentation. The uterine tubes have four distinct segments.

1. The **interstitial segment** is the region where the oviducts penetrate the wall of the uterus.

2. The **isthmus** is the most proximal two-thirds of the uterine tubes.

3. The **ampullary segment** is an expanded segment distal to the isthmus.

4. The **infundibulum** is the most distal portion of the uterine tubes. It is an expanded, funnel-shaped segment with finger-like projections called **fimbria**.

C. Histology

1. **Mucosa.** The lumen of uterine tubes is lined by a simple columnar epithelium of **ciliated cells** and **secretory cells**.
   a. In the distal portions of the tubes (i.e., fimbria, infundibulum, ampulla), ciliated cells far outnumber secretory cells. In proximal portions of the tubes (i.e., isthmus, interstitial segment), secretory cells outnumber ciliated cells.
   b. The mucosal epithelium rests on a well-developed basement membrane.
   c. A thin lamina propria is present under the luminal epithelium, but no distinct submucosal compartment exists.
   d. The mucosa has an elaborate system of projections in the fimbria, a labyrinthine network in the infundibulum, and longitudinally arranged folds in the isthmus and interstitial segment (Figures 23-5 and 23-6).

2. **Muscularis.** External to the mucosa, the uterine tubes have a muscularis of smooth muscle that varies in thickness.
   a. The muscularis has an inner layer of smooth muscle fibers encircling the lumen and an outer layer of longitudinal smooth muscle fibers.
   b. The muscularis is relatively thin in the distal oviducts and becomes thicker proximally.
   c. The most proximal portion of the isthmus contains an inner layer of longitudinal fibers that merges into the uterine myometrium.
   d. The boundaries between the different layers of smooth muscle are indistinct.

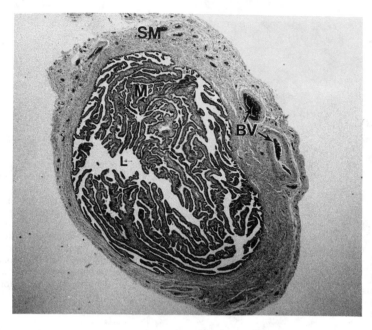

Figure 23-5. Light micrograph of the uterine tube isthmus. The wall contains several layers of smooth muscle (*SM*) and numerous blood vessels (*BV*). The mucosa (*M*) has numerous folds. The lumen (*L*) of the tube is visible as well.

3. Adventitia. The outer layer of the uterine tubes consists of a thin layer of connective tissue coated by a serosal layer of mesothelium.

IV. THE UTERUS

A. **Uterine endometrium** is the luminal layer of the uterus.

1. **General characteristics**
   a. The endometrium consists of an **endometrial epithelium** resting on a basement membrane and connective tissue stromal cells, blood vessels, and lymphatics.
   b. Most endometrial epithelial cells are tall columnar secretory cells; a few are ciliated cells.
   c. The endometrium has deep surface invaginations called glands.
   d. During a woman's reproductive life, the endometrium undergoes cyclic changes associated with the phases of the ovarian cycle.

2. **Endometrial changes during the follicular phase**
   a. The endometrium has two portions. The **functionalis** portion is sloughed monthly during **menstruation**. The **basalis** portion persists as a source of cells to renew the endometrium after menstruation.
   b. After menstruation, deep endometrial epithelial cells spread over and heal broken blood vessels, and bleeding ceases. Estrogen from the developing follicles stimulates mitoses in epithelial and stromal cells. Consequently, the endometrium thickens and endometrial glands elongate.
   c. Endometrial glands grow extensively and blood vessels and stromal cells beneath the epithelial basement membrane proliferate through mitotic activity.

3. **Endometrial changes during the luteal phase**
   a. After ovulation, the ovarian corpus luteum secretes progesterone, which stimulates further development of endometrial glands, stromal elements, and blood vessels.
      (1) Endometrial glands become secretory under the influence of progesterone. During the early luteal phase, their lumina broaden and distend with secretion products and they become more coiled and branched.
      (2) Endometrial blood vessels elongate and become distended and engorged.
   b. Toward the end of the luteal phase, progesterone secretion by the corpus luteum begins to trail off.
      (1) At this time, the endometrium becomes ischemic and secretory activity stops.

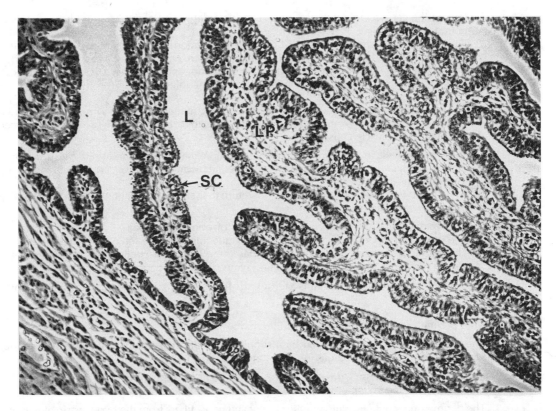

Figure 23-6. Light micrograph of the mucosa of the uterine tube. Simple columnar epithelium (*SC*) lines all surfaces facing the lumen (*L*). The lamina propria (*LP*) is a loose areolar connective tissue beneath the epithelium.

    (2) The entire functionalis becomes thinner due to dehydration, stromal cells becoming more closely packed together, and a reduction in the width of the lumina of endometrial glands.

    (3) Transient interruptions in blood supply due to constricted endometrial blood vessels cause degeneration of the entire endometrial functionalis and initiate the shedding of blood, endometrial remnants, mucus, and endometrial blood vessels in the menses.

  c. During menstruation, most of the endometrial epithelium (including stromal cells, blood, and blood vessels) is sloughed from the uterus and expelled through the cervix. Typically, the menses last for 4 or 5 days and then a new endometrial cycle begins.

4. Endometrial changes after fertilization. Successful fertilization after ovulation produces a **zygote**.

  a. The conceptus reaches the uterine cavity 3 days after ovulation (approximately day 17 of the ovarian cycle). During the next 2 days, a blastocyst forms and then attaches to and burrows into the endometrium during implantation.

    (1) Certain cells at the periphery of the developing conceptus differentiate into the trophoblast, which produces hCG.

    (2) hCG stimulates the corpus luteum of pregnancy to continue secreting progesterone and suppresses ovulation.

  b. The endometrial sloughing of the menses does not occur. Instead, endometrial vascularity and glandular proliferation and secretion continue to increase as the **gestational endometrium** develops to support early development of the conceptus.

  c. Trophoblastic cells in the conceptus invade the endometrium and induce a **decidual cell reaction** in endometrial stromal cells. This reaction appears to limit further invasion.

5. Cyclic changes in endometrial epithelial cells. The ultrastructure of endometrial epithelial cells undergoes changes related to phases of the ovarian cycle.

  a. During the follicular phase, tall columnar epithelial cells have an apical Golgi apparatus and microvilli, small mitochondria, and basal oval nuclei. They do not contain cytoplasmic glycogen granules.

b. During the early secretory phase, mitochondria enlarge and glycogen granules become an abundant cytoplasmic inclusion.

c. During the late secretory phase, just before menstruation, the apices of cells become filled with glycogen and sometimes rupture, releasing glycogen into the glands.

d. In the late secretory phase, lysosomes become abundant and there is evidence of cell degeneration (e.g., myelin figures and lipid-filled vacuoles).

B. Uterine myometrium is a thick muscular layer containing numerous smooth muscle cells, a rich blood supply, and impressive lymphatic drainage. Broad bands of connective tissue cells lie between the layers of smooth muscle cells.

1. **Smooth muscle cells** in the myometrium of the nonpregnant uterus are about 250 μm long.
   a. During pregnancy, smooth muscle cells proliferate by mitosis and become much longer (up to 5 mm) and much thicker.
   b. After delivery, smooth muscle cells do not return to their nonpregnant morphology, resulting in uterine enlargement.

2. The myometrium has three poorly defined **layers of smooth muscle**.
   a. The outer and inner layers tend to run at oblique angles to the long axis of the uterus.
   b. The middle layer tends to run perpendicular to the long axis, circling the uterus. This layer carries numerous large blood vessels and lymphatics.

3. The peripheral myometrium contains a few **elastic fibers**. During pregnancy, these fibers become more abundant.

C. Uterine perimetrium is the outer layer of the uterine wall.

1. The uterine perimetrium is covered by a simple squamous mesothelium, which is continuous with the serosal lining of the uterine tubes and broad ligament.

2. Large coiled blood vessels traverse the broad ligament to the perimetrium and penetrate the myometrium to the endometrium.

3. The perimetrium contains sympathetic ganglia and nerve fibers from the hypogastric plexus and parasympathetic fibers from sacral nerves. Branches of these nerves supply uterine blood vessels, musculature, and endometrial glands.

D. The cervix is the lowest portion of the uterus. It encircles the **cervical canal,** which connects the uterine cavity to the vagina.

1. The mucosa of the cervix (**endocervix**) is continuous with the uterine endometrium and is lined by an epithelium predominantly composed of tall columnar cells that can secrete mucus and scattered ciliated cells that beat toward the vagina.
   a. The apical portion of endocervical epithelial cells has a marked distention. The basal portion of the cell contains a nucleus and rough endoplasmic reticulum, a well-developed Golgi apparatus just above the nucleus, and numerous membrane-bound mucous vacuoles.
   b. These cells are very similar to surface cells in the gastric mucosa and goblet cells in the intestines and respiratory system.
   c. In the cervical os, the epithelium changes from columnar to unkeratinized stratified squamous (Figure 23-7).

2. In histologic sections, the cervical mucosa appears to be glandular.
   a. When viewing the endocervix **en face** from the cervical canal, it is apparent that the mucosa is invaginated into longitudinal and oblique folds called **plicae palmatae**.
   b. Plicae palmatae are particularly prominent along the middle dilated portion of the cervical canal and diminish gradually until they disappear at the cervical opening into the vagina.

3. The cervix produces a **cervical mucus** that is mostly water, but also contains glycoproteins, glucose, and ions. The cervical mucus formed near the time of ovulation differs from the cervical mucus formed after ovulation.
   a. At ovulation, cervical mucus is more hydrated and easier for spermatozoa to penetrate.
   b. After ovulation, when the corpus luteum is formed, cervical mucus is less hydrated and, thus, more difficult for spermatozoa to penetrate.
   c. During pregnancy, a plug of cervical mucus forms to prevent bacteria in the vagina from entering the uterus and attacking the fetus and fetal membranes.

4. The cervical os is encircled by a thick, circular band of smooth muscle fibers.

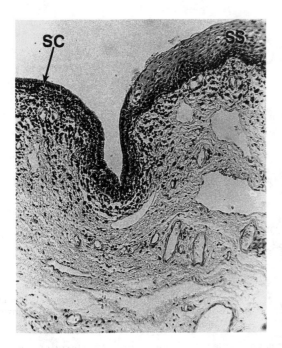

Figure 23-7. Light micrograph of the transitional zone of the cervix, where the simple columnar epithelium (*SC*) of the endocervix becomes the stratified squamous epithelium (*SS*) of the ectocervix.

V. THE VAGINA AND EXTERNAL GENITAL ORGANS

A. The vagina. The vaginal wall contains a mucosa, a muscularis, and an outer layer of connective tissue.

  1. Mucosa (Figure 23-8). The vaginal mucosa consists of a thick unkeratinized stratified squamous epithelium. (Some apical epithelial cells may accumulate keratohyalin granules, but they never become fully keratinized like epidermal epithelial cells.)

    a. Epithelial cells. Vaginal epithelial cells contain abundant glycogen, especially at midcycle when estrogen secretion is high.

      (1) Glycogen can be degraded and converted to lactic acid by certain benign vaginal microorganisms.

      (2) The acid pH of the vagina probably limits the growth of pathogenic organisms.

      (3) Late in the ovarian cycle, as estrogen levels decline, glycogen is less abundant and vaginal pH increases, increasing the likelihood of vaginal infections.

    b. The **lamina propria** beneath the vaginal epithelium and its basement membrane contains connective tissue fibroblasts; many papillae, which anchor the vaginal epithelium to underlying connective tissues; and many lymphocytes, solitary lymph nodules, and neutrophils.

      (1) Neutrophils become particularly numerous near menstruation.

      (2) Many of these leukocytes move into the vagina and can be seen with desquamated epithelial cells in vaginal smears.

      (3) The lamina propria contains numerous blood vessels, but no glands. During sexual arousal, the blood vessels engorge, the vaginal mucosa reddens, and lubricating transudates accumulate in the vagina.

  2. Muscularis. The vaginal muscularis consists of an outer and inner layer of smooth muscle cells and skeletal muscle fibers that form a sphincter around the outer vaginal opening.

    a. The outer layer is a longitudinal group of smooth muscle cells that merge with the outer longitudinal smooth muscle cells of the cervix and uterus.

    b. The inner layer of smooth muscle cells more or less encircles the vagina.

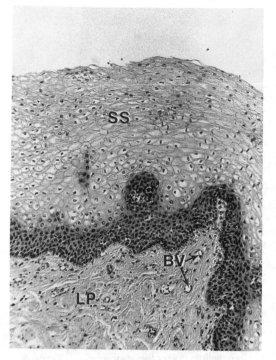

Figure 23-8. Light micrograph of the vaginal mucosa, showing the unkeratinized stratified squamous epithelium (*SS*) and blood vessels (*BV*) in the vascularized lamina propria (*LP*).

3. **Connective tissue.** The entire muscularis is surrounded by a layer of connective tissue embedded with numerous elastic fibers.

B. External genitalia

1. The vaginal orifice is flanked by two narrow flaps of tissue called the **labia minora**. The space bounded by the labia minora is called the **vestibule**.

   a. The labia minora are covered by hairless stratified squamous epithelium containing numerous sebaceous glands and more melanin pigmentation than the vagina.

   b. Minor vestibular glands secrete mucus around the clitoris and urethral orifice.

   c. The lateral walls of the vestibule contain larger accessory **vestibular glands (Bartholin's glands),** which are homologous to the bulbourethral glands **(Cowper's glands)** in males, and secrete a milky lubricant during sexual arousal.

2. The **clitoris** is located at the superior border of the vestibule where the labia minora meet. It is an erectile tissue richly supplied with blood vessels and contains numerous sensory nerve endings for touch and pressure (e.g., **Meissner's corpuscles**).

3. The labia minora are flanked laterally by the **labia majora,** two flaps of skin covered by keratinized stratified squamous epithelium. The labia majora of sexually mature women have pubic hairs, sebaceous glands, and apocrine sweat glands.

STUDY QUESTIONS

Directions: The groups of questions below consist of lettered choices followed by several numbered items. For each numbered item select the **one** lettered choice with which it is **most** closely associated. Each lettered choice may be used once, more than once, or not at all.

Questions 1–6

Match each description below with the most appropriate hormone.

(A) Estrogen
(B) Progesterone
(C) Luteinizing hormone
(D) Follicle-stimulating hormone
(E) Chorionic gonadotropin

1. Secreted in large quantities by the corpus luteum during the luteal phase

2. Secreted by developing follicles during the follicular phase; stimulates endometrial glandular growth

3. Corpus luteum secretion that stimulates endometrial glandular secretion during the luteal phase

4. Secreted by gonadotropes; stimulates granulosa cell mitosis

5. Secretory surge at midcycle causes ovulation

6. Secreted by the developing placenta; suppresses ovulation

Questions 7–10

Match each statement below with the most appropriate structure or structures of the female reproductive system.

(A) Uterus
(B) Fallopian tubes
(C) Both
(D) Neither

7. The mucosal epithelium contains more ciliated cells than secretory cells

8. The lamina propria contains loose areolar connective tissue

9. Exhibits functional regulation by hormones during the menstrual cycle

10. Contains glands that become distended with secretory products during the luteal phase of the menstrual cycle

Questions 11–15

Match each statement below with the appropriate lettered structure in the electron micrograph.

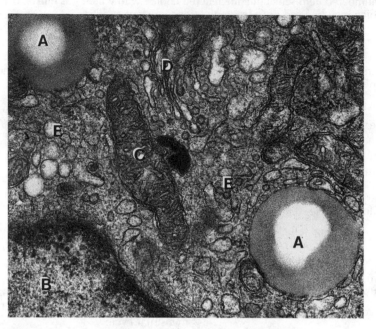

Micrograph courtesy of Dr. B. Gulyas, National Institute of Child Health and Development.

11. Source of energy-rich compounds used in steroid biosynthesis

12. Vacuole containing stored cholesterol esters for steroid biosynthesis

13. Bounded by a double unit membrane; contains most of the cell's genetic material

14. Golgi apparatus

15. Membranous labyrinth; location of many steroid biosynthesis enzymes

ANSWERS AND EXPLANATIONS

1–6. The answers are: 1-B, 2-A, 3-B, 4-D, 5-C, 6-E. [*I C 2; II C 1, 2; IV A 2–4*] Estrogen is a steroid secreted by granulosa cells during the follicular phase of the menstrual cycle. It stimulates the growth of endometrial glands.

Progesterone is a steroid secreted by the corpus luteum after ovulation. It stimulates endometrial gland secretory activity.

Luteinizing hormone (LH) is secreted by gonadotropes in the adenohypophysis. LH secretion rises sharply at midcycle, causing ovulation.

Follicle-stimulating hormone (FSH) is secreted by gonadotropes in the adenohypophysis. It stimulates proliferation of granulosa cells during the follicular phase of the menstrual cycle.

Human chorionic gonadotropin (hCG) is a glycoprotein hormone secreted by the syncytiotrophoblast of the developing placenta. It prevents menstruation and ovulation during pregnancy.

7–10. The answers are: 7-B, 8-C, 9-C, 10-A. [*III C; IV A*] The uterus and fallopian tubes have an epithelium consisting of tall columnar secretory cells (predominant in the uterus) and ciliated cells (predominant in the fallopian tubes). These cells exhibit functional changes during the menstrual cycle. For example, endometrial secretory cells begin secreting during the luteal phase, and ciliary beating in the fallopian tubes becomes more pronounced near ovulation. The lamina propria of both consists of a loose areolar connective tissue infiltrated with leukocytes.

11–15. The answers are: 11-C, 12-A, 13-B, 14-D, 15-E. [*II C 2 b (1) (b); Figure 23-4*] This is an electron micrograph of a granulosa lutein cell in a functional corpus luteum. (*A*) is a lipid droplet, (*B*) is the nucleus, and (*C*) is a mitochondrion. Notice that the cristae of the mitochondrion have a peculiar tubular arrangement. Although the functional significance of this particular arrangement is not known, it is common to mitochondria in many steroidogenic cells. (*D*) is the Golgi apparatus. The smooth endoplasmic reticulum (*E*) contains several enzymes for steroid biosynthesis and, therefore, is always prominent in steroidogenic cells such as granulosa lutein cells.

24
The Male
Reproductive System

I. INTRODUCTION

A. Components

1. The male reproductive system includes the following components:
   a. Two testes
   b. The epididymis
   c. The ductus deferens
   d. Seminal vesicles
   e. The prostate and bulbourethral glands
   f. Ejaculatory ducts
   g. The urethra and penis

2. The **scrotum** contains the testes and the small excurrent ducts that connect the testes to the epididymis.

3. The epididymis connects to the ductus deferens. The seminal vesicles are a pair of diverticula of the ductus deferens.

4. The male urinary and reproductive systems meet at the prostate gland. The ejaculatory ducts, which drain the ductus deferens, pierce the prostate gland and unite with the urethra, which drains urine from the bladder. Beyond this point, the urethra conveys both male gametes and urine.

5. The urethra passes into the penis and opens onto the body surface at the **fossa navicularis** in the head of the penis.

B. Functions. The male reproductive system produces **spermatozoa** (male gametes), maintains them, and then conveys them into the female reproductive tract. Developmental maturation and normal reproductive competence of the male reproductive system depends on testicular androgens.

1. **Spermatogenesis** occurs in the **seminiferous tubules** within the testes. **Tubuli recti, rete testis,** and **efferent ductules** convey spermatozoa into the epididymis. From the epididymis, spermatozoa enter the ductus deferens, which is a conduit that conveys spermatozoa to the prostate gland.

2. **Formation of ejaculate.** The seminal vesicles and prostate gland produce most of the noncellular components of ejaculate. This viscous material has the important task of conveying spermatozoa from the male reproductive tract, through the vagina, and into the uterus.

3. **Release of ejaculate.** Ejaculate is released from the penis into the vagina near the cervix, increasing the likelihood that spermatozoa will penetrate the cervical mucus and enter the main lumen of the uterus.

II. THE TESTES

A. Microscopic anatomy

1. **Tunica albuginea and testicular lobules**
   a. The **tunica albuginea** (Figure 24-1) is the connective tissue that surrounds the testes.

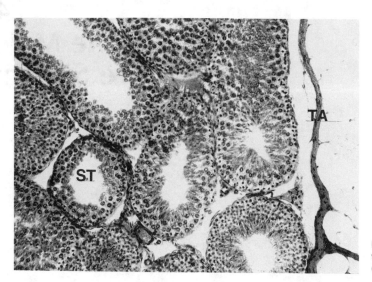

Figure 24-1. Light micrograph of the testis, showing the tunica albuginea (*TA*) and seminiferous tubules (*ST*).

     (1) The tunica albuginea is thickest on the posterior aspect of the testicle, forming a mass of connective tissue called the **mediastinum testis**.
     (2) Efferent ducts bearing gametes, testicular blood vessels, lymphatic vessels, and nerves enter the testis through the mediastinum testis.
  b. Connective tissue trabeculae project from the mediastinum into the testicular parenchyma, dividing the testis into the several hundred **testicular lobules**.

2. Seminiferous tubules. Each testicular lobule has as many as four seminiferous tubules. Spermatogenesis occurs in these tubules.
  a. Seminiferous tubules are closed continuous loops that empty into straight sections of the seminiferous tubules **(tubuli recti)**.
     (1) The combined length of the seminiferous tubules is about 400 m, and they produce 100 million spermatozoa each day.
     (2) Some seminiferous tubules form cul-de-sacs rather than looping back to the tubuli recti; others branch or anastomose with neighboring tubules.
     (3) Tubuli recti empty into an anastomosing network of channels called the **rete testis**.
  b. Seminiferous tubules contain cells of the gamete-producing line and **Sertoli cells,** which support the gamete-producing cells. They are bounded by numerous cellular and acellular barriers, including several layers of **myoid cells**.
     (1) Myoid cells have many of the characteristics of smooth muscle cells.
     (2) Histologists believe that myoid cells cause the peristaltic waves of contraction that pass along the seminiferous tubules and propel spermatozoa toward the epididymis.
     (3) The deep myoid cell layers include layers of collagen fibers, elastic fibers, and the basal lamina of the seminiferous epithelium.

3. Seminiferous epithelium. The seminiferous epithelium is a continuous layer of Sertoli cells joined by prominent junctional complexes. Compartments between the lateral boundaries of adjacent Sertoli cells contain spermatogonia resting on the basal lamina, primary spermatocytes, secondary spermatocytes, spermatids, and spermatozoa (Figure 24-2).

4. Interstitial tissue. The space between seminiferous tubules is occupied by interstitial tissue that is a continuation of the tunica albuginea. Interstitial tissue contains connective tissue fibroblasts and collagenous fibrils that hold the spermatogenic tissue together. It also contains macrophages, lymphocytes, mast cells, and Leydig cells.
  a. Leydig cells (Figure 24-3) are the most important functional cells in the interstitium.
     (1) Leydig cells synthesize and secrete male sex steroids called androgens. **Testosterone** is the most abundant androgen released by Leydig cells.
     (2) Leydig cells exist in small clusters of cells. These cells have a diameter of 15–20 μm and a central round nucleus with peripheral heterochromatin and one or more prominent nucleoli.

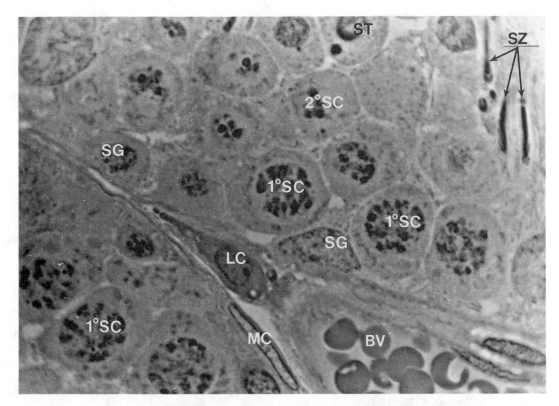

Figure 24-2. Light micrograph of the seminiferous epithelium, showing spermatogonia (*SG*), primary spermatocytes (*1°SC*), secondary spermatocytes (*2°SC*), spermatids (*ST*), and mature spermatozoa (*SZ*). Myoid cells (*MC*), Leydig cells (*LC*), and a blood vessel (*BV*) also are visible.

  (3) Leydig cell cytoplasm contains a well-developed Golgi apparatus close to the nucleus. The Golgi apparatus helps elaborate an abundance of smooth endoplasmic reticulum, which contains the enzymes for testosterone synthesis.
  (4) The cytoplasm also contains lysosomes, peroxisomes, and conspicuous **crystals of Reinke**. The function of crystals of Reinke is not known.
  (5) Leydig cells contain many mitochondria with tubular rather than lamellar cristae.
  b. Interstitial tissue contains numerous small lymphatic vessels and abundant interstitial fluid. The testosterone produced by Leydig cells enters the interstitial fluid, where it has a local effect on spermatogenesis.
  c. Androgens enter the systemic circulation through the interstitial fenestrated capillaries near Leydig cells. Once in the systemic circulation, androgens affect target organs in the body.

B. Sertoli cells. The seminiferous epithelium contains a continuous layer of tall columnar epithelial cells called **Sertoli cells,** which support the spermatogenic cell lines. These cells rest on the seminiferous epithelial basal lamina and reach the lumen.

1. **Structure**
   a. Sertoli cells have a large basal nucleus with numerous indentations, an abundance of euchromatin, and two or more prominent nucleoli.
   b. Sertoli cells have a conspicuous smooth endoplasmic reticulum, many conventional mitochondria, a well-developed Golgi apparatus, lipid droplets, microfilaments, and scattered lamellae of rough endoplasmic reticulum.

2. **Function**
   a. Sertoli cells have a trophic function in spermatogenesis.
   b. Sertoli cells help phagocytose and destroy cells that die during spermatogenesis and residual bodies shed by spermatids during spermiogenesis.
   c. Sertoli cells secrete fluids that create a hospitable environment for spermatozoa in the seminiferous tubules.

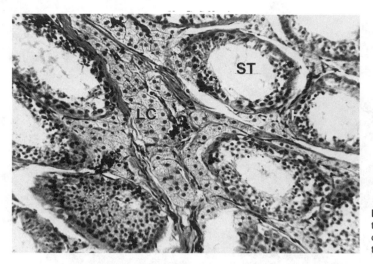

Figure 24-3. Light micrograph of the testis, showing the interstitial Leydig cells (*LC*) between the seminiferous tubules (*ST*).

    (1) Follicle-stimulating hormone (FSH) and testosterone stimulate Sertoli cells to secrete **androgen binding protein**.

    (2) Very high concentrations of testosterone are required to support the spermatogenic cycle of the seminiferous epithelium.

    (3) Androgen binding protein concentrates testosterone in the microenvironment of the seminiferous epithelium.

    d. Sertoli cells also secrete the protein **inhibin,** which provides feedback that helps inhibit FSH production by adenohypophysial gonadotropes either by directly affecting the hormone-producing gonadotropes or by affecting the secretion of hypothalamic **gonadotropin releasing hormone (GnRH)**.

C. Functional compartmentalization of the seminiferous epithelium

    1. An extensive system of **tight junctions** between Sertoli cells effectively divides the entire seminiferous epithelium into a basal compartment and an adluminal compartment.

        a. The **basal compartment** extends apically from the basal lamina to the tight junctions. Macromolecular tracers introduced into the blood diffuse through the basal lamina between the Sertoli cells and bathe the entire basal compartment, including spermatogonia contained therein, but do not pass through or around the junctional complexes.

        b. The **adluminal compartment** extends from the tight junctions to the lumen.

    2. Prior to, and just after DNA synthesis, spermatogonia are in the basal compartment. All later stages of spermatogenesis (primary spermatocytes to spermatozoa) occur in the adluminal compartment.

    3. The mechanism that allows spermatocytes to pass into the adluminal compartment may involve proteolytic enzymes such as **plasminogen activator** (secreted by Sertoli cells) causing transient openings in these junctions.

    4. Spermatogenesis and the concomitant appearance of new antigens on spermatozoa and in semen begins during puberty.

        a. During childhood, the seminiferous epithelium contains only Sertoli cells and spermatogonia. This condition is established very early in development.

        b. A fetus can recognize antigens on spermatogonia as self because the antigens are present during the developmental stage when the fetus gains the immunologic ability to discriminate self from non-self.

        c. The surface antigens of spermatozoa and antigenic molecules in seminal fluid are not present until later in development; therefore, they are identified by the immune system as non-self (foreign) antigens.

    5. The tight junctions probably isolate the spermatozoa from the circulatory system and thus form a **blood-testis barrier**.

6. Also, the tight junctions may maintain high concentrations of Sertoli cell secretions such as androgen binding protein in the lumen of the tubule.

III. HORMONAL CONTROL OF THE MALE REPRODUCTIVE SYSTEM

A. Hormones

1. **Testosterone** is a key testicular secretion.
   a. High concentrations of testosterone are needed to maintain spermatogenesis and the functional activity of the epididymis, ductus deferens, seminal vesicles, prostate gland, and bulbourethral glands.
   b. Testosterone is necessary for the development and maintenance of male secondary sexual characteristics.

2. **Luteinizing hormone (LH)** formerly was called interstitial cell-stimulating hormone (ICSH).
   a. LH is produced by basophilic gonadotropes in the adenohypophysis.
   b. Hypothalamic Gn-RH stimulates gonadotropes to secrete LH.
   c. LH binds to specific receptors on Leydig cells and stimulates cyclic adenosine 3′,5′-monophosphate (cAMP) formation and protein kinase activity.

3. **Follicle-stimulating hormone (FSH)**
   a. FSH is produced by adenohypophysial basophilic gonadotropes.
   b. Histologists believe it binds to specific receptors on Sertoli cells and activates an adenyl cyclase.
   c. High levels of cytoplasmic cAMP activate certain protein kinases, causing an increase in protein synthesis in Sertoli cells.
   d. FSH may stimulate Sertoli cells to synthesize androgen binding protein, which is needed to initiate (and probably to maintain) spermatogenesis.

4. **Other hormones.** Sertoli cells elaborate androgen binding protein, which accumulates in the seminiferous tubules and helps establish the high local concentrations of testosterone needed for spermatogenesis; and inhibin, which may inhibit the secretion of FSH.

B. Hormonal interaction

1. Testosterone secretion produces negative feedback that inhibits the production of LH.

2. Inhibin secretion produces negative feedback that inhibits the production of FSH.

3. LH, and possibly prolactin, stimulate Leydig cells to secrete testosterone.

4. FSH stimulates Sertoli cells to secrete inhibin.

IV. TESTICULAR EXCURRENT DUCTS

A. Tubuli recti and rete testis. The seminiferous tubules empty into tubules with an epithelium that changes rapidly from one containing the spermatogenic cell line to an epithelium consisting only of Sertoli cells.

1. Short **tubuli recti** join the seminiferous tubules to an anastomosing network of channels in the mediastinum testis called the **rete testis**.

2. The channels of the rete testis are lined by a simple cuboidal epithelium. Many of these cells have a flagellum that circulates luminal fluids.

B. Efferent ductules. The rete testis empties into 10–15 **efferent ductules,** which convey the fluids and cells secreted by the testis into the **epididymis**.

1. Efferent ductules are gathered into a smaller number of conical structures called the **coni vasculosi**.
   a. The broad bases of coni vasculosi face the epididymis; the tapered apices point toward the rete testis.
   b. Coni vasculosi are contained within the connective tissues that ensheathe the epididymis.

2. The **epithelial lining** of efferent ductules has alternating groups of cuboidal nonciliated cells and taller columnar ciliated cells. Consequently, transverse sections of this epithelium appear to have a height that varies from one point to another.

   a. The cuboidal cells have apical microvilli, contain lysosomes, and endocytose tracers instilled in the lumen. These features are consistent with the theory that these cells perform an absorptive function.

   b. Ciliated cell cilia beat actively, propelling spermatozoa and residual testicular fluids toward the epididymis.

   c. Smooth muscle fibers beneath the basement membrane surround the efferent ductules and help propel their contents toward the epididymis.

C. Epididymis. Efferent ductules empty into a single, long convoluted duct called the epididymis.

1. The epididymis is the organ that stores spermatozoa after they leave the testes. It also secretes materials that may be important for the maintenance and maturation of spermatozoa.

2. Unraveled, the epididymis is about 6 m long; however, due to its extensive coiling, it is usually 7–8 cm long when palpated in the adult male.

3. The head of the epididymis contains so many coils that it looks glandular. The body has fewer coils, and the tail, which empties into the **ductus deferens** (vas deferens), is nearly straight.

4. The epididymis has a tall **pseudostratified columnar epithelium** containing numerous tall columnar **principal cells** and scattered shorter pyramidal **basal cells** (Figure 24-4).

   a. Principal cells extend from the basement membrane to the lumen of the epididymal epithelium.

   (1) Principal cells have numerous long, modified microvilli called **stereocilia,** coated pits and vesicles in the apical cytoplasm, and numerous lysosomes and multivesicular bodies.

   (2) Principal cells endocytose the luminal contents and probably recover much of the fluid that is exuded from the testes.

   (3) The well-developed basal rough endoplasmic reticulum and apical Golgi apparatus of principal cells produce a glycoprotein secretion that coats spermatozoa.

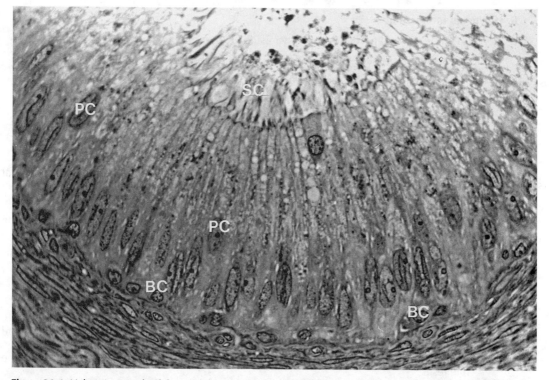

Figure 24-4. Light micrograph of the epididymal epithelium, showing principal cells (*PC*) with stereocilia (*SC*) and basal cells (*BC*).

   b. Basal cells are wedged between the principal cells and rest on the basement membrane of
   the epididymal epithelium. They may be an undifferentiated stem cell population that
   replaces effete principal cells.

5. The epididymis has an external investment of **smooth muscle** that is thick at the head and
   thinner at the tail.
   a. The smooth muscle in the upper sections of the epididymis causes peristaltic contractions
   that propel spermatozoa toward the ductus deferens. This portion of the epididymis has no
   extrinsic motor innervation.
   b. The more robust smooth muscle layers have a complex sympathetic innervation that causes
   the strong contractions during ejaculation that expel stored spermatozoa.

D. Ductus deferens (Figure 24-5). The entire length of the ductus deferens has conspicuous mucosal
folds.

1. The ductus deferens is lined by a **pseudostratified epithelium** that is quite similar to the
   epididymal epithelium but not as tall. For example, like epididymal epithelium, it has columnar
   cells with abundant stereocilia.

2. The epithelium rests on a **thin lamina propria** and is surrounded by three thick layers of **smooth
   muscle**. The inner and outer layers run along the lumen, and the middle layer runs around the
   lumen.

3. The lower, **ampullary portion** of the ductus deferens has numerous mucosal folds, a secretory
   columnar epithelium, and thinner layers of smooth muscle.

4. The ductus deferens has a glandular diverticulum called the **seminal vesicles** and passes into the
   prostatic urethra as paired **ejaculatory ducts**.
   a. The ejaculatory ducts are lined by a simple or pseudostratified columnar epithelium that
   gradually merges with the transitional epithelium of the prostatic urethra.
   b. The mucosa of each ejaculatory duct folds into the lumen.

V. THE SEMINAL VESICLES, PROSTATE GLAND, AND BULBOURETHRAL GLANDS

A. Seminal vesicles. The seminal vesicles are paired diverticula of the ampullary section of the ductus
deferens.

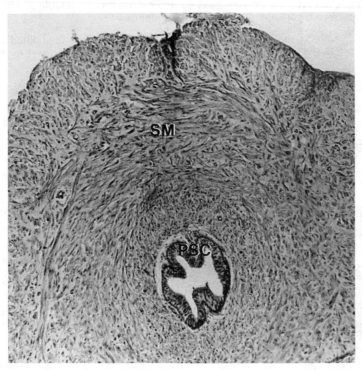

Figure 24-5. Light micrograph of the
ductus deferens. The mucosa has a
somewhat irregular shape and is
lined by a pseudostratified columnar
epithelium (*PSC*). The wall contains
three layers of smooth muscle (*SM*).

1. The **mucosa** has primary, secondary, and, occasionally, tertiary folds, which produce labyrinthine cavities in the walls of the seminal vesicles, all of which communicate with the central lumen of the gland (Figure 24-6).

2. The **epithelium** contains cuboidal cells or low columnar secretory cells containing granules of secretion product and yellowish lipochrome pigments, interspersed with shorter basal cells in an arrangement similar to the arrangement of cells in the epithelia of the epididymis and ductus deferens (Figure 24-7).

3. Seminal vesicles secrete a viscous, light-yellow material that contains mucoproteins, **prostaglandins,** and fructose.
   a. Prostaglandins may help induce smooth muscle contraction in the female reproductive tract, which is an important mechanism for transporting spermatozoa distally toward the uterine tubes.
   b. Fructose probably is a nutrient for spermatozoa.

4. The seminal vesicles have abundant smooth muscle, which contracts to expel their contents.

B. The prostate gland surrounds the **prostatic urethra** at the base of the urinary bladder.

1. The male urinary system and reproductive system meet in the prostate gland.
   a. The urethra carries urine from the urinary bladder to the prostate gland.
   b. The ejaculatory ducts carry spermatozoa from the epididymis and secretions from the seminal vesicles to the prostate gland.

2. The prostate is composed of 30–50 **main prostatic glands** with ducts symmetrically disposed to the left and right of the **seminal colliculus (verumontanum)**—a mound of fibromuscular tissue that imparts a C shape to the prostatic urethra.
   a. The main prostatic glands produce most of the secretory material released from the prostate gland.
   b. The epithelium of the main prostatic glands is a low columnar pseudostratified epithelium composed of numerous secretory cells with apical microvilli and a smaller number of basal cells.
   c. The lumina of main prostatic glands may contain solidified secretion products called **prostatic concretions**. The number of prostatic concretions present increases with age.

3. Smaller **submucosal glands** surround the prostatic urethra and extend a considerable distance into the submucosa. The mucosa contains short **periurethral glands**.

4. The **prostatic utricle** is a cul-de-sac diverticulum of the prostatic urethra within the colliculus. The ejaculatory ducts pass through the colliculus posterior to the utricle and connect to the prostatic urethra.

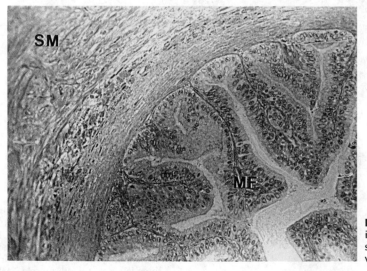

Figure 24-6. Light micrograph showing the mucosal folds (*MF*) and smooth muscle (*SM*) in the seminal vesicles.

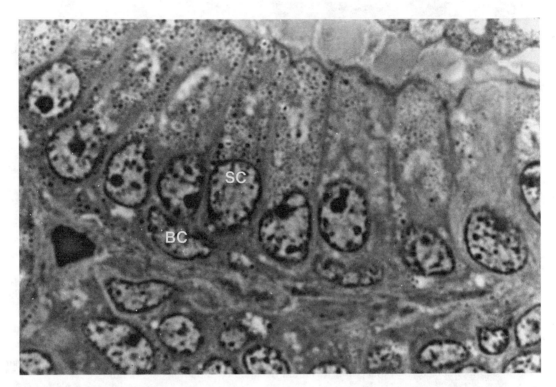

Figure 24-7. Light micrograph of the tall granule-laden secretory cells (*SC*) and short basal cells (*BC*) in the epithelium of seminal vesicles.

5. The periglandular portions of the prostatic stroma are dense masses of connective tissue cells and numerous smooth muscle cells. The smooth muscle contracts to help expel the glandular secretions during ejaculation.

6. Prostatic secretions are opalescent and have a slightly acidic pH of 6.0. They contain acid phosphatase, several proteolytic enzymes (e.g., **fibrinolysin**), zinc, and citric acid. Spermatozoa are contained in seminal fluid because they cannot otherwise survive the normally acidic milieu of the vagina.

C. **Bulbourethral (Cowper's) glands** are small glands at the base of the penis with ducts that join the penile urethra. They have glandular acini consisting of simple columnar cells that secrete a clear, slippery mucoid material during sexual arousal.

VI. THE PENIS

A. **Erectile tissues.** The penis has three masses of erectile tissue surrounding the penile urethra.

1. The urethra lies in a single ventral midline **corpus spongiosum,** which ends with a distal expanded head, called the glans penis (Figure 24-8).

2. Above the corpus spongiosum, a pair of **corpora cavernosa** run almost the entire length of the penis.
   a. The corpora cavernosa are the primary erectile tissue of the penis. This tissue is a complex anastomosing network of vascular channels and, when this tissue engorges with blood, the penis becomes erect.
   b. The flow of blood into the penis exceeds the outflow during erection, perhaps due to compression of venous drainage channels.

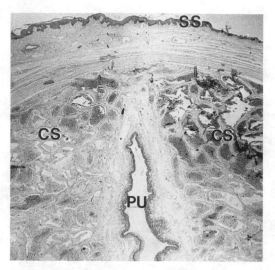

Figure 24-8. Light micrograph of the penile urethra (*PU*) passing through the corpus spongiosum (*CS*). A stratified squamous keratinized epithelium (*SS*) covers the penis.

B. Connective tissue and skin

1. The corpora of the penis are surrounded by a dense fibrous band of connective tissue called the **tunica albuginea**.

2. Dermal connective tissue and skin cover the tunica albuginea. The dermis of the glans contains numerous **Meissner's corpuscles,** which account for the great tactile sensitivity of the penis.

3. Up to the fossa navicularis, the penile urethra is lined by stratified or pseudostratified columnar epithelium. Beyond this point, it is lined by stratified squamous epithelium.

4. Diverticula of the penile urethra called the **glands of Littré** contain stratified columnar epithelium with small nests of pyramidal mucus-secreting cells.

STUDY QUESTIONS

Directions: Each question below contains five suggested answers. Choose the **one best** response to each question.

1. Which of the following statements best characterizes the testis?

(A) Functional compartmentalization of the seminiferous epithelium depends on tight junctions
(B) The seminiferous epithelium contains only proliferative cells
(C) The seminiferous epithelium contains numerous capillaries
(D) The interstitial tissue contains few capillaries
(E) Parts of the seminiferous epithelium lack Sertoli cells

2. Which of the following statements concerning the ductus deferens is true?

(A) Its wall does not contain smooth muscle
(B) It conveys spermatozoa to the prostate gland with ciliary currents
(C) It has an adventitia covered by a mesothelium
(D) Its lumen is lined by tall columnar cells with long apical stereocilia
(E) It carries mature spermatozoa to the epididymis

3. Which of the following statements best characterizes the epididymis?

(A) It has motile cilia
(B) It has a simple columnar epithelium
(C) It has long, nonmotile microvilli
(D) Its wall lacks smooth muscle
(E) It is a lobulated gland

4. Which of the following statements best characterizes the seminal vesicles?

(A) They are lobulated glands
(B) They have a columnar epithelium
(C) They have abundant mural smooth muscle
(D) They contribute nothing to the ejaculate
(E) They are inferior to the prostate gland

5. Each of the following statements concerning testosterone is true EXCEPT

(A) it is synthesized by Leydig cells
(B) it is required for spermatogenesis
(C) it is required for male secondary sexual characteristics
(D) it is excluded from the adluminal compartment of the seminiferous epithelium
(E) it is concentrated by the action of androgen binding protein

6. Each of the following statements concerning LH is true EXCEPT

(A) it is secreted by gonadotropes
(B) it is required for testosterone production
(C) it is secreted by adenohypophysial acidophils
(D) its secretion is influenced by GnRH
(E) a hypothalamic secretion influences secretion of LH

7. Each of the following statements concerning the seminiferous epithelium is true EXCEPT

(A) before puberty, it contains spermatids
(B) it consists of Sertoli cells and spermatogenic cells
(C) it has a robust basement membrane
(D) its basal compartment contains spermatogonia
(E) it is surrounded by smooth muscle-like myoid cells

Directions: The groups of questions below consist of lettered choices followed by several numbered items. For each numbered item select the **one** lettered choice with which it is **most** closely associated. Each lettered choice may be used once, more than once, or not at all.

Questions 8–11

Match each characteristic below with the most appropriate testicular duct or ducts.

(A) Epididymis
(B) Ductus deferens
(C) Both
(D) Neither

8. Lined by pseudostratified epithelium with stereocilia

9. All parts contain mature spermatozoa

10. Stimulated by testosterone

11. Has three distinct layers of mural smooth muscle

Questions 12–15

Match each description below with the appropriate lettered structure in the illustration.

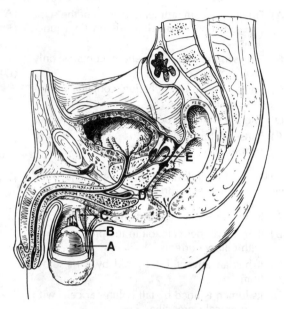

12. Contains a pseudostratified columnar epithelium; its secretion product is involved in sperm maturation

13. Contains Leydig cells in interstitial tissue

14. A diverticulum of the ductus deferens that has a secretory pseudostratified epithelium

15. A tubuloalveolar gland; its secretion is a clear viscous lubricant

ANSWERS AND EXPLANATIONS

1. The answer is A. [*II C 1*] Each testis contains seminiferous tubules for spermatogenesis and interstitial tissue rich in Leydig cells for steroidogenesis. The seminiferous epithelium is avascular, like all epithelia. The interstitial tissue has a rich supply of capillaries, like all endocrine tissue. Interstitial capillaries nourish the seminiferous epithelium. The seminiferous epithelium contains a continuous layer of Sertoli cells, which are joined by tight intercellular junctions. These junctions divide the epithelium into a basal compartment that contains spermatogonia and an adluminal compartment that contains spermatocytes, spermatids, and spermatozoa.

2. The answer is D. [*IV D*] The wall of the ductus deferens contains abundant smooth muscle. Peristaltic contractions of this smooth muscle help propel spermatozoa; ciliary currents are not involved. The ductus deferens carries capacitated sperm away from the epididymis. It is lined by a pseudostratified epithelium partially composed of tall columnar cells with stereocilia. The adventitia is not covered by mesothelium.

3. The answer is C. [*IV C*] The epididymis is a long, tortuously coiled duct lined by a pseudostratified columnar epithelium of tall cells with apical stereocilia. These apical modifications are not like motile cilia, which have a central core consisting of microtubules. They are long, highly modified, nonmotile microvilli. The wall of the epididymis contains abundant smooth muscle.

4. The answer is C. [*V A 4*] The seminal vesicles are paired, sac-like diverticula of the ductus deferens, which lie superior to the prostate gland. Although they have a glandular appearance, the seminal vesicles are really sacs lined by a pseudostratified epithelium that secretes a substantial amount of fluid into the ejaculate. Like the epididymis and ductus deferens, the seminal vesicles contain abundant mural smooth muscle.

5. The answer is D. [*II A 4 a (1); III A 1*] Testosterone is the primary male sex steroid. It is an androgenic steroid synthesized by interstitial Leydig cells in the testes. It is required for spermatogenesis, functional differentiation of the excurrent ducts of the testes, and the development of male secondary sexual characteristics. Androgen binding protein synthesized by Sertoli cells assures that a high concentration of testosterone is present in the adluminal compartment of the seminiferous epithelium.

6. The answer is C. [*III A 2, B 3*] Luteinizing hormone (LH) has a key role in the function of the male reproductive system. LH stimulation of Leydig cells is required for testosterone secretion. LH is a secretion product of the basophilic gonadotropes of the adenohypophysis. Certain cells in the hypothalamus release gonadotropin releasing hormone (GnRH), which stimulates gonadotropes in the adenohypophysis to release LH.

7. The answer is A. [*II A 2, 3; B, C 1, 4*] The seminiferous epithelium lines the seminiferous tubules in the testis. It consists of a population of Sertoli cells and a spermatogenic cell line. Sertoli cells span the thickness of the seminiferous epithelium. Sertoli cell characteristics, including the distribution of their intercellular junctions, divide the seminiferous epithelium into a basal compartment containing only spermatogonia and an adluminal compartment containing primary and secondary spermatocytes, spermatids, and spermatozoa. The differentiation of spermatogonia into spermatids does not begin until puberty. Thus, the seminiferous epithelium does not contain spermatids prior to puberty. Like all epithelial layers, the basal surface of the seminiferous epithelium is bounded by a basement membrane.

8–11. The answers are: 8-C, 9-B, 10-C, 11-B. [*III A 1; IV C, D*] Both the epididymis and ductus deferens are stimulated by testosterone and are lined by a pseudostratified epithelium. Spermatozoa are mature by the time they enter the tail of the epididymis; however, they are not mature in the head of the epididymis. There is abundant smooth muscle in the wall of both the epididymis and the ductus deferens; however, only the ductus deferens has three distinct layers of smooth muscle.

12–15. The answers are: 12-B, 13-A, 14-E, 15-D. [*II A 2; IV C, D; V A, C*] The testis (*A*) contains seminiferous tubules for spermatogenesis and interstitial tissue rich in testosterone-secreting Leydig cells.

The epididymis (*B*) conveys sperm from the seminiferous tubules to the ductus deferens (*C*). It is lined by a pseudostratified columnar epithelium with tall stereocilia. The epididymis modifies the fluids from the seminiferous tubules and functions in sperm maturation.

The seminal vesicle (*E*) is a sac-like diverticulum of the ductus deferens. It has a secretory pseudostratified epithelium.

The bulbourethral gland (*D*) lies at the base of the penis. During sexual arousal, it secretes a clear viscous lubricant.

I. INTRODUCTION

A. Function

1. The entire outer surface of the body, with the exception of the cornea and conjunctiva, is covered by skin.

2. This complex covering is essential to life. It serves as a barrier against infection, a hydrophobic layer that prevents the loss of vital fluids, and a mechanism that alters the core body temperature (i.e., by regulating blood flow through capillary beds in the skin and by sweating).

B. Components

1. The skin is composed of a stratified squamous keratinized epithelium called the **epidermis** and a dense irregular connective tissue called the **dermis**. Epidermis is derived from the ectoderm and dermis is derived from the mesoderm.

2. Some skin locations produce **hair** and **nails,** which are protective acellular appendages. These structures are formed entirely from the epidermis and are subsequently influenced by the dermis that is closely associated with hair follicles and nail beds.

3. **Cells** in the basal layer of the skin constantly proliferate and differentiate into keratinized cells, which eventually dry out and desquamate from the skin surface.

4. **Glands** in the skin produce sweat to cool the body and sebum to lubricate and condition the skin.

5. Skin has **color,** which varies somewhat among different ethnic and racial populations due to the action of neural crest derivatives called **melanocytes**.

II. EPIDERMIS (Figure 25-1)

A. Microscopic anatomy. The epidermis has four easily recognizable layers. They are presented here in order from deep to superficial.

1. **Stratum basale (stratum germinativum)**
   a. The stratum basale, the most basal layer of cells in the epidermis, rests on a thick basement membrane.
   b. Cells in the stratum basale are highly proliferative. They undergo repeated mitoses to produce a proliferative stem cell layer that remains on the basement membrane and yields the next generation of cells, and cells that differentiate into **keratinocytes**.
   c. Epidermal layering represents the progressive accumulation of keratin within the progeny of proliferative cells in the stratum basale.

2. **Stratum spinosum** (Figure 25-2)
   a. The thickness of the stratum spinosum varies from two to six cell layers.
   b. In the light microscope, small spine-like projections appear to join cells in this layer; however, examination by electron microscope reveals that the cells are connected by numerous desmosomes.

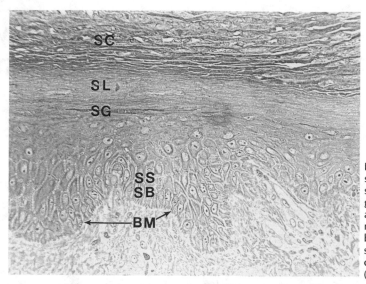

Figure 25-1. Light micrograph of a section through the epidermis. This stratified squamous epithelium begins at the basement membrane (*BM*) and extends beyond the top of the micrograph. Cells of the stratum basale (*SB*), stratum spinosum (*SS*), stratum granulosum (*SG*), stratum lucidum (*SL*), and stratum corneum (*SC*) are visible.

    c. Cells in the stratum spinosum contain a peculiar organelle called the **lamellated granule**. This small membrane-delimited ellipsoid granule is 0.3 mm in diameter and has lamellae running perpendicular to its long axis. Often, lamellated granules are closely associated with the Golgi apparatus.

    d. Some histologists discuss the stratum basale and stratum spinosum together as one layer called the **stratum malpighii**. Examination of cells in this region by electron microscopy reveals numerous mitochondria, melanin granules, abundant rough endoplasmic reticulum, and numerous bundles of 6-nm to 8-nm filaments.

3. Stratum granulosum

    a. The stratum granulosum is three to five layers of flattened, irregularly shaped cells containing numerous **keratohyalin granules**.

        (1) Examination by light microscopy reveals that these granules stain intensely with various basic dyes including hematoxylin.

        (2) Classical histologists believe the granules are an intermediate stage in the keratinization process.

    b. Electron microscopic investigation reveals that cells in the stratum granulosum are filled with small granules that have an irregular amorphous core and filaments that run up to or through the granules.

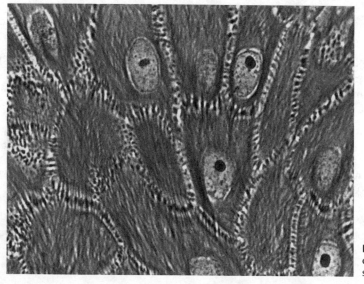

Figure 25-2. High-power light micrograph of the epidermal stratum spinosum.

    c. In addition, cells in this layer have lamellated granules that migrate from the Golgi apparatus toward the periphery of the cell. In some cases, lamellated granules discharge their contents into the intercellular space.

4. Stratum corneum

    a. The stratum corneum is the most apical layer of the epidermis, ranging in thickness from as few as five to ten cell layers to as many as several hundred cell layers. For example, the stratum corneum on the back of the hand or on the eyelid is extremely delicate, while the stratum corneum on the sole of the foot is extremely thick. Furthermore, the stratum corneum can become thicker if circumstances warrant (e.g., in the case of callus formation).

    b. In the light microscope, the stratum corneum appears as a poorly stained layer of dead and desiccated cells that flake easily from the epidermis. Where the stratum corneum is particularly thick, there is a thin transitional zone called the **stratum lucidum** between the stratum granulosum and the stratum corneum.

    c. In the electron microscope, cells in the stratum corneum appear to be dead or inactive.

        (1) Instead of the usual nucleus, mitochondria, and rough endoplasmic reticulum, these cells have masses of 6-nm to 8-nm filaments that are intermingled with masses of amorphous material.

        (2) The inner aspect of the plasma membrane is considerably thickened, and the intercellular clefts are filled by the discharged contents of the lamellated granules.

        (3) Desmosomes are present between cells of the stratum corneum. Typically, the outermost layers of this stratum are loosely attached to underlying layers and appear ready to separate from them.

B. Keratinization. Stratified squamous epithelium is distributed throughout the body, typically existing where friction is encountered (e.g., conjunctiva and anterior surface of the cornea, oral cavity, esophagus, vagina, and anal canal). The surface of the body is covered by this type of epithelium; however, it differs from all other stratified squamous epithelium in that it is keratinized.

    1. Keratinization is the process whereby living cells in the stratum basale differentiate into the daughter, keratin-filled dead cells of the stratum corneum. These dried cells filled with hydrophobic proteins limit water loss from the body and limit the entry of external noxious substances and microorganisms into the body.

        a. First, the cells that commit to this fatal differentiation are produced by repeated mitosis of proliferative cells in the stratum basale.

        b. Second, cells in the stratum basale and stratum spinosum become committed to producing large amounts of several proteins. This occurs when certain select genes become activated to synthesize messenger RNA (mRNA) and eventually to synthesize proteins in the filaments and keratohyalin granules that are present in a later stage of differentiation.

        c. Third, so many proteins accumulate within the cells that they kill the cells and exclude all other organelles.

    2. The death of an epidermal cell is faster, but no less inevitable, than the death of a highly differentiated red blood cell or pyramidal neuron in the cerebrum. All three cell types become highly differentiated and lose the ability to proliferate, thus ensuring a shortened life span.

    3. In contrast, some cells do not differentiate but remain proliferative throughout the life of the organism. Cells in the epidermal stratum basale, spermatogonia, hematogenous cells, and cells lining intestinal crypts are proliferative throughout life.

    4. Epidermal stratification is a representation of the different stages in the keratinization process. Completely proliferative, unkeratinized cells exist near the basement membrane and nonproliferative, completely keratinized cells comprise the outermost layers. Between these extremes are layers in the intermediate stages of keratin accumulation.

C. Keratin biochemistry

    1. Keratin contains some sulphur cross-linked by disulfide bridges. These bonds make keratin a very stable protein that can be dissolved only by strong acid.

    2. Also, the proteins derived from keratohyalin granules are extensively cross-linked to each other and to the proteins of the keratin by disulfide bonds.

    3. This extensive cross-linking makes proteins of the stratum corneum very resistant to dissolution.

D. Unusual cell types in the epidermis. In addition to keratinizing cells, the epidermis contains two other unusual types of cells.

1. **Langerhans cells** are a cell population present throughout the epidermis. They also exist in the dermis and nearby lymph nodes. These cells stain poorly in routine histologic preparations.
   a. Langerhans cells are part of the immune system; they apparently help present antigens to helper T cells. They have the cell surface markers common to most B cells, some T cells, and macrophages and monocytes.
   b. They lack the tonofilaments and melanosomes present in other epidermal cells, and have an indented nucleus and characteristic racket-shaped **Birbeck granules** (function unknown).

2. **Merkel cells** are present only in the basal layer of the epidermis.
   a. They are innervated by myelinated nerve fibers and are most abundant in areas of greatest tactile sensitivity, such as the fingertips.
   b. They have irregular, lobulated nuclei and cytoplasm that is less electron-dense than adjacent epidermal keratinocytes.
   c. They contain electron-dense, membrane-bound granules that are morphologically similar to the catecholamine-containing granules of the adrenal medulla. Histologists believe they are APUD (amine precursor uptake and decarboxylation) cells.

III. GLANDS

A. Sweat glands are epithelial modifications distributed throughout the body. They are specialized for the production of sweat, which cools the body by evaporation, and other complex secretions.

1. **Apocrine sweat glands**
   a. Cuboidal and columnar eosinophilic secretory cells deep in these glands produce a milky white secretion, which is expelled from the glands after adrenergic stimulation of myoepithelial cells.
   b. The ducts of apocrine sweat glands empty into hair follicles.
      (1) In lower forms of animal life, apocrine glands are distributed uniformly in the body surface and have an important role in sexual behavior and in marking territorial zones.
      (2) In humans, they are restricted to the axilla, the area around the genitalia, and the nipples of the mammary glands. Their role in humans is poorly understood.
   c. Secretion from apocrine glands begins after puberty under the influence of sex steroids. Initially, the milky secretion is odorless; later, it develops an odor due to the action of cutaneous bacteria.
   d. The ceruminous glands in the external auditory meatus and glands of Moll in the eyelid are modified apocrine sweat glands.
   e. Recent research indicates that apocrine sweat glands use a merocrine mechanism of secretion.

2. **Eccrine sweat glands**
   a. Eccrine sweat glands are distributed throughout the surface of the body. They are present everywhere except on the lips, penis, clitoris, and labia minora. They are elaborately coiled glandular structures connected to the surface of the body by coiled ducts.
   b. The glandular portion of eccrine sweat glands is composed of myoepithelial cells. The glands also contain large eosinophilic clear cells, which appear to secrete sodium, and smaller basophilic dark cells, which appear to produce a glycoconjugate-rich material.
   c. Sweat contains sodium, water, chloride, and other substances. It is formed by active pumping of sodium out of clear cells, passive diffusion of water out of clear cells, and a resorption of sodium in the sweat ducts, which makes the sweat hypotonic.
   d. Eccrine glands respond to cholinergic stimulation and, under certain circumstances, to adrenergic stimulation. The composition of the sweat produced varies according to the stimuli.
   e. Sweating begins whenever heat radiation from the body is inadequate and the core body temperature rises. As the water in sweat evaporates, the body cools.

B. Sebaceous glands (Figure 25-3). All sebaceous glands are associated with hair follicles and are found on most body surfaces. A short duct connects the gland to the shaft of a hair follicle.

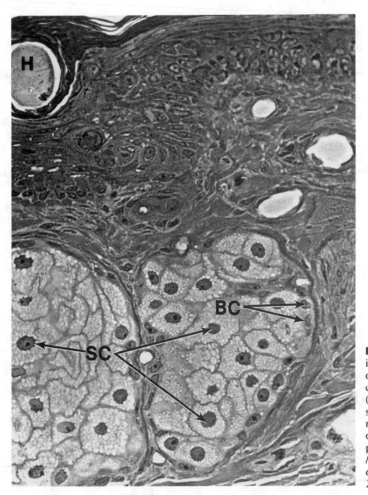

Figure 25-3. Light micrograph showing two large sebaceous glands. Most of the cells are lipid-laden secretory cells (*SC*); however, some basal cells (*BC*) are visible. A section of a hair shaft (*H*) also is visible, but the remainder of the pilosebaceous unit is out of this plane of section. (Reprinted from Johnson KE: *Histology: Microscopic Anatomy and Embryology*. New York, John Wiley, 1982, p 253.)

1. The gland is composed of two types of cells: an outer layer of stem cells, called basal cells, and a central group of cells that accumulate lipid droplets in their cytoplasm.
   a. The largest cells in the central portion of a sebaceous gland acinus are full of lipid droplets.
   b. These lipid-laden cells burst and become part of the secretion product. This type of secretion is called **holocrine secretion**.

2. Sebum contains abundant triglycerides and free fatty acids. The secretion of sebum is partially controlled by sex hormones. The high lipid content of sebum suggests that it may condition the skin or that it may be a part of the hydrophobic barrier that exists in the epidermis.

IV. HAIR

A. General characteristics

1. Hairs are appendages of the skin. They are produced by the deep epithelial ingrowths that form hair follicles.

2. Hair and nails are largely composed of hard keratin, a substance that is chemically similar to the soft keratin in the skin but is more extensively cross-linked by disulfide bridges.

B. Hair follicles

1. Hair follicles are deep ingrowths of the epithelium of the skin that terminate as bulbous cores indented by a connective tissue papilla. The papilla has blood vessels that carry nutrients to the forming hair. The bulbous portion contains a group of cells called the **germinal matrix,** which produces the hair in a process similar to the process used by the stratum germinativum to produce the stratum corneum.

2. The **external root sheath** is the portion of the epithelial ingrowth that connects the germinal matrix to the skin surface.
   a. Near the surface of the skin, the external root sheath resembles an invagination of the stratified squamous epithelium and exhibits the same strata.
   b. Moving down the external root sheath toward the germinal matrix, the layering of the skin is less apparent. Eventually, only a stratum germinativum-like layer exists, mingled in the jumbled mass of cells in the germinal matrix.

3. **Proliferative cells** in the germinal matrix produce two structures: hair and a cellular collar that surrounds the hair shaft between the shaft and the external root sheath.
   a. The tubular collar is called the internal root sheath and is made of soft keratin. The soft keratin is formed from trichohyalin granules; structures similar to keratohyalin granules.
   b. The internal root sheath extends only partway up the hair shaft; above this collar a naked shaft is exposed.

4. **Melanocytes** in the germinal matrix add melanin granules to the growing hair. Hair color varies from light blond to black, depending on the amount of melanin deposited in the hard keratin of the hair shaft.

5. The **ducts** of sebaceous glands and apocrine sweat glands empty secretions into the hair follicle.

V. MELANOCYTES AND MELANIN

A. **General characteristics.** Melanocytes synthesize melanin, the pigment responsible for coloration of skin, hair, and the iris. Melanocytes are derived from the neural crest and, therefore, originate in the ectoderm.

B. **Melanocyte structure and function**

1. Melanocytes rest on the apical side of the epidermal basement membrane, usually between cells of the stratum basale, and produce long processes that are pushed around and between the basal cells of the epidermis.

2. Melanocytes have an enzyme called **tyrosinase,** which catalyzes one step in the reaction that converts tyrosine into the polymeric oxidation product called melanin.

3. Melanocytes can be distinguished from surrounding cells in the stratum basale by the dopa reaction. (In essence, these cells produce extraordinary amounts of melanin when supplied with 3,4-dihydroxyphenylalanine, or dopa, due to the action of their tyrosinase. Then, they mark themselves by accumulating a dark reaction product.)

4. Melanocytes synthesize melanin in membrane-delimited melanin granules (melanosomes). These granules pass from the melanocytes into the cells of the stratum basale, presumably by exocytosis from the melanocyte and phagocytosis by keratinocytes.

C. **Melanin function**

1. The nuclei of cells in the stratum basale are shielded from ionizing (i.e., mutagenic) ultraviolet solar radiation by an apical cap of melanin granules. The melanin content of these cells increases rapidly in response to exposure to the sun, causing tanning.

2. Racial and ethnic differences in skin color are caused by several factors, including the number of melanocytes present and the types and sizes of melanin granules.

VI. DERMIS

A. **Microscopic anatomy**

1. The dermis is a dense, irregular connective tissue that lies beneath the epidermis. The epidermal-dermal boundary is the **epidermal basement membrane**.

2. Dermis contains a series of peg-like folds called **dermal papillae,** which increase the contact area between the epidermis and dermis and thus strengthen the adhesion between these layers. Skin in the palms and soles of the feet is subjected to profound and constant abrasion. These locations have the most complex contact between the epidermis and dermis.

3. The dermis is divided into two **layers,** based on differences in the texture and arrangement of collagen fibrils.
   a. The **papillary layer** is associated with the papillae. Here, collagen fibers are arranged in an irregular mesh.
   b. In the **reticular layer,** which is deep to the papillary layer, collagen fibers are arranged in coarser bundles that crisscross one another.
   c. The boundary between these two layers is indistinct; however, in histologic sections, it is relatively easy to differentiate the papillary and the reticular areas.

4. A thick layer of **subcutaneous connective tissue** (hypodermis) containing considerable adipose tissue exists deep to the dermis.

B. Extracellular matrix

1. The extracellular matrix of the dermis is rich in **Type I collagen** and contains a lesser amount of **Type III collagen.** Type I collagen is composed of two $\alpha1(I)$ chains and one $\alpha2(I)$ chain. Type III collagen is composed of three $\alpha1(III)$ chains. These two types of collagen also have other minor biochemical differences.

2. Also, the extracellular matrix is rich in **elastic fibers** and the glycosaminoglycans hyaluronic acid, chondroitin sulfate, and dermatan sulfate.

C. Cellular elements

1. The most important cellular element of the dermis is the **fibroblast,** which synthesizes and secretes the complex extracellular matrix of the dermis. The dermis also contains mast cells, macrophages, and other formed elements of the blood.

2. Small cutaneous lesions may result in localized infections and a concomitant increase in the number of phagocytic neutrophils in the dermis.

D. Blood supply. The blood supply of the skin comes from the dermis and is somewhat complex due to its role in thermoregulation.

1. The **rete cutaneum** is a deep network of branches of the main arteries that supply the skin. This network is located just deep to the dermis in the subcutaneous adipose tissue.

2. Some branches of the rete cutaneum project toward the epidermis, forming a second **rete subpapillare** at the junction between the papillary and reticular layers of the dermis.

3. Small arterioles in the rete subpapillare branch to dermal papillae, where the extensive capillary beds are bypassed by direct arteriovenous shunts.
   a. When the body is overheated, the blood supply to the papillary capillaries increases and sweating commences. The evaporation of the water in sweat cools the blood slightly, reducing body temperature.
   b. In contrast, when the body is chilled, blood is directed away from the capillary beds through the arteriovenous shunts to conserve body heat.

E. Dermal variation and skin thickness

1. Thick skin, such as skin on the sole of the foot, has a thick epidermis with a thick stratum corneum and a relatively thin and compact dermis with numerous large dermal papillae.

2. Thin skin, such as skin on the forearm, has a thin epidermis with a thin stratum corneum and a relatively thick and more diffuse dermis that lacks large dermal papillae.

STUDY QUESTIONS

Directions: The question below contains five suggested answers. Choose the **one best** response to the question.

1. Which of the following statements best describes the functional importance of melanocytes?

(A) Melanocytes are phagocytic cells
(B) Melanocytes perform an immunologic function
(C) Melanocytes secrete collagen in the dermis
(D) Melanocytes synthesize materials capable of absorbing ultraviolet radiation
(E) Melanocytes present antigens to helper T cells

Directions: The groups of questions below consist of lettered choices followed by several numbered items. For each numbered item select the **one** lettered choice with which it is **most** closely associated. Each lettered choice may be used once, more than once, or not at all.

Questions 2–5

Match each description below with the most appropriate cell type.

(A) Keratinocyte
(B) Fibroblast
(C) Melanocyte
(D) Langerhans cell
(E) Merkel cell

2. Present in the epidermis and dermis; probably derived from bone marrow monocytes; does not have tonofilaments

3. Present in the epidermis; its cytoplasm has numerous tonofilaments attached to desmosomes

4. Has a fusiform or stellate shape; contains abundant rough endoplasmic reticulum and a well-developed Golgi apparatus for the synthesis of dermal extracellular matrix

5. Innervated by myelinated fibers; contains catecholamine-like granules and is thought to be a sensory element of the epidermis

Questions 6–11

Match each description below with the appropriate lettered structure in the following micrograph.

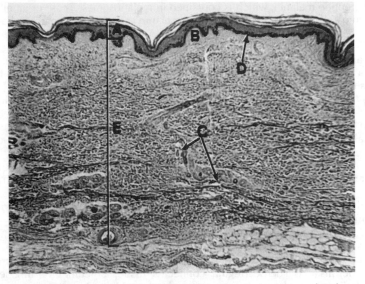

Reprinted from Johnson KE: *Histology: Microscopic Anatomy and Embryology.* New York, John Wiley, 1982, p 256.

6. Blood vessels and fibroblasts lie deep to this location; the epithelium is superficial to this location

7. A dense, irregular connective tissue rich in fibroblasts and collagen

8. Important in temperature homeostasis

9. An avascular layer of proliferative cells nourished by diffusion from nearby connective tissue

10. The basement membrane of the epidermis

11. Increases the contact area between the epidermis and the dermis; more prominent in thick skin

ANSWERS AND EXPLANATIONS

1. The answer is D. [*V B, C*] Melanocytes are cells on the apical side of the epidermal basement membrane. They are present in the dermis but pass melanin granules into epidermal cells. Melanocytes synthesize the polypeptide melanin, which is contained in melanosomes. These organelles are passed from melanocytes into cells of the stratum basale. The melanin granules form an umbrella-like cap over the nuclei of cells in the stratum basale, shielding the DNA from the ionizing radiation in sunlight.

2–5. The answers are: 2-D, 3-A, 4-B, 5-E. [*II A 1, B, D 1, 2; V B, C; VI C 1*] Skin contains several types of cells; each has a unique function in the skin. The keratinocyte is an epidermal cell in the process of keratinization. It is formed by proliferative cells in the stratum basale.

The fibroblast is the most important cell type in the dermis. It synthesizes the extracellular matrix of the dermis.

Melanocytes are located in the epidermis. They produce melanin granules that are passed into cells in the stratum basale. The melanin granules form an umbrella-like cap over the nuclei of cells in the stratum basale, shielding the DNA from the ionizing radiation in sunlight.

Langerhans cells are present in the epidermis. They are bone marrow–derived cells of the mononuclear phagocyte system and perform an immune function. Langerhans cells also exist in the dermis and nearby lymph nodes.

The epidermis also contains Merkel cells. Merkel cells are innervated by myelinated nerve fibers, contain catecholamine-like granules, and perform a sensory function in the skin.

6–11. The answers are: 6-D, 7-E, 8-C, 9-A, 10-D, 11-B. [*I B; II A 1; III A; VI A 1, 2*] This is a light micrograph of skin. Skin is composed of an epithelial epidermis (*A*), which rests on a basement membrane (*D*). The dermis (*E*) is a dense irregular connective tissue, sometimes forming dermal papillae (*B*). The dermis underlies the epidermis. Eccrine sweat glands and their ducts (*C*) are a common feature of skin.

26
The Urinary System

I. INTRODUCTION

A. Components. The urinary system is composed of the **kidneys, ureters, urinary bladder,** and **urethra**.

B. Functions. The urinary system produces, stores, and eliminates **urine** after it produces and modifies a **urinary filtrate** consisting of a large volume of hypotonic blood filtrate rich in serum proteins.

   1. The urinary filtrate is produced in the **renal glomerulus** and then passes into a complex network of tubules where the protein, water, and ionic composition of the filtrate undergoes extensive modification. The urine that passes from the urinary system is hypertonic due to this modification.

   2. Urine secreted by the kidneys is collected by the **renal pelvis** (the funnel-shaped upper end of the ureters) and is conveyed through the ureters into the urinary bladder.

   3. Urine is stored in the urinary bladder until the bladder becomes completely distended, at which time the urine is eliminated from the body through the urethra by involuntary contractions of the smooth muscle cells in the bladder.

II. THE KIDNEYS

A. Macroscopic features

   1. **The renal cortex and medulla**
      a. The kidneys have an outer red **renal cortex** and pyramidal masses of yellowish **renal medulla**.
      b. The apices of the **renal pyramids** project into the minor calyces, and their bases project toward the cortex.

   2. **Renal lobes.** Each medullary pyramid and its connected cap of cortical tissue comprise a **renal lobe**.
      a. The cortical portion of each lobe contains many **nephrons** and the pyramids contain numerous **collecting tubules**.
      b. Each lobe contains many **medullary rays,** which project from the pyramid into the cortex. Medullary rays are cortical tissue comprised of small blood vessels and loops of Henle and are considered part of the cortex.
      c. A medullary ray and the nephrons that drain into it comprise a **renal lobule**.
      d. **Renal columns** of cortical tissue occupy the space between renal lobes.

B. Uriniferous tubules (Figure 26-1) are the basic functional units of the renal parenchyma. They perform all of the important functions that lead to urine production and elimination. Each uriniferous tubule consists of a **nephron,** which produces and modifies the urinary filtrate, and a **collecting tubule,** which concentrates urine and conveys it into the minor calyces.

   1. **Nephrons** are very elaborate tubes with a closed, ballooned proximal end called **Bowman's capsule**.
      a. A nephron is composed of a capillary tuft called the **renal glomerulus,** Bowman's capsule, a **proximal convoluted tubule,** a **loop of Henle,** and a **distal convoluted tubule,** and includes the **juxtaglomerular apparatus** (a structure at the vascular pole of the renal corpuscle).

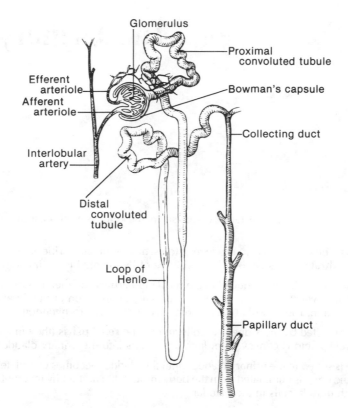

Figure 26-1. Illustration of the major components of the uriniferous tubule.

b. Details of nephron structure vary in different parts of the kidney. For example, cortical nephrons have a short loop of Henle that is restricted to the cortex. In contrast, juxtamedullary nephrons have a long loop of Henle that begins in the cortex and extends down into the medulla.

2. **Collecting tubules** receive urine from the distal convoluted tubules and carry it toward the apex of the renal pyramids. The apices sometimes are called renal papillae.
   a. Expanded portions of the collecting tubules, called the **papillary ducts,** pass through the papillae and exit into the renal sinus through papillae at the **area cribrosa**. The area cribrosa has many openings—one for each papillary duct.
   b. Near the distal convoluted tubules, collecting ducts are lined by a simple cuboidal epithelium. The epithelial cells become columnar as the ducts pass toward the area cribrosa. The papillary ducts are lined by a simple columnar epithelium (Figure 26-2).

C. Nephron histology

1. **Renal glomerulus**
   a. The renal glomerulus is an anastomosing network of **fenestrated capillaries** that receives high-pressure blood from the renal arteries.
      (1) Endothelial cells in the fenestrated capillaries are joined to each other by firm tight junctions.
      (2) The fenestrations allow many of the constituents of the blood, including macromolecules, to escape from the bloodstream.
      (3) The capillaries are surrounded by a basement membrane.
   b. The capillary tuft fits neatly into a deep cup-like invagination of Bowman's capsule, which has two epithelial cell layers that are reflected upon one another.

2. **Bowman's capsule.** Bowman's capsule epithelium has a **visceral layer** intimately associated with the capillary tuft and a **parietal layer** that is continuous with the epithelial cells of the proximal convoluted tubule (Figure 26-3).

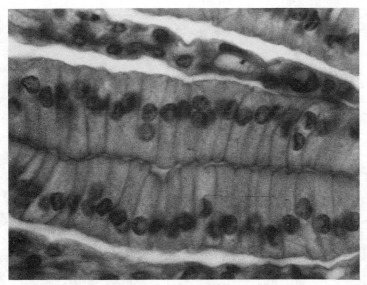

Figure 26-2. Light micrograph of a papillary duct lined by a simple columnar epithelium. (Reprinted from Johnson KE: *Histology: Microscopic Anatomy and Embryology.* New York, John Wiley, 1982, p 356.)

   a. Visceral layer

      (1) The visceral layer of Bowman's capsule is a layer of specialized epithelial cells called **podocytes**.

         (a) Podocytes are shaped somewhat like octopi. They have large **primary processes,** which have many secondary and tertiary branches.

         (b) The ends of these branches have minute processes called **pedicles** (foot processes), which form complex interdigitations with the pedicles on branches of adjacent cells (Figure 26-4).

         (c) The interdigitations form numerous small slits between cells. Small diaphragms, called **filtration slit membranes,** bridge the gap between adjacent pedicles.

      (2) The visceral layer of Bowman's capsule is adjacent to a prominent basement membrane that presses tightly against the basement membrane of the adjacent capillary, forming the extraordinarily thick **glomerular basement membrane**. This basement membrane is a selective filter for macromolecules in blood (Figure 26-5).

         (a) High molecular weight proteins in the blood pass through the capillary fenestrations but are blocked from passing into the urinary space by the basement membrane.

         (b) Intermediate molecular weight proteins cross the basement membrane but cannot cross the filtration slit membranes.

         (c) Low molecular weight proteins cross both barriers and enter the urine that is filtering into the urinary space.

   b. Parietal layer. The parietal layer of Bowman's capsule is a simple squamous epithelium, which is continuous with the visceral layer of podocytes (at the vascular pole) and the cuboidal epithelial cells of the proximal convoluted tubules (at the urinary pole).

3. Proximal convoluted tubules (Figure 26-6)

   a. Proximal convoluted tubules have a simple cuboidal epithelium with a prominent apical brush border consisting of many long microvilli.

      (1) The cytoplasm of these epithelial cells is markedly acidophilic because it contains many mitochondria.

      (2) The lateral borders of adjacent epithelial cells are convoluted and interdigitate, much like pieces in a jigsaw puzzle. This modification creates an extensive compartment for ion pumping.

      (3) Mitochondria in this region are elongated and arranged radially in relation to the lumen, so that the proximal convoluted tubule often is striated.

   b. An enormous volume of glomerular filtrate is produced daily, and the proximal convoluted tubules absorb about 80% of the filtrate immediately after it is formed.

      (1) Sodium ions are recaptured from the glomerular filtrate by ATPase activity in the membranes of tubule cells and are pumped into the lateral compartments. Chloride ions and water passively follow sodium ions. Carbohydrates and amino acids are recovered in the proximal convoluted tubules as well.

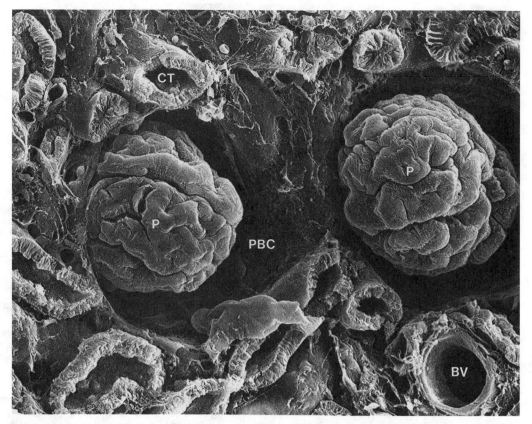

Figure 26-3. Scanning electron micrograph of a fractured portion of the renal cortex. Two glomeruli are visible, each covered by a layer of podocytes (*P*). The parietal layer of Bowman's capsule (*PBC*) is covered by a simple squamous epithelium. A blood vessel (*BV*) and a convoluted tubule (*CT*) also are visible. (Micrograph courtesy of Mr. F. Lightfoot, Department of Anatomy, George Washington University.)

    (2) The apices of tubule cells have an actively endocytotic mechanism that recovers most urinary filtrate protein. At the base of the deep surface invaginations between microvilli, small membrane-bound packets of luminal contents are continuously pinched off and internalized by the cytoplasm. These small vesicles fuse with each other to form larger vacuoles, which fuse with lysosomes.

    (3) Resorbed proteins pass through the epithelium into the bloodstream, where they are used again.

4. Loop of Henle
  a. The proximal tubule has a long, thin, straight section that leads into the thin descending limb of the **loop of Henle**. Epithelial cells here are similar to those in the convoluted section; however, they are not as tall and their mitochondria, although abundant, are not as elongated as mitochondria in cells in the convoluted section (Figure 26-7).
  b. The straight portion of the proximal tubule leads into the thin limb of the loop of Henle. This portion of the nephron is lined by a squamous epithelium that is slightly thicker than the endothelium of a capillary. These epithelial cells have a nucleus that projects into the lumen of the small tubules.
  c. The portion of the loop of Henle that ascends toward the cortex has an ascending thin limb that becomes an ascending thick limb. The ascending thick limb passes into a **distal convoluted tubule**.

5. Distal convoluted tubules
  a. Epithelial cells in distal convoluted tubules are cuboidal. They are taller than epithelial cells in the loop of Henle but not as tall as epithelial cells in the proximal convoluted tubules.

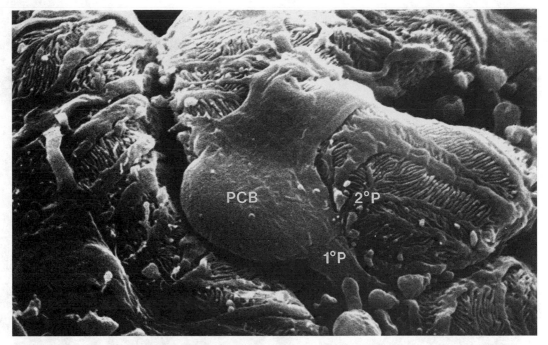

Figure 26-4. Scanning electron micrograph showing a podocyte cell body (*PCB*) with primary processes (*1°P*), secondary processes (*2°P*), and interdigitating foot processes (pedicles). (Reprinted from Johnson KE: *Histology: Microscopic Anatomy and Embryology.* New York, John Wiley, 1982, p 354.)

    (1) Epithelial cells in the distal convoluted tubules have fewer apical microvilli than cells in the proximal convoluted tubules and they have fewer mitochondria; consequently, their cytoplasm is less acidophilic.

    (2) The mitochondria often are arranged in basal rows, giving distal convoluted tubules a striated appearance similar to the appearance of proximal convoluted tubules.

  b. Distal convoluted tubules pass the renal corpuscle and join the collecting tubules. As they pass the renal corpuscle, they contribute to the **juxtaglomerular apparatus**.

D. **The juxtaglomerular apparatus** is involved in events that regulate the volume and composition (salt concentration, fluid balance) of the blood; consequently, it is involved in blood pressure regulation.

  1. **Microscopic anatomy.** The juxtaglomerular apparatus is located at the vascular pole of the renal corpuscle and is composed of three different groups of cells.

    a. **Macula densa.** As the distal convoluted tubules pass the renal corpuscle near its vascular pole, their cells become tall, thin, and columnar, forming a group of cells called the macula densa. These cells have a basal polarization and monitor the chemical composition of fluid in the distal convoluted tubules.

    b. **Juxtaglomerular cells** are modified smooth muscle cells. They are most common in the wall of the afferent arteriole but may be present in the wall of the efferent arteriole as well. These cells contain cytoplasmic granules rich in the enzyme **renin**.

    c. **Mesangial cells** are pericyte-like cells.

  2. **Function.** When the blood pressure falls, juxtaglomerular cells release renin. This causes an increase in blood pressure by way of a homeostatic mechanism that involves blood pressure and ionic composition, the juxtaglomerular apparatus, and hormone secretion by the adrenal cortex.

    a. Renin is an enzyme that cleaves a decapeptide called **angiotensin I** from the blood protein **angiotensinogen**.

    b. Another proteolytic enzyme (located in the lungs) converts angiotensin I into the octapeptide **angiotensin II**. This polypeptide profoundly affects the zona glomerulosa of the adrenal cortex, causing it to release the mineralocorticoid **aldosterone**.

    c. Aldosterone is a steroid that acts directly on the distal convoluted tubules, causing them to stimulate sodium ion reabsorption.

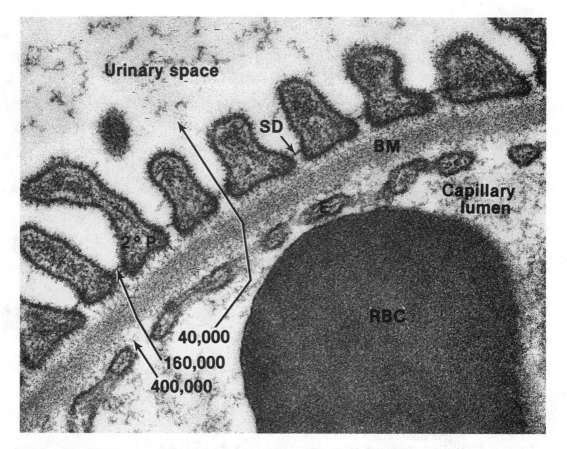

Figure 26-5. Electron micrograph of the glomerular basement membrane (*BM*). The capillary lumen is lined by a fenestrated endothelium (*E*), and the urinary space is lined by a complex barrier formed by podocyte secondary processes (*2°P*) and slit diaphragms (*SD*). Note that 400,000 molecular weight substances do not cross the glomerular basement membrane; 160,000 molecular weight substances do not cross the slit diaphragm; and that 40,000 molecular weight substances enter the glomerular filtrate. (Reprinted from Weiss L, Greep RO: *Histology.* New York, McGraw-Hill, 1977.)

   d. Angiotensin II production causes an increase in the volume of the extracellular fluid and is a potent vasoconstrictor.

III. HISTOLOGY OF THE URETERS AND PROXIMAL URINARY PASSAGES

   A. Transitional epithelium. Urine exiting the renal papillae passes through the minor calyces, the major calyces, the renal pelvis, and then into the ureters. The calyces, pelvis, and ureters are lined by **transitional epithelium,** a characteristic epithelium that exists only in the urinary system.

   1. Transitional epithelium is a stratified epithelium that rests on a basement membrane. The upper luminal cells have a pillowy appearance that distinguishes transitional epithelium from stratified squamous epithelium. The deeper cells are more uniform.

   2. The apical cells in transitional epithelium have an elaborately scalloped apical border and numerous **fusiform vesicles** in their cytoplasm.
      a. These fusiform vesicles and apical scallops may be reserves of surface membrane that can be added to the apical cell surface or removed from it as the bladder distends and empties.
      b. The plasma membranes of apical cells in the scallops and in the fusiform vesicles have a peculiar structure with intramembranous protein particles arranged in hexagonal arrays.
      c. These modifications have an important function in preventing damage to epithelial cells of the urinary passages as they are bathed by deleterious constituents of urine.

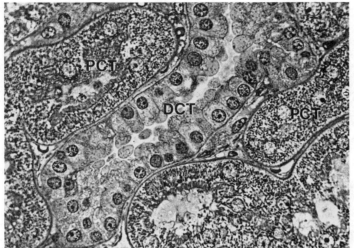

Figure 26-6. Light micrograph of the renal cortex showing a distal convoluted tubule (*DCT*) and several proximal convoluted tubules (*PCT*).

3. The transitional epithelium of the calyces, pelvis, and ureters is similar in appearance, although it increases in thickness from the calyces, through the pelvis and ureters, to the bladder.

4. Transitional epithelium varies morphologically according to the degree of distension of the urinary bladder.

B. Lamina propria. The lamina propria beneath the transitional epithelium contains collagen fibers and fibroblasts. No distinct submucosa exists in the tubular portions of the urinary system.

C. Muscularis externa

1. The prominent muscularis consists of many smooth muscle cells gathered into bundles and surrounded by connective tissue elements (Figure 26-8).

2. The calyces, pelvis, and upper two-thirds of the ureters have two poorly demarcated layers of smooth muscle fibers. The inner layer is predominantly oriented longitudinally; the outer layer is primarily circular.

3. The lower third of the ureters contains an outer third layer of smooth muscle fibers oriented longitudinally.

4. Peristaltic contraction of the smooth muscle propels urine toward the urinary bladder.

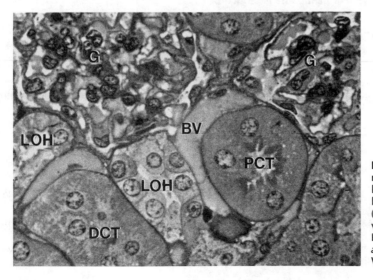

Figure 26-7. Light micrograph of the renal cortex, showing a distal convoluted tubule (*DCT*), proximal convoluted tubule (*PCT*), loop of Henle (*LOH*), glomerulus (*G*), and a blood vessel (*BV*). (Reprinted from Johnson KE: *Histology: Microscopic Anatomy and Embryology.* New York, John Wiley, 1982, p 355.)

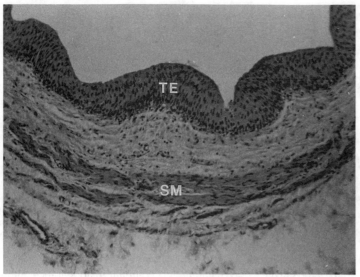

Figure 26-8. Light micrograph of the ureter, which is lined by transitional epithelium (*TE*) and contains smooth muscle (*SM*) in its wall. (Reprinted from Johnson KE: *Histology: Microscopic Anatomy and Embryology.* New York, John Wiley, 1982, p 359.)

5. The ureters enter the bladder at an angle as the inner longitudinal smooth muscle bundles mix with the muscularis of the bladder. This oblique insertion causes closure of the ureters when the bladder is distended, preventing urinary reflux toward the kidneys.

D. Adventitia. The calyces, renal pelvis, and ureters are surrounded by a thin adventitial connective tissue layer.

IV. THE URINARY BLADDER AND URETHRA

A. The urinary bladder

1. Transitional epithelium. The urinary bladder is lined by transitional epithelium that is similar to but thicker than the transitional epithelium in the upper urinary passages. This transitional epithelium also rests on a basement membrane.
  a. When the bladder is empty, the transitional epithelium is many cell layers thick. The epithelial cells, especially apical cells, are quite rounded.
  b. As the bladder fills, the rounded cells flatten and the number of cell layers between the lumen and the basement membrane is reduced.

2. Lamina propria. The lamina propria of the bladder is slightly folded when the bladder is empty. These folds disappear as the bladder fills.

3. Muscularis
  a. The muscularis has three different layers; however, due to the odd shape of the bladder, it is difficult to describe the orientation of the bundles of muscle fibers.
  b. A prominent circular bundle of smooth muscle tissue surrounds the urethral outlet, forming the internal sphincter of the bladder.

4. Adventitia. The upper portion of the bladder projects into the peritoneal cavity. Here, its adventitia is covered by a mesothelium. Elsewhere, the bladder has an adventitia but no mesothelium.

B. The male urethra is about 20 cm long. It has a prostatic portion (4 cm), a membranous portion (1 cm), and a penile portion (15 cm).

1. The **prostatic urethra** is lined by transitional epithelium. Numerous glands branch from it into the prostate gland.
  a. Ejaculatory ducts enter the prostatic urethra and convey gametes from the testes to the lower portion of the urethra.

    b. The prostatic urethra has a highly vascularized lamina propria under the transitional epithelium and an inner longitudinal and an outer circular layer of smooth muscle fibers. The circular layer of smooth muscle is continuous with the thick layer of muscle tissue that constitutes the internal sphincter of the bladder.

2. The **membranous urethra** penetrates the urogenital diaphragm and is lined by pseudostratified or stratified columnar epithelium. It is surrounded by the skeletal muscle fibers that form the external sphincter of the bladder.

3. The **penile urethra** passes through the corpus spongiosum of the penis. Along most of its length, it is lined by pseudostratified columnar epithelium; however, this lining is highly variable.

    a. Stratified columnar epithelium is present near the membranous urethra and stratified squamous epithelium is present toward the **fossa navicularis,** which is lined exclusively by stratified squamous epithelium.

    b. Ducts from the **bulbourethral glands** (Cowper's glands) enter the penile urethra. It is also associated with some small mucoid glands (glands of Littré).

C. The female urethra

1. The female urethra is about 4 cm long. Along most of its length, it is lined by a stratified squamous epithelium that is continuous with, and quite similar to, the vaginal epithelium.

2. The upper portion of the female urethra has either transitional epithelium or pseudostratified columnar epithelium.

STUDY QUESTIONS

Directions: Each question below contains five suggested answers. Choose the **one best** response to each question.

1. Each of the following statements concerning the kidneys is true EXCEPT

(A) renal pyramids empty into the major calyces
(B) papillary ducts empty into the minor calyces through the area cribrosa
(C) each major calyx connects to more than one minor calyx
(D) cortical tissue is present between pyramids in the renal columns
(E) some nephrons have medullary components

2. Each of the following statements concerning the ureters is true EXCEPT

(A) they are lined by transitional epithelium
(B) they have several layers of smooth muscle in the muscularis externa
(C) they have a thin serosal layer
(D) they enter the bladder at an oblique angle
(E) they are retroperitoneal

3. Each of the following statements concerning the renal calyces is true EXCEPT

(A) renal calyces are lined by pseudostratified epithelium
(B) renal calyces have smooth muscle in their walls
(C) renal pyramids empty into minor calyces
(D) minor calyces empty directly into the major calyces
(E) major calyces convey urine from the minor calyces to the renal pelvis

4. Each of the following statements concerning the juxtaglomerular apparatus is true EXCEPT

(A) mesangial cells are modified pericytes
(B) the wall of the distal convoluted tubule contains macula densa cells
(C) renin secretion decreases blood pressure
(D) the wall of the afferent arteriole contains juxtaglomerular cells
(E) juxtaglomerular cells contain granules of renin

Questions 5 and 6

The following two questions pertain to the micrograph below.

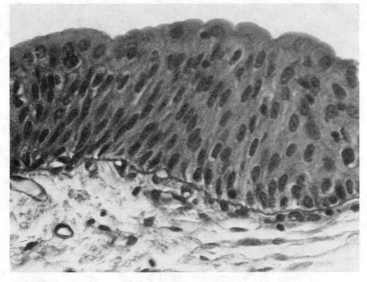

Reprinted from Johnson KE: *Histology: Microscopic Anatomy and Embry-ology.* New York, John Wiley, 1982, p 42.

5. Which of the following epithelia is pictured in the micrograph?

(A) Simple cuboidal epithelium
(B) Simple squamous epithelium
(C) Transitional epithelium
(D) Stratified squamous epithelium
(E) Pseudostratified epithelium

6. This epithelium would be present in each of the following locations EXCEPT

(A) collecting tubules
(B) minor calyces
(C) major calyces
(D) ureter
(E) urinary bladder

Directions: The groups of questions below consist of lettered choices followed by several numbered items. For each numbered item select the **one** lettered choice with which it is **most** closely associated. Each lettered choice may be used once, more than once, or not at all.

Questions 7–9

Match each description below with the most appropriate renal tubular structures.

(A) Proximal convoluted tubules
(B) Distal convoluted tubules
(C) Both
(D) Neither

7. Contain a few apical microvilli and a few mitochondria

8. Contain intensely acidophilic cells due to an abundance of mitochondria; have a prominent apical microvillous brush border

9. Lined by transitional epithelium

Questions 10–14

Match each description below with the appropriate lettered structure in the micrograph.

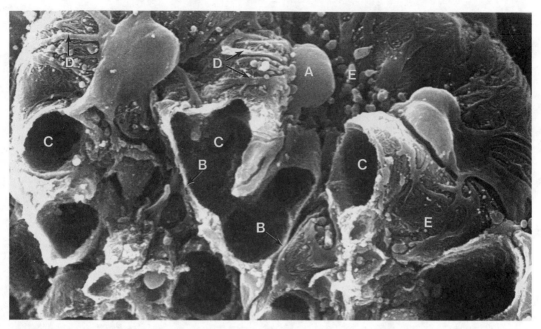

Micrograph courtesy of Dr. M. J. Koering, Department of Anatomy, George Washington University. Reprinted from Johnson KE: *Histology: Microscopic Anatomy and Embryology*. New York, John Wiley, 1982, p 364.

10. Vascular channel that carries red blood cells

11. Initial portion of the protein-sieving mechanism of the kidney

12. Podocyte primary process

13. Component characterized by transcellular fenestrations without diaphragms

14. An epithelial cell on the visceral portion of Bowman's capsule

ANSWERS AND EXPLANATIONS

1. The answer is A. [*II A, B*] Renal pyramids empty into minor calyces, which, in turn, join major calyces. Minor calyces and renal pyramids are far more numerous than major calyces. The papillary ducts drain into minor calyces at the area cribrosa. Juxtamedullary nephrons have long loops of Henle that reach into the renal medulla.

2. The answer is C. [*III A–D*] The ureters are long tubes that convey urine from the renal pelvis to the urinary bladder. The ureters enter the urinary bladder at an oblique angle, which serves as a kind of sphincter. They are lined by transitional epithelium, have several layers of smooth muscle in their walls, and are coated by a thin adventitial layer. Ureters are retroperitoneal; therefore, the adventitial layer has no mesothelial coating and is not correctly designated a serosal covering.

3. The answer is A. [*III A*] Renal calyces are lined by transitional epithelium, not pseudostratified epithelium, and have some smooth muscle in their walls. Minor calyces receive papillary ducts draining renal pyramids. Minor calyces empty into the major calyces, which then empty into the renal pelvis.

4. The answer is C. [*II D 2*] The juxtaglomerular apparatus helps regulate blood pressure and blood composition. Renin is a proteolytic enzyme secreted by juxtaglomerular cells in the wall of the afferent arteriole in the juxtaglomerular apparatus. Renin secretion increases blood pressure. Renin cleaves angiotensinogen into the smaller polypeptide angiotensin I, which is subsequently cleaved into the still smaller polypeptide angiotensin II. Angiotensin II causes the secretion of aldosterone, an adrenocortical steroid that stimulates sodium ion reabsorption. Macula densa cells are present in the wall of the distal convoluted tubule. Mesangial cells are pericyte-like cells and are part of the juxtaglomerular apparatus.

5 and 6. The answers are: 5-C, 6-A. [*II B 2; III A*] This is a micrograph of transitional epithelium. The apical cells have the pillowy appearance characteristic of transitional epithelium. This type of epithelium is restricted to the urinary passages in the minor and major calyces, renal pelvis, ureters, urinary bladder, and upper portion of the urethra. The collecting tubules and papillary ducts are lined by simple columnar epithelium.

7–9. The answers are: 7-B, 8-A, 9-D. [*II C 3, 5*] The proximal convoluted tubule attaches to Bowman's capsule at the urinary pole. It is lined by a simple cuboidal epithelium. Epithelial cells in the proximal convoluted tubule are strongly acidophilic because they contain many mitochondria. They also have a dense apical brush border of microvilli.

The distal convoluted tubule attaches the ascending loop of Henle to collecting tubules. It is lined by a simple cuboidal epithelium, with fewer apical microvilli and fewer mitochondria than are present in the proximal convoluted tubule. Transitional epithelium exists only in the urinary passages between the minor calyces and the upper part of the urethra.

10–14. The answers are: 10-C, 11-B, 12-D, 13-C, 14-A. [*II C 1, 2*] This is a scanning electron micrograph of a fractured renal glomerulus. Podocytes (*A*) have numerous primary processes (*D*), which are interdigitated in a complex fashion. Blood flows through the capillaries of the renal glomerulus (*C*), and a filtrate of blood passes through the glomerular basement membrane (*B*) into the urinary space (*E*) of Bowman's capsule.

I. INTRODUCTION

A. Components of the eye (Figure 27-1)

1. The **cornea** is the transparent anterior portion of the capsule of the eye. Light passes through the cornea and then through a variable diaphragm called the **iris**.

2. The **pupil** is the aperture of the iris. Contraction of smooth muscle fibers automatically changes the aperture in reaction to the level of light in the environment, in the subject of inspection, or both.

3. After passing through the anterior chamber and the iris, light passes through the **lens,** traverses the **vitreous body** (vitreous humor), and strikes the **retina**. Here, light evokes an action potential in cells that synapse with the **cones** and **rods** (photoreceptors).

4. The **optic nerve** carries nerve impulses to the brain.

B. Functions of the eye. The eye receives a variety of visual stimuli that can have a significant effect on human life and behavior. It is an elegantly constructed transducer that converts light into electrical impulses and transmits these electrical impulses to the brain for processing. The eye also has a mechanism that forms images.

C. Components of the ear (Figure 27-2)

1. The **external ear** is composed of the **auricle** and the **external auditory meatus**.

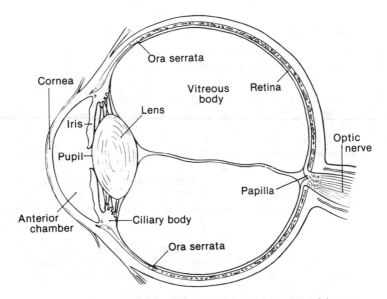

Figure 27-1. Illustration of the gross relationships of the parts of the eye.

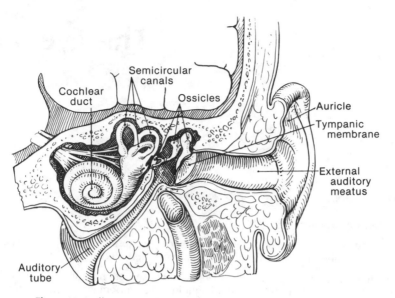

Figure 27-2. Illustration of the gross relationships of the parts of the ear.

2. The **middle ear** contains the **auditory ossicles** and connects to the **auditory tube,** which, in turn, connects to the nasopharynx.

3. The **tympanic membrane** separates the external ear from the middle ear.

4. The **internal ear** contains the **cochlea,** which contains the **organ of Corti** (a specialized mechanoreceptor formed by a highly modified epithelium). The internal ear also contains a **vestibular apparatus** with several types of neuroepithelial cells.

D. Functions of the ear. The ear contains complex mechanoreceptors. Motion of these mechanoreceptors is converted into electrical impulses, which are transmitted to the brain. The brain interprets the impulses as sound, a sense of the body's position in space, or a motion of the head.

II. THE EYE

A. The sclera and cornea

1. The **sclera** is a tough, fibrous layer of connective tissue that forms in the outer layer of the eye. Tendons of the oculomotor muscles are inserted into the sclera. This connective tissue is rich in randomly arranged collagenous fibers, fibroblasts, and elastic fibers.

2. The **cornea** is a highly modified, clear anterior portion of the outer capsule of the eye. It is slightly thicker than the sclera and has a smaller radius of curvature. The cornea's high refractive index (1.38) and small radius of curvature make it extremely important in image formation.
   a. Corneal layers
   (1) The **corneal epithelium** (Figure 27-3) is a stratified squamous epithelium, 5–7 cell layers thick.
      (a) Corneal epithelium contains several free nerve endings that, when stimulated, cause the blinking reflex (i.e., the eyelids close, protecting the cornea).
      (b) Corneal epithelium also has a remarkable capacity to heal and regenerate. Minor wounds are closed by cell migration; larger wounds are healed by mitosis in the basal cell layers and by the production of new cells.
   (2) The corneal epithelium rests on **Bowman's membrane,** a clear, acellular layer that is 6–9 mm thick. Bowman's membrane is not a true membrane but a composite of the basement membrane of the corneal epithelium and the outer layer of the underlying substantia propria.
   (3) The **substantia propria** (stroma) is the thickest corneal layer, comprising about 90% of the corneal thickness.

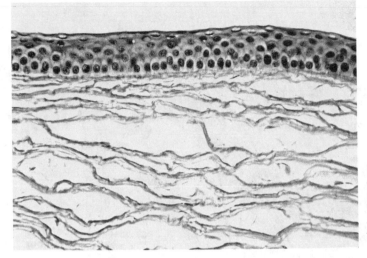

Figure 27-3. Light micrograph of the corneal epithelium. The underlying stroma is abnormally distended as a fixation artifact. (Reprinted from Johnson KE: *Histology: Microscopic Anatomy and Embryology.* New York, John Wiley, 1982, p 380.)

(a) The substantia propria is a mixture of fibroblastic cells, collagen fibers, and an amorphous ground substance rich in the glycosaminoglycans chondroitin sulfate and keratan sulfate.

(b) This layer does not contain blood vessels.

(4) **Descemet's membrane** is an extremely thick basement membrane containing a peculiar kind of hexagonally arranged collagen, which is especially prominent in older people. Descemet's membrane probably is the basement membrane of, and is secreted by, the corneal endothelium.

(5) **Corneal endothelium** is a simple squamous epithelium.

   b. **Corneal transparency** is due to the regular arrangement of its fibrous elements. The fibers probably are kept in close register by glycosaminoglycans in the cornea. The embryonic cornea is not transparent because stromal fibroblasts have yet to secrete glycosaminoglycans and accumulate them around the collagen fibers.

   c. **Corneal blood supply.** The cornea is avascular. The center of the cornea receives nutrition by diffusion from the aqueous humor; the periphery receives nutrition from blood vessels in the limbus (corneal-scleral junction).

3. The **limbus** (Figure 27-4). The cornea and the sclera are continuous structures and are part of the outer layer of the eye. They join at the limbus. The anatomy of this region is complex, and the area is quite significant clinically.

   a. Structures that regulate the outflow of aqueous humor from the eye are located here. Failure to regulate the outflow of aqueous humor can cause an increase in the intraocular pressure, which is a characteristic feature of a severe common disease called **glaucoma**.

   b. A cavity containing the trabecular meshwork exists at the junction of the cornea and iris. This complex network of anastomosing connective tissue cords is covered by an endothelium that is continuous with the corneal endothelium.

   c. The **canal of Schlemm** is deep to the trabecular meshwork. These two structures comprise an apparatus that regulates the outflow of aqueous humor from the anterior chamber.

B. The vascular layer of the eye

1. The **choroid layer** is a thin pigmented layer just beneath the sclera. It is composed of three layers: an outer layer of arterioles and venules, a middle capillary layer, and an inner layer called **Bruch's membrane**.

2. The **ciliary body** occupies the space between the edge of the visual retina (**ora serrata** or **ora terminalis**) and the edge of the lens.

   a. Numerous long, radially arranged ciliary processes project from the thick ciliary body toward the lens.

   b. The main mass of the ciliary body, exclusive of the ciliary processes, is composed of muscles of accommodation called the **ciliary muscles**.

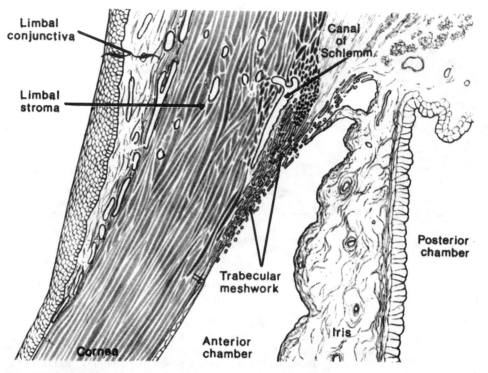

Figure 27-4. Diagram of the limbus of the human eye. (Reprinted from Hogan MJ, Alvarado JA, Weddell JE: *Histology of the Human Eye.* Philadelphia, WB Saunders, 1971.)

      (1) Contraction of the ciliary muscles reduces tension on the ligaments attached to the lens, and the lens becomes more convex. This allows the lens to focus images of nearby objects on the retina.

      (2) To remember this mechanism, recall that eye fatigue during prolonged periods of reading results from smooth muscle contraction.

   c. The surface facing the central cavity of the eye is covered by a heavily pigmented layer of the retina. This layer lacks photoreceptors and extends anteriorly from the ora serrata as the **ciliary epithelium**.

      (1) Ciliary epithelium covers the ciliary body with two layers: an outer layer of unpigmented cells and a deep layer of pigmented cells. Both layers rest on a basement membrane.

      (2) Ciliary epithelium in the posterior chamber secretes the aqueous humor. Aqueous humor nourishes the lens and other inner structures of the eye and then flows over the lens, through the pupil, through the trabecular meshwork into the anterior chamber, and then through the canal of Schlemm.

3. The iris

   a. The main mass of the iris consists of a loose, highly pigmented connective tissue mass containing many blood vessels.

      (1) The iris has a ciliary margin at the ciliary body and a pupillary margin at the pupil.

      (2) The posterior surface of the iris rests on the lens.

   b. The iris and lens form the boundary between the anterior and posterior chambers of the eye.

   c. The iris has color. The hue of color depends on the quantity and arrangement of melanin and the thickness of the connective tissue lamellae.

   d. The **pupil** (the iris aperture) varies in size with the amount of light in the environment.

      (1) Cells derived from the anterior pigmented layer of the iris lose pigment and differentiate into contractile cells that form the myoepithelium of the pupillary muscles.

      (2) Two groups of myoepithelial cells exist. One group is arranged concentrically around the pupil. Contraction of these cells reduces the aperture of the iris. The second group is arranged radially around the pupil. Contraction of these cells increases the aperture of the iris.

C. Refractive media. The refractive apparatus of the eye includes the cornea and anterior chamber of the eye (see II A 2) and the lens and vitreous humor.

1. The **lens** is a transparent biconvex structure located immediately posterior to the pupil. It is approximately 7 mm in diameter and 3.7–4.5 mm thick, depending on the degree of contraction or relaxation of ciliary muscles.

  a. The lens consists of lens fibers, which are highly modified epithelial cells.

  (1) Each lens fiber is a hexagonal prismatic structure that is 8–12 μm wide, about 2 μm thick, and 7–10 mm long. Each fiber contains a nucleus but is relatively devoid of organelles and intracellular inclusions.

  (2) The plasma membranes of adjacent lens fibers are separated by 15-mm wide intercellular clefts and are elaborately interdigitated.

  b. The lens is held in position by a series of fibers called the **ciliary zonule**. These fibers arise from the epithelium of the ciliary body. In the electron microscope, they appear as small hollow tubular structures that differ from microtubules.

2. The **vitreous body** (vitreous humor) is 99% water and fills the cavity between the lens and the retina.

  a. Most of the nonaqueous portion is composed of hyaluronic acid.

  b. The vitreous body contains a random network of collagen fibers that lack the usual collagen periodicity. The vitreous body also contains hyalocytes and has high concentrations of collagen and hyaluronate at its periphery. The hyalocytes may secrete the collagen and hyaluronate.

  c. The remnant of the embryonic hyaloid artery forms the hyaloid canal through the middle of the vitreous body.

D. The retina has two main components: a pigmented retina next to the choroid layer and a neural retina next to the vitreous body.

1. The **pigmented retina** is a simple layer of cells rich in melanin granules. Its epithelium has an extensive network of tight junctions between cells.

  a. This tightly joined layer of cells probably is a barrier between the blood and the neural retina.

  b. The pigmentation prevents extraneous light from stimulating the retina and absorbs light after it passes through the neural retina.

  c. The basement membrane of the pigmented retina is part of Bruch's membrane.

2. The **neural retina** (Figure 27-5) consists of many layers and two types of photoreceptor neurons called rods and cones.

  a. Rods and cones

  (1) Rods are long and thin and connect to numerous other cells. They are responsible for night vision, with their peak function occurring in low levels of illumination.

  (2) Cones are shorter and fatter than rods, and they connect to fewer other cells. They are responsible for visual discrimination and color vision.

  (3) Common features

  (a) Both rods and cones have an outer segment consisting of multiple layers of stacked plasma membranes. This segment is rich in the photoreceptive substances **rhodopsin** (in rods) and **iodopsin** (in cones).

  (b) Ellipsoids are inner segments of each rod and cone that contain a concentration of mitochondria.

  b. Layers of the neural retina

  (1) The **outer nuclear layer** is the thickest retinal layer and is composed of the cell bodies and nuclei of the rods and cones. Müller's cells separate individual rods and cones, form contacts with them at the **outer limiting membrane** (not a true membrane, but a layer where numerous cells contact each other), and then extend radially between the ellipsoids.

  (2) The **outer plexiform layer** and the **synaptic layer** are composed of nerve processes and synapses between rods, cones, and bipolar cells.

  (3) The **inner nuclear layer** is a layer of bipolar cells that connects rods and cones to a ganglion cell layer. This layer contains the nuclei of the Müller's cells, horizontal cells, amacrine cells (large unipolar cells), and a few ganglion cells. The inner nuclear layer integrates the function of rods and cones and performs summation of rod impulses.

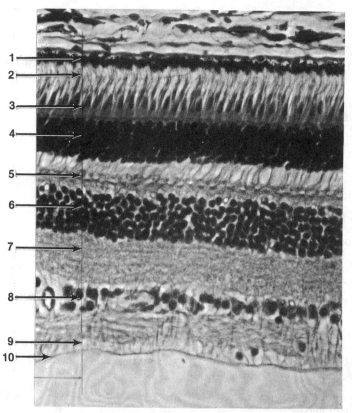

Figure 27-5. Light micrograph showing the ten layers of the retina. They are: (*1*) pigmented epithelium, (*2*) rod and cone outer segments, (*3*) outer limiting membrane, (*4*) outer nuclear layer, (*5*) outer plexiform layer, (*6*) inner nuclear layer, (*7*) inner plexiform layer, (*8*) ganglion cell layer, (*9*) optic nerve fiber layer, and (*10*) inner limiting membrane. (Reprinted from Johnson KE: *Histology: Microscopic Anatomy and Embryology.* New York, John Wiley, 1982, p 385.)

   (4) The **inner plexiform layer** contains synapses between bipolar cells and ganglion cells, as well as a few glial cells and small blood vessels.

   (5) The **ganglion cell layer** is the most anterior layer of cells in the retina. Dendrites of cells in this layer synapse with bipolar cells, and their axons are bundled to form the optic nerve.

   (6) The **optic nerve fiber layer** is composed of the many axons of the ganglion cells.

   (7) The **inner limiting membrane** is a basement membrane and the most anterior layer of the retina.

  3. The **fovea centralis** is the region of the retina that is responsible for maximal visual acuity.

    a. Although the cellular composition of the retina varies considerably from one region to another, most regions of the light-sensitive portions of the retina contain a preponderance of rods. The fovea centralis (macula lutea), however, contains no rods; instead, it has a high concentration of tall thin modified cones.

    b. Cones in the fovea synapse one-to-one with bipolar cells. Other retinal layers are pushed away from the fovea to minimize structural interference with entering light rays.

III. ACCESSORY ORGANS OF THE EYE.
The conjunctiva, eyelids, meibomian glands, and lacrimal glands are accessory organs that protect the eyes.

A. Conjunctiva

  1. Structure

    a. The anterior surface of the eyeball (the cornea) is covered by a stratified squamous epithelium that is reflected onto the posterior surface of the eyelid as the **conjunctiva**.

    b. The conjunctiva and the corneal epithelium are continuous at the conjunctival fornices.

    c. Apical cells in the conjunctiva are tall conical or cylindrical cells; basal cells are cuboidal.

    d. As the conjunctiva passes over the margin of the eyelid, the conjunctival mucous membrane gives way to epidermis.

2. Function. The conjunctiva helps clean the surface of the cornea. Mucous secretions also moisten the conjunctiva and cornea and help prevent the evaporation of tears.

B. Eyelids

1. Structure

   a. The inside of the eyelids is coated by conjunctival epithelium; the outside is coated by thin epidermal epithelium.

   b. The eyelids contain numerous coarse hairs (eyelashes), highly modified sebaceous glands (the tarsal or meibomian glands), and sweat glands with straight ducts (Moll's glands).

2. Function.

The eyelids protect and clean the cornea. Oily secretions from the meibomian glands mix with mucus from the conjunctival glands to form a thin film on tears, which helps prevent the evaporation of tears.

C. Lacrimal glands

1. Structure.

The lacrimal glands are situated in the superior lateral portion of the orbit. Each gland is composed of many serous acini that drain into the orbit through 10–20 ducts.

2. Function.

The lacrimal glands produce tears, which cleanse and lubricate the cornea.

IV. THE EAR

A. The external ear

1.

The **auricle** (pinna) contains abundant elastic cartilage and is covered by skin with hairs, sebaceous glands, and a few eccrine sweat glands. The elastic cartilage extends partway down toward the external auditory meatus, which penetrates the temporal bone.

2. External auditory meatus.

The epithelium that lines the meatus is similar to the skin of the pinna but also contains especially large sebaceous glands. The meatus has a highly modified type of apocrine sweat glands called **ceruminous glands**.

   a. Secretions from ceruminous glands are expressed by myoepithelial cells.

   b. Ducts in ceruminous glands are quite large and may enter hair follicles with sebaceous glands or empty directly onto the surface of the external auditory meatus.

   c. Cerumen repels insects and other vermin entering the ear.

B. The middle ear

1. Tympanic cavity

   a. The tympanic cavity is an irregular cavity in the temporal bone. Its lateral boundary is the tympanic membrane, and its medial boundary is the bony wall of the inner ear. Anteriorly, it is continuous with the auditory tube.

   b. The tympanic cavity is lined by a simple squamous epithelium, except near the tympanic membrane and the beginning of the auditory tube, where it is lined by a cuboidal epithelium or columnar epithelium with cilia.

2.

The **ossicles,** which include the **malleus, incus,** and **stapes,** are supported by miniature ligaments and are covered by reflections of the simple squamous epithelium that lines the middle ear.

3. Tympanic membrane

   a. The lateral side of the tympanic membrane is covered by a thin layer of the stratified squamous epithelium of the external auditory meatus (without glands and hairs); its medial side is covered by a thin simple squamous epithelium.

   b. **Shrapnell's membrane** is a thin connective tissue domain between the epithelial layers, which has a layer of radially arranged collagenous fibers and a layer of circularly arranged fibers, elastic fibers, and fibroblasts. The anterosuperior quadrant does not contain connective tissue.

4. Auditory tube

   a. The auditory tube courses anteromedially to an opening in the nasopharynx. The portion near the middle ear is surrounded by the temporal bone; the portion near the nasopharynx is partially surrounded by a spiral of elastic cartilage.

b. The mucosa has a low ciliated columnar epithelium in the temporal bone. Near the nasopharynx, this gives way to a pseudostratified layer with tall columnar ciliated cells, some goblet cells, and mucous glands.

c. The lamina propria in the medial portion is thicker and may be extensively infiltrated with lymphocytes, which may form discrete nodules called **tubal tonsils** (Gerlach's tonsils).

d. The auditory tube usually is closed, but the pharyngeal orifice becomes patent during yawning and swallowing, thereby equalizing the pressure differential between the tympanic cavity and the environment outside the ear.

C. The internal ear (Figure 27-6) occupies a complex cavity in the petrous portion of the temporal bone called the **osseous labyrinth,** which is divided into two major cavities called the **vestibule** and the **cochlea**. The vestibule contains the saccule, utricle, and three semicircular canals. The cochlea contains the **organ of Corti**.

1. Cochlea

a. The cochlea is a spiral-shaped cavity with an axis formed by a pillar of bone called the **modiolus**. The cochlea has a broad base and tapers up, like a cone.

b. The broad base of the modiolus opens into the cranial cavity at the internal acoustic meatus.

(1) Here, afferent nerve processes from the cochlear division of the eighth cranial nerve pass through numerous tiny openings.

(2) The cell bodies of these nerve processes are arranged together in the spiral ganglion. Dendrites of these cells innervate the hair cells of the internal ear, and axons project into the central nervous system.

c. The cochlea is divided into two chambers by a bony shelf called the **osseous spiral lamina** and a membrane called the **basilar membrane** (membranous spiral lamina).

d. The cochlea is subdivided further by the **vestibular membrane** (Reissner's membrane), which extends between the spiral lamina and the wall of the cochlea.

e. The cochlea has three compartments: the upper passage is called the **scala vestibuli,** the lower passage is called the **scala tympani,** and the intermediate passage is called the **cochlear duct** (scala media).

(1) The **ductus reuniens** connects the cochlear duct to the vestibular apparatus.

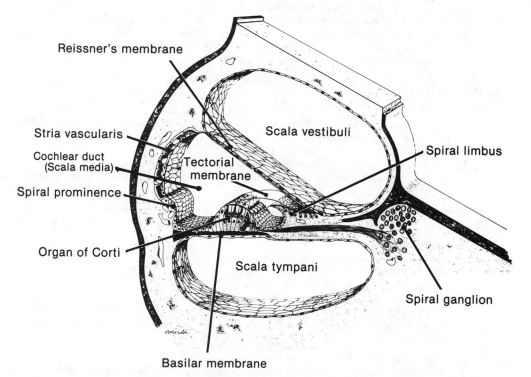

Figure 27-6. Diagram of the internal ear. (Reprinted from Bloom W, Fawcett DW: *A Textbook of Histology,* 10th ed. Philadelphia, WB Saunders, 1975, p 976.)

(2) The scala vestibuli and scala tympani are perilymphatic spaces. The former ends at the fenestra ovalis and the latter ends at the fenestra rotunda. The scala vestibuli and scala tympani connect at the apex of the cochlear duct in a small opening called the **helicotrema**.

2. Vestibular apparatus
  a. Utricle and saccule
   (1) The utricle connects to three semicircular canals, which are filled with endolymph and contain three dilations called **ampullae**.
     (a) Each ampulla has a patch of neuroepithelium called the **crista**, which is composed of hair cells.
     (b) The cristae ampullares have an extracellular material called the **cupula**, which rests on hair cells.
   (2) Both the utricle and saccule have a macula of neuroepithelium.
     (a) The macula utriculi and macula sacculi have an extracellular otolithic membrane with embedded (calcified) otoliths (Figure 27-7).
     (b) Hair cells in the maculae are stimulated by movements of otoliths and otolithic membrane which, in turn, are generated by movements of the head.
  b. The **endolymphatic sac** has a columnar epithelium that contains some cells with microvilli and numerous apical pinocytotic vesicles. Macrophages and neutrophils cross the epithelium with ease. Endolymphatic fluid and cellular debris of the endolymph probably are removed at the endolymphatic sac.

3. The **cochlear duct** is a highly specialized diverticulum of the saccule, which contains the organ of Corti. It probably is more histologically complex than any other area in the body.
  a. The vestibular membrane has two back-to-back layers of flattened cells. One layer of cells faces and lines the scala vestibuli. The other layer faces and lines the roof of the cochlear duct.
  b. On the outer wall of the cochlear duct, the inner layer of cells of the vestibular membrane becomes continuous with a stratified epithelium called the **stria vascularis**.
   (1) Basal cells in the stria vascularis have deep basal infoldings and numerous mitochondria.
   (2) The stria vascularis is involved in the maintenance of the unusual ionic composition of the endolymph.
  c. The epithelium of the cochlear duct reflects from the stria vascularis onto the basilar membrane. Cells of Claudius and cells of Böttcher are present in the epithelium before it reaches the organ of Corti.

4. The **organ of Corti** (Figure 27-8) is composed of many different types of cells. It has six kinds of supporting cells and two kinds of hair cells.
  a. Supporting cells
   (1) The supporting cells are tall and slender and contain conspicuous tonofibrils. Their apical surfaces contact each other, hair cells, or both to form a continuous surface called the **reticular membrane**.

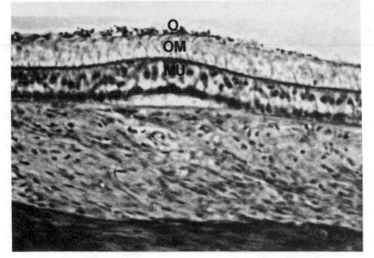

Figure 27-7. Light micrograph of the macula utriculi (*MU*), with its otolithic membrane (*OM*) and otoliths (*O*). (Reprinted from Johnson KE: *Histology: Microscopic Anatomy and Embryology.* New York, John Wiley, 1982, p 397.)

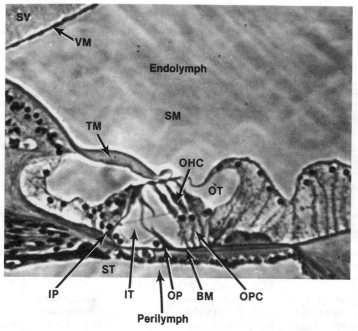

Figure 27-8. Light micrograph of the organ of Corti. This complex structure rests on a basilar membrane (*BM*) in the scala media (*SM*) between the scala vestibuli (*SV*) and the scala tympani (*ST*). The tectorial membrane (*TM*) contacts outer hair cells (*OHC*) and outer phalangeal cells (*OPC*). The outer tunnel (*OT*) and inner tunnel (*IT*) also are visible. Inner pillar (*IP*) and outer pillar (*OP*) cells border the inner tunnel. The vestibular membrane (*VM*) is the boundary between the scala vestibuli and the scala media. (Reprinted from Johnson KE: *Histology: Microscopic Anatomy and Embryology.* New York, John Wiley, 1982, p 398.)

(2) The inner tunnel in the middle of the organ of Corti rests on the basilar membrane. The floor and wall of the inner tunnel are composed of inner pillar cells, which contact inner hair cells, and outer pillar cells, which contact outer hair cells.

(3) The inner border of the organ of Corti is made of **border cells;** the outer border is made of **cells of Hensen**.

(4) **Phalangeal cells** rest on the basilar membrane. Each phalangeal cell has a cup-like indentation in its apical surface. The inferior third of a hair cell rests in this depression.

  (a) Tunnels containing afferent and efferent nerve processes pass through phalangeal cells and synapse with the hair cells.

  (b) The apical portion of phalangeal cells expand into umbrella-like phalangeal processes, which contact the apical portions of the hair cells.

b. Hair cells

(1) Both inner hair cells and outer hair cells are associated with phalangeal cells and are contacted by afferent and efferent nerve endings.

(2) Each inner and outer hair cell has long apical microvilli called **stereocilia** and an apical centriole.

  (a) Outer hair cells have approximately 100 stereocilia arranged in a W pattern.

  (b) Inner hair cells have fewer stereocilia that are arranged in a straight line.

c. Tectorial membrane

(1) In the inner angle of the scala media, the connective tissue covering the osseous spiral lamina forms a crest called the **spiral limbus**.

(2) The epithelium covering the spiral limbus secretes the tectorial membrane, which projects away from the spiral limbus into the scala media.

(3) The tips of the hair cells are embedded in the tectorial membrane.

D. Vibrations in the basilar membrane

1. Perilymph

  a. The vibrations of the tympanic membrane are transmitted through the auditory ossicles to the fenestra ovalis and, thus, to the perilymph of the scala tympani. Vibrations in the perilymph cause vibrations in the basilar membrane.

  b. The basilar membrane contains about 20,000 basilar fibers. These fibers project away from the bony modiolus and have free ends that vibrate like reeds in a mouth organ.

2. The **basilar fibers** near the base of the cochlea are blunt and short. Near the apex of the cochlea (the helicotrema), basilar fibers are longer and more slender.

    a. The short fat fibers vibrate in resonance with high frequency sounds; the long thin fibers vibrate in resonance with low frequency sounds.

    b. The difference in the mechanical properties of the basilar fibers have an interesting result. Movements in the perilymph cause a wave in the basilar membrane that is dampened where basilar fibers vibrate in resonance.

        (1) The maximum amplitude of the displacement of the basilar membrane varies with the frequency of the sound stimulus: high frequency sounds cause a maximum displacement of the basilar membrane near the base of the cochlea; low frequency sounds cause a maximum displacement of the basilar membrane furthest from the base of the cochlea, near the helicotrema.

        (2) The sensation of different sounds results from the different ways the sounds stimulate hair cells. It should be noted that the tips of the hair cells are in contact with the tectorial membrane, which is fixed with respect to the moving hair cells.

STUDY QUESTIONS

Directions: Each question below contains five suggested answers. Choose the **one best** response to each question.

1. Each of the following statements concerning corneal layers is true EXCEPT

(A) the epithelium is stratified squamous
(B) the endothelium has a basement membrane
(C) the stroma is acellular
(D) the stroma contains collagen
(E) Bowman's membrane is anterior to Descemet's membrane

2. Which of the following statements best describes the functional histology of the sclera?

(A) It is covered by an epithelium
(B) It contains more blood vessels than the choroid layer
(C) It is richly pigmented
(D) It has few fibroblasts and collagen fibers
(E) Its anterior component is modified to form the cornea

Directions: The groups of questions below consist of lettered choices followed by several numbered items. For each numbered item select the **one** lettered choice with which it is **most** closely associated. Each lettered choice may be used once, more than once, or not at all.

Questions 3–10

Match each description below with the most appropriate membrane.

(A) Reticular membrane
(B) Tectorial membrane
(C) Basilar membrane
(D) Vestibular membrane
(E) Tympanic membrane

3. Secreted from spiral limbus and makes contact with tips of hair cells

4. Boundary between scala media and scala tympani

5. Boundary between scala media and scala vestibuli

6. Boundary between external ear and middle ear

7. Conducts traveling wave of deformation

8. Formed from phalangeal cell processes and apices of hair cells

9. Covered by stratified squamous epithelium on lateral surface

10. Remains nearly stationary when stapes moves

Questions 11–16

Match each description of a retinal layer with the letter of the appropriate layer in the micrograph below.

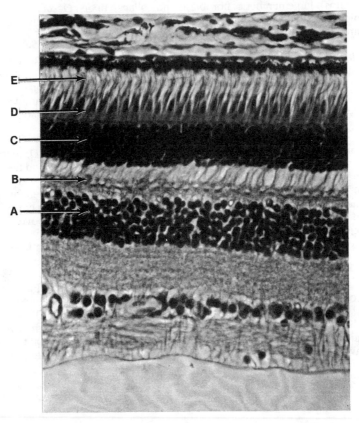

Reprinted from Johnson KE: *Histology: Microscopic Anatomy and Embryology.* New York, John Wiley, 1982, p 385.

11. Contains axons of rods and cones

12. Contains nuclei of rods and cones

13. Contains outer segments of rods and cones

14. Contains bipolar cell bodies

15. Receives light before all other (lettered) layers

16. Caused by rows of contacts between Müller's cells and photoreceptors

ANSWERS AND EXPLANATIONS

1. The answer is C. [*II A 2 a*] The first of the five corneal layers is a stratified squamous epithelium. The epithelium rests on the second layer, Bowman's membrane. The third corneal layer, the stroma, has no blood supply but contains a large number of fibroblasts and collagen fibers. Descemet's membrane, the fourth layer, is a thick basement membrane. The final corneal layer, the endothelium, is a simple squamous epithelium.

2. The answer is E. [*II A 1*] The sclera is the outermost coating of the eye. It is a dense, connective tissue layer rich in fibroblasts and collagen fibers, but it does not have extensive pigmentation. The oculomotor muscles are inserted into the sclera. It is opaque in most areas, but its anterior component is the transparent cornea. The anterior surface of the cornea is covered by a stratified squamous unkeratinized epithelium. No epithelium covers the rest of the sclera. The choroid layer has many more capillaries than the sclera.

3–10. The answers are: 3-B, 4-C, 5-D, 6-E, 7-C, 8-A, 9-E, 10-B. [*I C 3; IV B, C, D; Figure 27-8*] The reticular membrane is a planar structure formed by the phalangeal cell processes and the apices of hair cells. The tectorial membrane contacts the tips of hair cells and remains essentially motionless because it is fixed to the skull. The basilar, vestibular, and tympanic membranes all move when the stapes moves. The basilar membrane is part of the boundary between the scala media and the scala tympani. It contains the basilar fibers and conducts a traveling wave of deformation when the stapes moves. The vestibular membrane is the boundary between the scala media and the scala vestibuli. The tympanic membrane is the boundary between the external auditory meatus and the middle ear. On its lateral surface, the tympanic membrane is lined by a thin stratified squamous epithelium.

11–16. The answers are: 11-B, 12-C, 13-E, 14-A, 15-A, 16-D. [*II D 2; Figure 27-5*] In this micrograph, (A) is the layer that contains the outer segments of rods and cones. Contacts between Müller's cells and photoreceptors occur at the outer limiting membrane (B). The outer nuclear layer (C) contains the nuclei of rods and cones. The axons of rods and cones are contained in the outer plexiform layer (D). Bipolar cell bodies are contained in the inner nuclear layer (E). Light first strikes the retina at the bottom of this micrograph and travels through all the retinal layers before striking the outer segments of the photoreceptors.

Challenge
Exam

Introduction

One of the least attractive aspects of pursuing an education is the necessity of being examined on what has been learned. Instructors do not like to prepare tests, and students do not like to take them.

However, students are required to take many examinations during their learning careers, and little if any time is spent acquainting them with the positive aspects of tests and with systematic and successful methods for approaching them. Students perceive tests as punitive and sometimes feel that they are merely opportunities for the instructor to discover what the student has forgotten or has never learned. Students need to view tests as opportunities to display their knowledge and to use them as tools for developing prescriptions for further study and learning.

A brief history and discussion of the National Board of Medical Examiners (NBME) examinations are presented in this introduction, along with ideas concerning psychological preparation for the examinations. Also presented are general considerations and test-taking tips as well as how practice exams can be used as educational tools. (The literature provided by the various examination boards contains detailed information concerning the construction and scoring of specific exams.)

National Board of Medical Examiners Examinations

Before the various NBME exams were developed, each state attempted to license physicians through its own procedures. Differences between the quality and testing procedures of the various state examinations resulted in the refusal of some states to recognize the licensure of physicians licensed in other states. This made it difficult for physicians to move freely from one state to another and produced an uneven quality of medical care in the United States.

To remedy this situation, the various state medical boards decided they would be better served if an outside agency prepared standard exams to be given in all states, allowing each state to meet its own needs and have a common standard by which to judge the educational preparation of individuals applying for licensure.

One misconception concerning these outside agencies is that they are licensing authorities. This is not the case; they are examination boards only. The individual states retain the power to grant and revoke licenses. The examination boards are charged with designing and scoring valid and reliable tests. They are primarily concerned with providing the states with feedback on how examinees have performed and with making suggestions about the interpretation and usefulness of scores. The states use this information as partial fulfillment of qualifications upon which they grant licenses.

Students should remember that these exams are administered nationwide and, although the general medical information is similar, educational methodologies and faculty areas of expertise

The author of this introduction, Michael J. O'Donnell, holds the positions of Assistant Professor of Psychiatry and Director of Biomedical Communications at the University of New Mexico School of Medicine, Albuquerque, New Mexico.

differ from institution to institution. It is unrealistic to expect that students will know all the material presented in the exams; they may face questions on the exams in areas that were only superficially covered in their classes. The testing authorities recognize this situation, and their scoring procedures take it into account.

The Exams

The first exam was given in 1916. It was a combination of written, oral, and laboratory tests, and it was administered over a 5-day period. Admission to the exam required proof of completion of medical education and 1 year of internship.

In 1922, the examination was changed to a new format and was divided into three parts. Part I, a 3-day essay exam, was given in the basic sciences after 2 years of medical school. Part II, a 2-day exam, was administered shortly before or after graduation, and Part III was taken at the end of the first postgraduate year. To pass both Part I and Part II, a score equalling 75% of the total points available was required.

In 1954, after a 3-year extensive study, the NBME adopted the multiple-choice format. To pass, a statistically computed score of 75 was required, which allowed comparison of test results from year to year. In 1971, this method was changed to one that held the mean constant at a computed score of 500, with a predetermined deviation from the mean to ascertain a passing or failing score. The 1971 changes permitted more sophisticated analysis of test results and allowed schools to compare among individual students within their respective institutions as well as among students nationwide. Feedback to students regarding performance included the reporting of pass or failure along with scores in each of the areas tested.

During the 1980s, the ever-changing field of medicine made it necessary for the NBME to examine once again its evaluation strategies. It was found necessary to develop questions in multidisciplinary areas such as gerontology, health promotion, immunology, and cell and molecular biology. In addition, it was decided that questions should test higher cognitive levels and reasoning skills.

To meet the new goals, many changes have been made in both the form and content of the examination. These changes include reduction in the number of questions to approximately 800 on Parts I and II to allow students more time on each question, with total testing time reduced on Part I from 13 to 12 hours and on Part II from 12.5 to 12 hours. The basic science disciplines are no longer allotted the same number of questions, which permits flexible weighing of the exam areas. Reporting of scores to schools include total scores for individuals and group mean scores for separate discipline areas. Only pass/fail designations and total scores are reported to examinees. There is no longer a provision for the reporting of individual subscores to either the examinees or medical schools. Finally, the question format used in the new exams, now referred to as Comprehensive (Comp) I and II, is predominantly multiple-choice, best-answer.

The New Format

New question formats, designed specifically for Comp I, are constructed in an effort to test the student's grasp of the sciences basic to medicine in an integrated fashion. The questions are designed to be interdisciplinary. Many of these questions are presented as a vignette, or case study, followed by a series of multiple-choice, best-answer questions.

The scoring of this exam also is altered. Whereas, in the past, the exams were scored on a normal curve, the new exam has a predetermined standard, which must be met in order to pass. The exam no longer concentrates on the trivial; therefore, it has been concluded that there is a common base of information that all medical students should know in order to pass. It is anticipated that a major shift in the pass/fail rate for the nation is unlikely. In the past, the average student could only expect to feel comfortable with half the test and eventually would complete approximately 67% of the questions correctly, to achieve a mean score of 500. Although with the standard setting method it is likely that the mean score will change and become higher, it is unlikely that the pass/fail rates will differ significantly from those in the past. During the first testing in 1991, there will not be differential weighing of the questions. However, in the future, the NBME will be researching methods of weighing questions based on both the time it takes to answer questions vis à vis their difficulty and the perceived importance of the information. In addition, the NBME is attempting to design a method of delivering feedback to the student that will have considerable importance in discovering weaknesses and pinpointing areas for further study in the event that a retake is necessary.

Since many of the proposed changes will be implemented for the first time in June 1991, specific information regarding actual standards, question emphasis, pass/fail rates, and so forth were unavailable at the time of publication. The publisher will update this section as information becomes available as we attempt to follow the evolution and changes that occur in the area of physician evaluation.

Materials Needed for Test Preparation

In preparation for a test, many students collect far too much study material only to find that they simply do not have the time to go through all of it. They are defeated before they begin because either they cannot get through all the material leaving areas unstudied, or they race through the material so quickly that they cannot benefit from the activity.

It is generally more efficient for the student to use materials already at hand; that is, class notes, one good outline to cover or strengthen areas not locally stressed and for quick review of the whole topic, and one good text as a reference for looking up complex material needing further explanation.

Also, many students attempt to memorize far too much information, rather than learning and understanding less material and then relying on that learned information to determine the answers to questions at the time of the examination. Relying too heavily on memorized material causes anxiety, and the more anxious students become during a test, the less learned knowledge they are likely to use.

Positive Attitude

A positive attitude and a realistic approach are essential to successful test taking. If concentration is placed on the negative aspects of tests or on the potential for failure, anxiety increases and performance decreases. A negative attitude generally develops if the student concentrates on "I must pass" rather than on "I can pass." "What if I fail?" becomes the major factor motivating the student to **run from failure rather than toward success**. This results from placing too much emphasis on scores rather than understanding that scores have only slight relevance to future professional performance.

The score received is only one aspect of test performance. Test performance also indicates the student's ability to use information during evaluation procedures and reveals how this ability might be used in the future. For example, when a patient enters the physician's office with a problem, the physician begins by asking questions, searching for clues, and seeking diagnostic information. Hypotheses are then developed, which will include several potential causes for the problem. Weighing the probabilities, the physician will begin to discard those hypotheses with the least likelihood of being correct. Good differential diagnosis involves the ability to deal with uncertainty, to reduce potential causes to the smallest number, and to use all learned information in arriving at a conclusion.

This same thought process can and should be used in testing situations. It might be termed **paper-and-pencil differential diagnosis**. In each question with five alternatives, of which one is correct, there are four alternatives that are incorrect. If deductive reasoning is used, as in solving a clinical problem, the choices can be viewed as having possibilities of being correct. The elimination of wrong choices increases the odds that a student will be able to recognize the correct choice. Even if the correct choice does not become evident, the probability of guessing correctly increases. Just as differential diagnosis in a clinical setting can result in a correct diagnosis, eliminating incorrect choices on a test can result in choosing the correct answer.

Answering questions based on what is incorrect is difficult for many students since they have had nearly 20 years experience taking tests with the implied assertion that knowledge can be displayed only by knowing what is correct. It must be remembered, however, that students can display knowledge by knowing something is wrong, just as they can display it by knowing something is right. **Students should begin to think in the present as they expect themselves to think in the future.**

Paper-and-Pencil Differential Diagnosis

The technique used to arrive at the answer to the following question is an example of the paper-and-pencil differential diagnosis approach.

> A recently diagnosed case of hypothyroidism in a 45-year-old man may result in which of the following conditions?

(A) Thyrotoxicosis
(B) Cretinism
(C) Myxedema
(D) Graves' disease
(E) Hashimoto's thyroiditis

It is presumed that all of the choices presented in the question are plausible and partially correct. If the student begins by breaking the question into parts and trying to discover what the question is attempting to measure, it will be possible to answer the question correctly by using more than memorized charts concerning thyroid problems.

- The question may be testing if the student knows the difference between "hypo" and "hyper" conditions.
- The answer choices may include thyroid problems that are not "hypothyroid" problems.
- It is possible that one or more of the choices are "hypo" but are not "thyroid" problems, that they are some other endocrine problems.
- "Recently diagnosed in a 45-year-old man" indicates that the correct answer is not a congenital childhood problem.
- "May result in" as opposed to "resulting from" suggests that the choices might include a problem that **causes** hypothyroidism rather than **results from** hypothyroidism, as stated.

By applying this kind of reasoning, the student can see that choice **A,** thyroid toxicosis, which is a disorder resulting from an overactive thyroid gland ("hyper") must be eliminated. Another piece of knowledge, that is, Graves' disease is thyroid toxicosis, eliminates choice **D.** Choice **B,** cretinism, is indeed hypothyroidism, but it is a childhood disorder. Therefore, **B** is eliminated. Choice **E** is an inflammation of the thyroid gland—here the clue is the suffix "itis." The reasoning is that thyroiditis, being an inflammation, may **cause** a thyroid problem, perhaps even a hypothyroid problem, but there is no reason for the reverse to be true. Myxedema, choice **C,** is the only choice left and the obvious correct answer.

Preparing for Board Examinations

1. **Study for yourself.** Although some of the material may seem irrelevant, the more you learn now, the less you will have to learn later. Also, do not let the fear of the test rob you of an important part of your education. If you study to learn, the task is less distasteful than studying solely to pass a test.

2. **Review all areas.** You should not be selective by studying perceived weak areas and ignoring perceived strong areas. This is probably the last time you will have the time and the motivation to review **all** of the basic sciences.

3. **Attempt to understand, not just to memorize, the material.** Ask yourself: To whom does the material apply? When does it apply? Where does it apply? How does it apply? Understanding the connections among these points allows for longer retention and aids in those situations when guessing strategies may be needed.

4. **Try to anticipate questions that might appear on the test.** Ask yourself how you might construct a question on a specific topic.

5. **Give yourself a couple days of rest before the test.** Studying up to the last moment will increase your anxiety and cause potential confusion.

Taking Board Examinations

1. In the case of NBME exams, be sure to **pace yourself** to use time optimally. Each booklet is designed to take 2 hours. You should check to be sure that you are halfway through the booklet at the end of the first hour. You should use all your alloted time; if you finish too early, you probably did so by moving too quickly through the test.

2. **Read each question and all the alternatives carefully** before you begin to make decisions. Remember the questions contain clues, as do the answer choices. As a physician, you would not make a clinical decision without a complete examination of all the data; the same holds true for answering test questions.

3. **Read the directions for each question set carefully.** You would be amazed at how many students make mistakes in tests simply because they have not paid close attention to the directions.

4. It is not advisable to leave blanks with the intention of coming back to answer the questions later. Because of the way board examinations are constructed, you probably will not pick up any new information that will help you when you come back, and the chances of getting numerically off on your answer sheet are greater than your chances of benefiting by skipping around. If you feel that you must come back to a question, mark the best choice and place a note in the margin. Generally speaking, it is best not to change answers once you have made a decision, unless you have learned new information. Your intuitive reaction and first response are correct more often than changes made out of frustration or anxiety. **Never turn in an answer sheet with blanks.** Scores are based on the number that you get correct; you are not penalized for incorrect choices.

5. **Do not try to answer the questions on a stimulus–response basis.** It generally will not work. Use all of your learned knowledge.

6. **Do not let anxiety destroy your confidence.** If you have prepared conscientiously, you know enough to pass. Use all that you have learned.

7. **Do not try to determine how well you are doing as you proceed.** You will not be able to make an objective assessment, and your anxiety will increase.

8. **Do not expect a feeling of mastery** or anything close to what you are accustomed to. Remember, this is a nationally administered exam, not a mastery test.

9. **Do not become frustrated or angry** about what appear to be bad or difficult questions. You simply do not know the answers; you cannot know everything.

Specific Test-Taking Strategies

Read the entire question carefully, regardless of format. Test questions have multiple parts. Concentrate on picking out the pertinent key words that might help you begin to problem solve. Words such as "always," "all," "never," "mostly," "primarily," and so forth play significant roles. In all types of questions, distractors with terms such as "always" or "never" most often are incorrect. Adjectives and adverbs can completely change the meaning of questions—pay close attention to them. Also, medical prefixes and suffixes (e.g., "hypo-," "hyper-," "-ectomy," "-itis") are sometimes at the root of the question. The knowledge and application of everyday English grammar often is the key to dissecting questions.

Multiple-Choice Questions

Read the question and the choices carefully to become familiar with the data as given. Remember, in multiple-choice questions there is one correct answer and there are four distractors, or incorrect answers. (Distractors are plausible and possibly correct or they would not be called distractors.) They are generally correct for part of the question but not for the entire question. Dissecting the question into parts aids in discerning these distractors.

If the correct answer is not immediately evident, begin eliminating the distractors. (Many students feel that they must always start at option A and make a decision before they move to B, thus forcing decisions they are not ready to make.) Your first decisions should be made on those choices you feel the most confident about.

Compare the choices to each part of the question. **To be wrong,** a choice needs to be incorrect for only part of the question. **To be correct,** it must be **totally** correct. If you believe a choice is partially incorrect, tentatively eliminate that choice. Make notes next to the choices regarding tentative decisions. One method is to place a minus sign next to the choices you are certain are incorrect and a plus sign next to those that potentially are correct. Finally, place a zero next to any choice you do not understand or need to come back to for further inspection. Do not feel that you must make final decisions until you have examined all choices carefully.

When you have eliminated as many choices as you can, decide which of those that are left has the highest probability of being correct. Remember to use paper-and-pencil differential diagnosis. Above all, be honest with yourself. If you do not know the answer, eliminate as many choices as possible and choose reasonably.

Vignette-Based Questions

Vignette-based questions are nothing more than normal multiple-choice questions that use the same case, or grouped information, for setting the problem. The NBME has been researching question types that would test the student's grasp of the integrated medical basic sciences in a more cognitively complex fashion than can be accomplished with traditional testing formats. These questions allow the testing of information that is more medically relevant than memorized terminology.

It is important to realize that several questions, although grouped together and referring to one situation or vignette, are independent questions; that is, they are able to stand alone. Your inability to answer one question in a group should have no bearing on your ability to answer subsequent questions.

These are multiple-choice questions, and just as is done with the single best answer questions, you should use the paper-and-pencil differential diagnosis, as was described earlier.

Single Best Answer–Matching Sets

Single best answer–matching sets consist of a list of words or statements followed by several numbered items or statements. Be sure to pay attention to whether the choices can be used more than once, only once, or not at all. Consider each choice individually and carefully. Begin with those with which you are the most familiar. It is important always to break the statements and words into parts, as with all other question formats. **If a choice is only partially correct, then it is incorrect.**

Guessing

Nothing takes the place of a firm knowledge base, but with little information to work with, even after playing paper-and-pencil differential diagnosis, you may find it necessary to guess at the correct answer. A few simple rules can help increase your guessing accuracy. Always guess consistently if you have no idea what is correct; that is, after eliminating all that you can, make the choice that agrees with your intuition or choose the option closest to the top of the list that has not been eliminated as a potential answer.

When guessing at questions that present with choices in numerical form, you will often find the choices listed in an ascending or descending order. It is generally not wise to guess the first or last alternative, since these are usually extreme values and are most likely incorrect.

Using the Challenge Exam to Learn

All too often, students do not take full advantage of practice exams. There is a tendency to complete the exam, score it, look up the correct answers to those questions missed, and then forget the entire thing.

In fact, great educational benefits can be derived if students would spend more time using practice tests as learning tools. As mentioned earlier, incorrect choices in test questions are plausible and partially correct or they would not fulfill their purpose as distractors. This means that it is just as beneficial to look up the incorrect choices as the correct choices to discover specifically why they are incorrect. In this way, it is possible to learn better test-taking skills as the subtlety of question construction is uncovered.

Additionally, it is advisable to go back and attempt to restructure each question to see if all the choices can be made correct by modifying the question. By doing this, four times as much will be learned. By all means, look up the right answer and explanation. Then, focus on each of the other choices and ask yourself under what conditions they might be correct? For example, the entire thrust of the sample question concerning hypothyroidism could be altered by changing the first few words to read:

> "Hyperthyroidism recently discovered in . . . ."
> "Hypothyroidism prenatally occurring in . . . ."
> "Hypothyroidism resulting from . . . ."

This question can be used to learn and understand thyroid problems in general, not only to memorize answers to specific questions.

The Challenge Exam that follows contains 119 questions and explanations. Every effort has been made to simulate the types of questions and the degree of question difficulty in the various licensure and qualifying exams (i.e., NBME Comp I and FLEX). While taking this exam, the student should attempt to create the testing conditions that might be experienced during actual testing situations.

Summary

Ideally, examinations are designed to determine how much information students have learned and how that information is used in the successful completion of the examination. Students will be successful if these suggestions are followed:

- Develop a positive attitude and maintain that attitude.
- Be realistic in determining the amount of material you attempt to master and in the score you hope to attain.

- Read the directions for each type of question and the questions themselves closely and follow the directions carefully.
- Guess intelligently and consistently when guessing strategies must be used.
- Bring the paper-and-pencil differential diagnosis approach to each question in the examination.
- Use the test as an opportunity to display your knowledge and as a tool for developing prescriptions for further study and learning.

NBME examinations are not easy. They may be almost impossible for those who have unrealistic expectations or for those who allow misinformation concerning the exams to produce anxiety out of proportion to the task at hand. They are manageable if they are approached with a positive attitude and with consistent use of all the information the student has learned.

Michael J. O'Donnell

QUESTIONS

Directions: Each of the numbered items or incomplete statements in this section is followed by answers or by completions of the statement. Select the **one** lettered answer or completion that is **best** in each case.

1. Which statement below best describes the papillae of the tongue?

(A) Fungiform papillae are the predominant papillae
(B) Fungiform papillae are located at the root of the tongue
(C) Circumvallate papillae are located all over the dorsal surface
(D) Circumvallate papillae contain taste buds
(E) Filiform papillae contain taste buds

2. Which activity below distinguishes cells in the zona glomerulosa of the adrenal cortex?

(A) Cortisol secretion
(B) Mineralocorticoid secretion
(C) Pregnenolone synthetase activity
(D) 17-α-Hydroxylase activity
(E) Responds to adrenocorticotropic hormone

3. Which one of the following statements is true concerning the presence of artifacts in histologic specimens?

(A) Artifacts are visible artificial structures produced by fixation
(B) Fixation artifacts exist in phase contrast microscopy specimens
(C) Glutaraldehyde fixation does not introduce artifacts
(D) Staining rarely introduces artifacts
(E) The cell nucleus is a fixation artifact

Questions 4 and 5

The epithelium pictured below was taken from a 24-year-old patient.

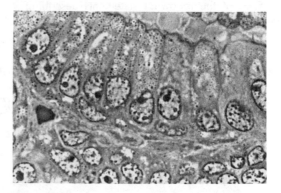

Reprinted from Johnson KE: *Histology: Microscopic Anatomy and Embryology.* New York, John Wiley, 1981, p 219.

4. Which of the following types of epithelia is pictured in the micrograph?

(A) Columnar
(B) Cuboidal
(C) Pseudostratified
(D) Transitional
(E) Stratified columnar

5. Which of the following locations contains the epithelium pictured in this micrograph?

(A) Respiratory system
(B) Vagina
(C) Oviduct
(D) Seminal vesicle
(E) Duodenum

6. Each of the following statements concerning granulopoiesis is true EXCEPT

(A) heterochromatinization of the nucleus occurs in association with specific granule accumulation
(B) metamyelocytes contain specific and azurophilic granules
(C) specific granules invariably outnumber azurophilic granules
(D) promyelocytes lack specific granules
(E) myeloblasts lack specific granules

7. Which statement below best describes the prostate gland?

(A) It functions normally without androgenic stimulation
(B) Its secretions are rich in proteolytic enzymes
(C) It is lined by a ciliated epithelium
(D) It stores spermatozoa
(E) It contributes little to the volume of the ejaculate

8. Each of the following statements describing parathyroid hormone is true EXCEPT

(A) it is a low molecular weight polypeptide hormone
(B) it is antagonistic to the effects of calcitonin
(C) it inhibits osteocytic osteolysis
(D) it promotes renal tubular resorption of calcium
(E) it stimulates renal tubular excretion of phosphate

9. The ciliary body is characterized by all of the following EXCEPT

(A) an epithelium that contains cones
(B) an epithelium that secretes aqueous humor
(C) an epithelium that is continuous with the retina
(D) ciliary muscles
(E) ciliary processes that suspend the lens

10. Each respiratory system component below has ciliated cells in its epithelium EXCEPT

(A) olfactory mucosa
(B) nasal cavity
(C) trachea
(D) bronchiole
(E) alveolus

11. Which of the following histologic techniques best identifies an acidophilic structure?

(A) Staining with the acid-phosphatase reaction
(B) Staining with toluidine blue
(C) Staining with eosin or orange G
(D) Obtaining a positive Feulgen reaction
(E) Obtaining a positive PAS reaction

12. Connective tissue contains all of the following components EXCEPT

(A) motile cells
(B) cells that commonly secrete collagen
(C) extracellular fibers
(D) amorphous ground substance
(E) a basal lamina

13. A macrophage has all of the following features EXCEPT

(A) an irregularly shaped nucleus
(B) many lysosomes
(C) many surface projections
(D) few microfilaments
(E) the ability to engage in active pinocytosis

14. Each statement below concerning T cells is true EXCEPT

(A) they appear similar to B cells in the light microscope
(B) they undergo functional maturation in the spleen
(C) they require thymosin to differentiate
(D) they are abundant in the thymus and spleen
(E) they have fewer surface immunoglobulins than B cells

15. Which of the following vessels or structures receives drainage directly from the central veins of the liver?

(A) Biliary apparatus
(B) Hepatic connective tissue
(C) Hepatic sinusoids
(D) Sublobular veins
(E) Portal veins

16. Adipose tissue contains all of the following components EXCEPT

(A) basement membranes
(B) collagen fibers
(C) fibroblasts
(D) amorphous ground substance
(E) elastic fibers

17. Proximal convoluted tubules (PCTs) are morphologically and functionally distinct from distal convoluted tubules (DCTs). Which statement below best characterizes this distinction?

(A) PCTs have simple cuboidal epithelium; DCTs have simple columnar epithelium
(B) PCTs have no microvilli
(C) PCTs have more mitochondria than DCTs
(D) DCTs have stereocilia
(E) DCTs do not have a basement membrane

18. Each statement below is characteristic of the dermis EXCEPT

(A) it contains blood vessels
(B) it contains chondroitin sulfate
(C) it contains an abundance of collagen and elastic fibers
(D) it contains an abundance of Merkel cells
(E) it is thicker in thin skin than in thick skin

Questions 19 and 20

The micrograph below is a histologic section of an organ.

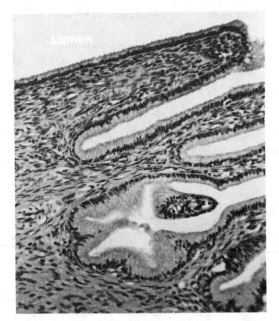

19. Which of the following types of epithelia lines the lumen of the organ in the micrograph?

(A) Simple columnar epithelium without goblet cells
(B) Pseudostratified ciliated columnar epithelium with goblet cells
(C) Simple cuboidal epithelium
(D) Simple columnar epithelium with goblet cells
(E) Transitional epithelium

20. Which one of the following structures or organs is pictured in this micrograph?

(A) Bronchus
(B) Renal pelvis
(C) Ureter
(D) Uterine tube
(E) Gallbladder

21. Each of the following statements concerning the extracellular matrix of hyaline cartilage is true EXCEPT

(A) it contains Type II collagen
(B) it contains many elastic fibers
(C) it contains chondroitin sulfate
(D) it contains glycoproteins
(E) it contains few reticular fibers

22. Each statement below is characteristic of the histologic organization of the auditory tube EXCEPT

(A) it communicates directly with the cavity containing the ossicles
(B) it is partially lined by pseudostratified epithelium
(C) it has mucous glands in its mucosa and lymphoid nodules in its lamina propria
(D) it contains ceruminous glands
(E) it is partially surrounded by elastic cartilage

23. Each statement below concerning testosterone is true EXCEPT

(A) it is made from cholesterol in Leydig cells
(B) the hypothalamus controls testosterone secretion
(C) it is required for normal prostatic secretion
(D) it inhibits LH and releasing hormone production
(E) low concentrations can maintain spermatogenesis

24. Which cell type below has an ultrastructure similar to the ultrastructure of chief cells?

(A) Pancreatic acinar cells
(B) Goblet cells
(C) Macrophages
(D) Erythrocytes
(E) Leydig cells

25. Which of the following structural features best differentiates the trachea from bronchi?

(A) Smooth muscle
(B) C-shaped rings of cartilage
(C) Scattered mucous cells in glands
(D) Goblet cells
(E) Ciliated cells

26. Each of the following enzymes or hormones is a component of the classical feedback loop that regulates blood thyroxine levels EXCEPT

(A) thyroperoxidase
(B) TSH
(C) TRH
(D) thyroxine
(E) thyroglobulin

27. Which of the following statements best characterizes the morphologic difference between dentin and enamel?

(A) Only dentin contains hydroxyapatite
(B) Dentin is harder than enamel
(C) Dentin has cellular processes; enamel does not
(D) Dentin is secreted by ameloblasts
(E) Enamel is secreted by odontoblasts

28. Each of the statements below concerning eccrine sweat glands is true EXCEPT

(A) they are abundant in the palm of the hand
(B) a coiled duct connects them to the body surface
(C) they produce copious secretion to cool the body
(D) they secrete a hypertonic solution of sodium, chloride, and urea
(E) their ducts contain a stratified cuboidal epithelium

29. Which of the following immune system components is part of the mononuclear phagocyte system?

(A) Killer T cells
(B) Memory B cells
(C) Suppressor T cells
(D) Effector T cells
(E) Macrophages

30. Which of the following properties best identifies a metachromatic structure?

(A) A positive PAS reaction
(B) A net positive charge
(C) Staining with eosin
(D) Staining purple with toluidine blue
(E) A high concentration of DNA

31. Each feature below is characteristic of neutrophils EXCEPT

(A) histamine-containing granules
(B) azurophilic granules
(C) nuclei with 3–5 lobes
(D) chemotactic activity
(E) phagocytotic activity

32. Each feature below is characteristic of the ultrastructure of cells in the adrenal medulla EXCEPT

(A) abundant RER
(B) large mitochondria with peculiar tubular cristae
(C) prominent perinuclear Golgi apparatus
(D) prominent granules that are 200 nm in diameter
(E) epinephrine-containing granules

33. Which epidermal layer below contains cells that show ^{3}H-thymidine incorporation?

(A) Stratum corneum
(B) Stratum lucidum
(C) Stratum granulosum
(D) Stratum spinosum
(E) Stratum basale

34. Each of the following statements concerning the layers of the cornea is true EXCEPT

(A) the stroma is rich in glycosaminoglycans
(B) the stroma contains fibroblasts and orthogonal layers of collagen
(C) Descemet's membrane rests on the endothelium
(D) the endothelium is a stratified epithelium
(E) the epithelium is continuous with the skin

35. Each of the following statements concerning thyroglobulin synthesis and secretion is true EXCEPT

(A) it is secreted with peroxidase into the lumen of the thyroid follicles
(B) its synthesis can be promoted by TSH
(C) it is iodinated in the RER and Golgi apparatus
(D) its polypeptides are assembled in the RER
(E) glycosylation occurs in part in the RER

36. Which of the following enteroendocrine cells stimulates the release of hydrochloric acid?

(A) G cells
(B) I cells
(C) S cells
(D) EC cells
(E) A cells

37. Each statement below is characteristic of epithelial cells in the small intestine EXCEPT

(A) they are joined by prominent apical zonulae occludentes
(B) the basement membrane usually is absent
(C) goblet cells have a basal nucleus
(D) Paneth cells contain apical granules of lysozyme
(E) sucrase is a component of the glycocalyx

38. Each of the following statements is characteristic of erythropoiesis EXCEPT

(A) polychromatic erythroblasts have more cytoplasmic RNA than basophilic erythroblasts
(B) polychromatic erythroblasts have less hemoglobin than orthochromatic erythroblasts
(C) reticulocytes lack a nucleus and can occur in normal peripheral blood
(D) orthochromatic erythroblasts have less hemoglobin than mature erythrocytes
(E) erythroblasts are 15 μm in diameter

39. Normally, erythrocytes are best described as

(A) more than 20% hemoglobin by weight
(B) less than 20% of the total volume of the blood
(C) loaded with mitochondria
(D) 3–6 μm in diameter
(E) circulating in the blood for less than 40 days

Questions 40–42

The micrograph below shows a rat epithelium resting on a field of connective tissue.

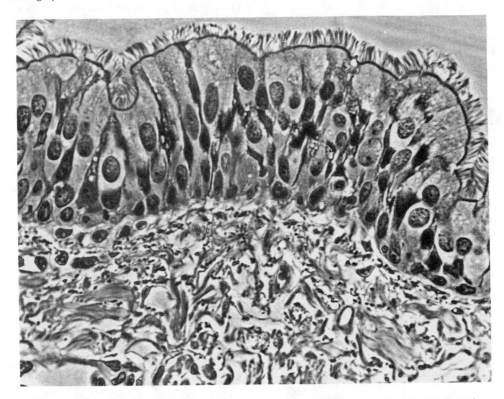

Reprinted from Johnson KE: *Histology: Microscopic Anatomy and Embryology.* New York, John Wiley, 1982, p 33.

40. Which of the following connective tissues is pictured in the micrograph?

(A) Dense regular
(B) Dense irregular
(C) Loose (areolar)
(D) Adipose
(E) Reticular

41. Which of the following epithelia is pictured in the micrograph?

(A) Simple columnar
(B) Simple cuboidal
(C) Stratified cuboidal
(D) Stratified squamous
(E) Pseudostratified ciliated columnar

42. Which of the following organs is lined by the epithelium in the micrograph?

(A) Trachea
(B) Stomach
(C) Gallbladder
(D) Epididymis
(E) Duodenum

43. A complex filtration system exists between the vascular and urinary space in the renal glomerulus. Which statement below best describes the behavior of high molecular weight (MW) tracers injected into the afferent blood supply of a renal glomerulus?

(A) Proteins with a MW greater than 400,000 cannot cross the fenestrated capillary
(B) Proteins with a MW greater than 160,000 cannot cross the fenestrated capillary
(C) Proteins with a MW greater than 160,000 cannot cross the podocyte slit diaphragm
(D) Proteins with a MW greater than 400,000 enter the glomerular filtrate
(E) Proteins with a MW greater than 160,000 enter the glomerular filtrate

44. Which statement below best characterizes the endosteum of bony tissue?

(A) It is a layer of osteocytes between the periosteum and cortical bone
(B) It is a layer of osteoblasts between the periosteum and cortical bone
(C) It is a layer of osteocytes between the marrow cavity and the inner shell of bone that lines the marrow cavity
(D) It is a layer of osteoblasts between the marrow cavity and the inner shell of bone that lines the marrow cavity
(E) It is a layer of osteoclasts between the marrow cavity and the inner shell of bone that lines the marrow cavity

45. Each of the following morphologic features is characteristic of the zona glomerulosa of the adrenal cortex EXCEPT

(A) they have fewer lipid droplets than parenchymal cells in the zona fasciculata
(B) they have mitochondria with tubular cristae
(C) the cell nucleus is elongated
(D) they have abundant SER
(E) they have fewer lipofuscin granules than parenchymal cells in the zona reticularis

46. Which of the following cells contains numerous microtubules arranged in an elongated axoneme?

(A) Neutrophil
(B) Spermatozoon
(C) Motor nerve axon
(D) Red blood cell
(E) Duodenal columnar absorptive epithelial cell

47. Which of the following cells is a glial cell of the peripheral nervous system?

(A) Schwann cell
(B) Fibrous astrocyte
(C) Ependymal cell
(D) Oligodendrocyte
(E) Purkinje cell

48. Each of the following statements concerning the cerebrospinal fluid is true EXCEPT

(A) it is secreted by cells of the choroid plexus
(B) it bathes the apical surface of ependymal cells
(C) it circulates in the subarachnoid space
(D) it is a blood filtrate
(E) it is absent from the central canal of the spinal cord

49. Which statement below best describes the microscopic anatomy of cardiac myocytes?

(A) They lack sarcomeres
(B) They are joined to each other by gap junctions
(C) They are anatomically syncytial
(D) They have A bands but lack I bands
(E) They have nuclei at the fiber periphery

50. Which one of the following cell types synthesizes testosterone?

(A) Sertoli cells
(B) Leydig cells
(C) Spermatogonia
(D) Primary spermatocytes
(E) Spermatids

51. Which statement below best characterizes the ultrastructure of the nuclear envelope?

(A) It is a single unit membrane
(B) Diaphragms across nuclear pores prevent macromolecules from crossing the nuclear envelope
(C) It is not continuous with the RER
(D) It has hexagonal nuclear pores
(E) It is a double unit membrane with pores in it

52. Which of the following body parts contains dense irregular connective tissue?

(A) Tendons
(B) Ligaments
(C) Mesenteries
(D) Lamina propria
(E) Dermis

53. The pituitary gland consists of an adenohypophysis and a neurohypophysis. Which of the cell types below exists only in the neurohypophysis?

(A) Gonadotropes

(B) Lactotropes

(C) Acidophils

(D) Pituicytes

(E) Chromophobes

54. The plasma membrane is the boundary between the cytoplasm and the extracellular environment. Each of the following statements characterize the plasma membrane EXCEPT

(A) it contains amphipathic phospholipids

(B) cholesterol reduces the stability of its hydrophobic domain

(C) its membrane proteins are amphipathic

(D) it contains numerous glycoproteins

(E) it contains amphipathic intramembranous proteins

Questions 55–57

Examine the electron micrograph below. Notice the apical surface projections filled with fibrous structures.

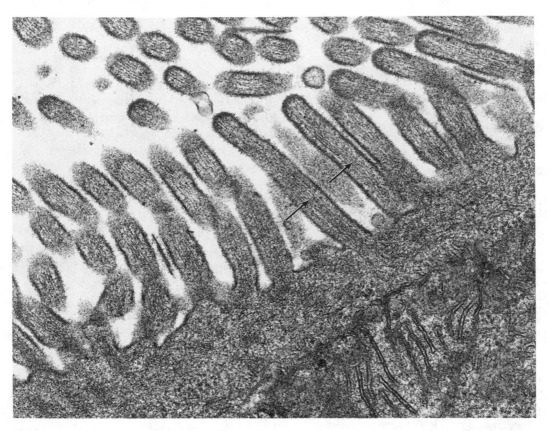

55. Which description below best describes the fibrous structures (*arrows*) in the micrograph?

(A) Actin-rich microfilaments

(B) Myosin-rich thick filaments

(C) Tubulin-rich microtubules

(D) Keratin-rich tonofilaments

(E) Desmin-rich intermediate filaments

56. Which of the following structures best describes the surface projections in the micrograph?

(A) Stereocilia

(B) Cilia

(C) Microvilli

(D) T tubules

(E) Flagella

57. Each location below has luminal epithelial cells with these apical surface projections EXCEPT

(A) choroid plexus
(B) proximal convoluted tubules
(C) distal convoluted tubules
(D) jejunum
(E) esophagus

Questions 58 and 59

Examine the micrograph below.

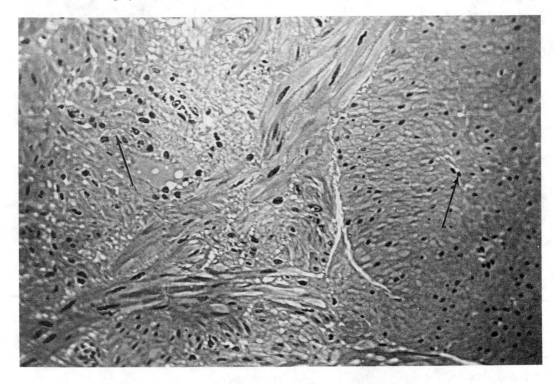

58. Which of the following tissues best describes the tissue marked by arrows in the micrograph?

(A) Dense irregular connective tissue
(B) Smooth muscle
(C) Bundles of Purkinje fibers
(D) Cardiac muscle
(E) Lamina propria

59. Which of the following body locations has the greatest amount of this tissue?

(A) Gastric muscularis externa
(B) Dermis
(C) Myocardium
(D) Endocardium
(E) Gastric mucosa

Questions 60–62

The micrograph below is taken from the male reproductive system.

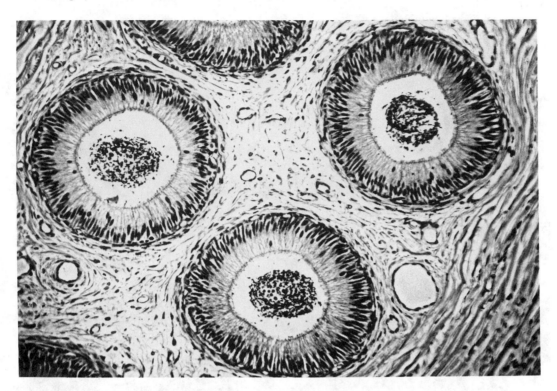

60. Which of the following organs is pictured in the micrograph?

(A) Testis
(B) Epididymis
(C) Vas deferens
(D) Seminal vesicle
(E) Bulbourethral gland

61. Which of the following epithelia best describes the epithelium of the organ in the micrograph?

(A) Simple cuboidal
(B) Simple columnar
(C) Stratified columnar
(D) Pseudostratified columnar with stereocilia
(E) Stratified squamous

62. Which phrase below best describes the function of the epithelium in the micrograph?

(A) Gametogenic
(B) Proliferative
(C) Secretory and absorptive
(D) Inactive
(E) Protective

Questions 63–65

This is a micrograph of a specialized type of human connective tissue.

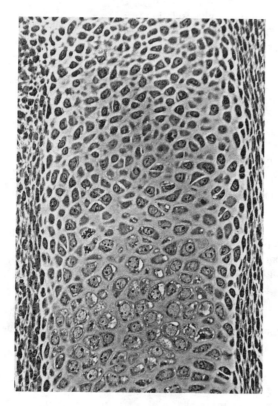

63. Which of the following connective tissues is pictured in the micrograph?

(A) Lamina propria (loose areolar connective tissue)
(B) Bone
(C) Hyaline cartilage
(D) Fibrocartilage
(E) Elastic cartilage

64. Which statement below best describes the connective tissue in the micrograph?

(A) Sparse cells and little extracellular matrix
(B) Many chondrocytes and an extracellular matrix rich in chondroitin sulfate
(C) Calcified extracellular matrix
(D) Many chondrocytes and elastic fibers
(E) Few chondrocytes and elastic fibers; many collagen fibers

65. Which of the following locations contains the connective tissue in the micrograph?

(A) Trachea
(B) Intervertebral disks
(C) Auricle and external auditory meatus
(D) Frontal bone of the skull
(E) Ligamentum flavum

66. Each statement below concerning the freeze-fracture-etch technique is true EXCEPT

(A) specimen fixation is minimized
(B) specimens receive a metal coating under atmospheric pressure
(C) glycerol infiltration minimizes ice crystal formation
(D) heavy metals make the fracture faces visible in the electron microscope
(E) specimens are rapidly frozen in liquid nitrogen

67. Each statement below concerning the use of autoradiography is true EXCEPT

(A) it can be used in light and electron microscopy
(B) it can reveal sites of DNA synthesis
(C) it is used to reveal sites of metabolic activity or cell movement
(D) it can be used with metabolically inactive cells
(E) it can be used to distinguish stem cells from their differentiated progeny

68. Which statement below best describes the ultrastructure of the Golgi apparatus?

(A) It contains membranes that are similar to the plasma membrane
(B) It has little functional interaction with the SER
(C) It has little functional interaction with the RER
(D) It has ribosomes bound to its membranes
(E) It contains mRNA for protein synthesis

69. Each statement below concerning zymogen granules is true EXCEPT

(A) they contain inactive enzymes
(B) their product is less mature than the product of condensing vacuoles
(C) they are surrounded by membranes
(D) they contain concentrated secretion product
(E) they fuse with the plasma membrane

70. Each statement below is a postulate of the signal hypothesis EXCEPT

(A) signal peptides are encoded in mRNA
(B) signal peptides interact strongly with RER receptors
(C) detachment factors release peptides from the RER
(D) signal peptidases cleave signal peptides
(E) polypeptides in the lumen of the RER are too large to escape

71. Mitochondria perform all of the following functions EXCEPT

(A) nucleic acid synthesis
(B) steroid synthesis
(C) ATP synthesis
(D) polysaccharide degradation
(E) electron transport

72. Lysosomes perform all of the following functions EXCEPT

(A) nucleic acid hydrolysis
(B) protein degradation
(C) protein glycosylation
(D) destruction of effete organelles
(E) destruction of bacteria

73. Each cell type below contains abundant microtubules EXCEPT

(A) neurons
(B) skeletal muscle cells
(C) ciliated epithelial cells
(D) spermatozoa
(E) cells in the stratum basale of skin

Directions: Each group of items in this section consists of lettered options followed by a set of numbered items. For each item, select the **one** lettered option that is most closely associated with it. Each lettered option may be selected once, more than once, or not at all.

Questions 74–77

Match each of the following functions with the appropriate lettered structure in this electron micrograph of a fibroblast.

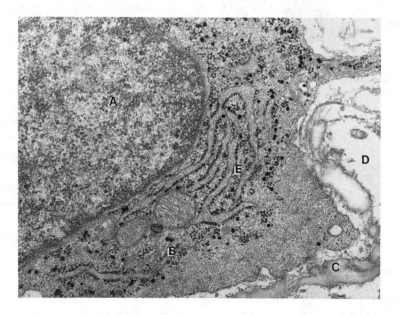

74. Produces ATP for assembling polypeptides into collagen

75. Polypeptide chains for collagen and for proteoglycans of the extracellular matrix are assembled here

76. The mRNA that directs protein synthesis is synthesized here

77. Collagen fibers are assembled and cross-linked here

Questions 78–80

Match each description below with the appropriate cell type.

(A) Granulosa cells
(B) Granulosa lutein cells
(C) Theca interna cells
(D) Theca lutein cells
(E) None of the above

78. Form after ovulation; secrete progesterone

79. Secrete glycoproteins

80. Secrete androgens

Questions 81–85

For each of the following descriptions of sites or structures of the pancreatic acinar cell, select the appropriate lettered component shown in the micrograph below.

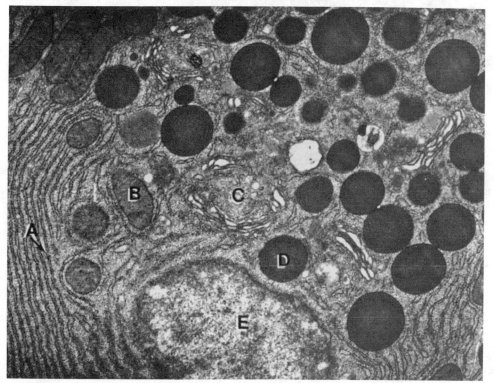

Courtesy of Dr. F. J. Slaby, Department of Anatomy, George Washington University.

81. Granules of secretion product

82. Membranous structure involved in processing secretion product for transport

83. Oxidative phosphorylation occurs here

84. Polypeptide chain synthesis occurs here

85. Glycosyltransferase enzymes are found here

Questions 86–90

Match each of the following descriptions with the most appropriate lettered structure or organelle in the micrograph below.

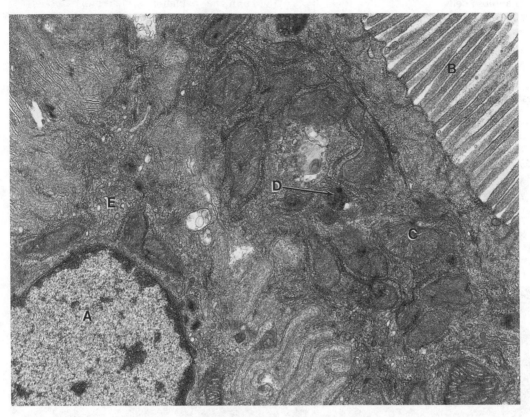

86. Plasma membrane modifications that increase surface area and contain many actin-containing microfilaments

87. Organelle bounded by a double unit membrane that synthesizes ATP

88. Organelle bounded by a double unit membrane containing pores large enough for macromolecule transport

89. Part of the lysosome system

90. Organelle that helps assemble absorbed lipids into micelles, which are then transported into the lateral compartments between cells

Questions 91–95

Match each of the following sarcomere regions with the appropriate lettered component of the micrograph below.

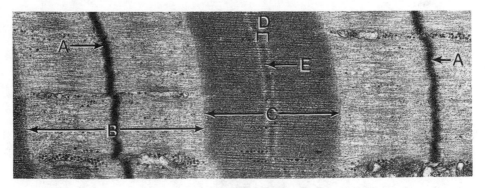

Courtesy of Dr. H.A. Padykula, Department of Anatomy, University of Massachusetts.

91. Z line

92. I band

93. H band

94. Middle of sarcomere

95. End of sarcomere

Questions 96–98

Match each component of an ileal villus described below with the appropriate lettered structure in the micrograph.

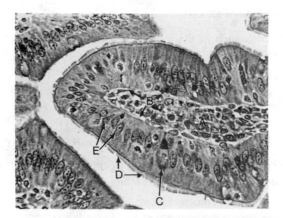

Reprinted from Johnson KE: *Histology: Microscopic Anatomy and Embryology.* New York, John Wiley, 1981, p 219.

96. These cells secrete protective mucus

97. Microvilli here increase the surface area available for absorption

98. Polymorphonuclear leukocytes, plasma cells, and lymphocytes occur here

Questions 99-103

Match each description below with the appropriate lettered organ in the diagram.

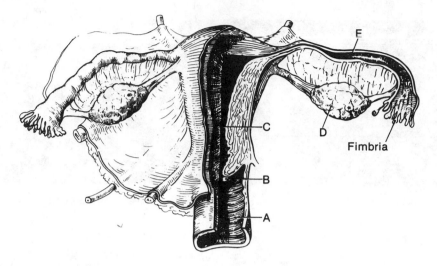

99. An organ covered by a simple cuboidal epithelium; its cortex contains stromal fibroblasts and gametogenic cells

100. An organ lined by a stratified squamous unkeratinized epithelium; it lacks glands

101. An organ with a zone of epithelial transition from simple columnar epithelium to stratified squamous unkeratinized epithelium

102. An organ with a simple columnar epithelium containing many ciliated cells and a few mucus-secreting cells

103. An organ with a simple columnar epithelium containing many mucus-secreting cells and a few ciliated cells; its glands become longer during the follicular phase of the menstrual cycle

Questions 104-108

Match each description below with the appropriate layer of the wall of the small intestine.

(A) Mucosa
(B) Muscularis mucosae
(C) Submucosa
(D) Muscularis externa
(E) Serosa/adventitia

104. In the duodenum, this layer contains many Brunner's glands

105. This layer contains ganglia of Auerbach's plexus

106. Plicae circulares are folds of the mucosa and this layer

107. The neurons of Meissner's plexus innervate the smooth muscle of this layer

108. The lamina propria in this layer contains lymphocytes and plasma cells

Questions 109-113

Match each of the following functions with the most appropriate tracheal epithelial cell.

(A) Tall columnar ciliated cell
(B) Goblet cell
(C) Basal cell
(D) Brush cell
(E) Granule cell

109. Secretes a viscous material for trapping debris

110. Moves mucus toward gastrointestinal tract

111. An undifferentiated type of cell that is capable of dividing

112. An APUD cell; its secretion products control mucus secretion

113. A tall columnar cell without mucus but with apical microvilli

Questions 114–117

Match each of the following descriptions with the appropriate lettered component in this micrograph of the trachea.

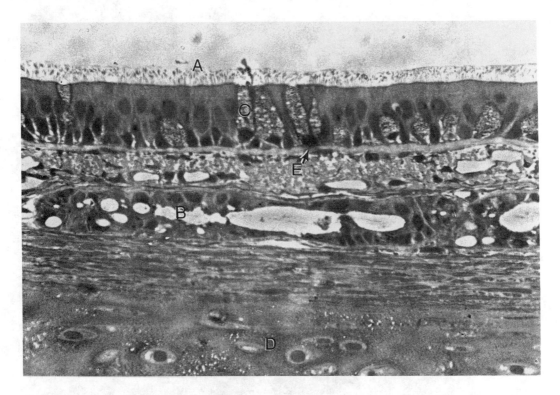

114. Basement membrane for the epithelium
115. Multicellular gland
116. Cell that secretes mucus
117. Flexible piece of avascular connective tissue that helps to maintain the tracheal lumen

Questions 118–121

Match each description below with the appropriate lettered structure in this micrograph of a chondrocyte.

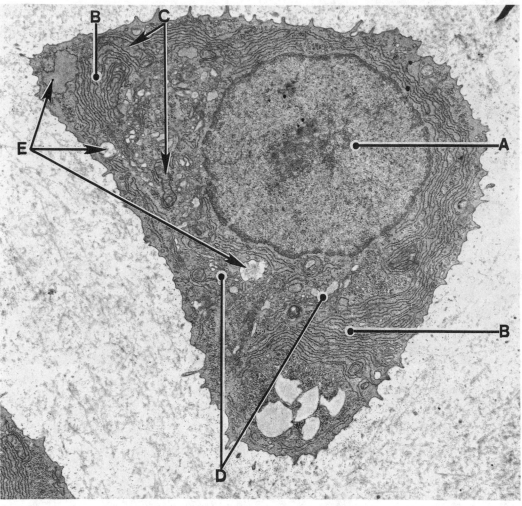

Courtesy of Dr. D. P. DeSimone, George Washington University.

118. Initial protein synthesis for proteoglycan aggregates occurs here
119. Produces ATP for synthetic activities of the cell
120. Contains hyaluronic acid, proteoglycan aggregates, and collagen molecules
121. Rich in glucuronyltransferases

Questions 122–126

Match each description of an inner ear component below with the appropriate lettered structure in the micrograph.

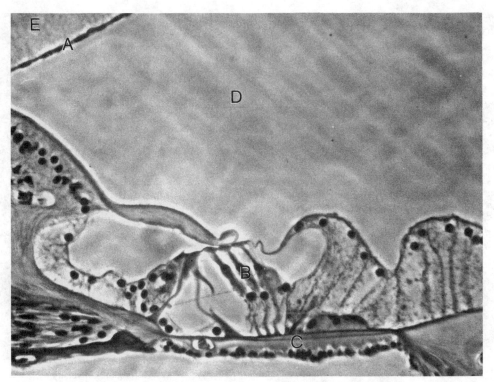

Reprinted from Johnson KE: *Histology: Microscopic Anatomy and Embryology.* New York, John Wiley, 1982, p 403.

122. Contains basilar fibers

123. Contains endolymph

124. Contains perilymph

125. Has apical stereocilia

126. Boundary between scala media and scala vestibuli

Questions 127–131

Match each description below with the appropriate lettered structure in this micrograph of a pancreatic acinar cell.

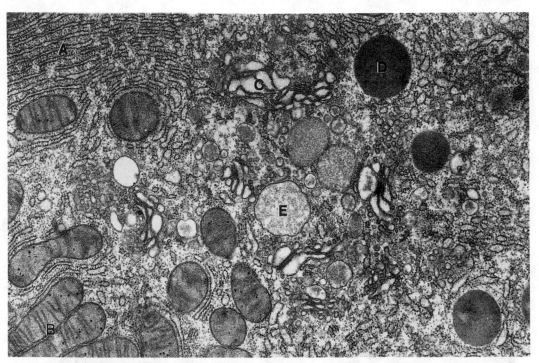

Courtesy of Dr. F. J. Slaby, Department of Anatomy, George Washington University.

127. Synthesis of nascent chains of hydrolytic enzymes begins here

128. Membrane-bound vesicle of enzymes, ready for secretion

129. Involved in glycosylation and packaging of secretion product

130. Contains unit membranes with attached ribonucleoprotein particles

131. Supplies ATP used in protein synthesis and contains enzymes for oxidative phosphorylation

Questions 132–136

Match each structure below with the appropriate lettered component in the electron micrograph of the renal cortex.

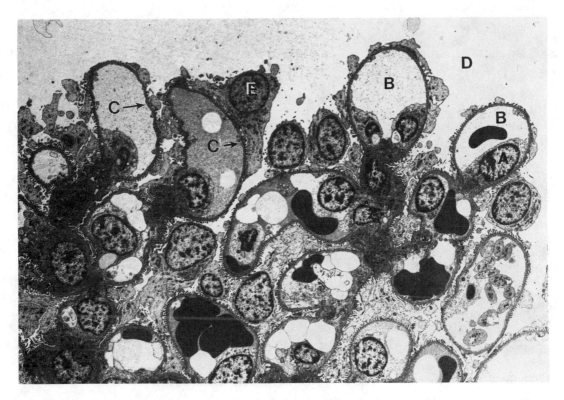

132. Urinary space

133. Podocyte

134. Vascular space

135. Vascular endothelial cell

136. Glomerular basement membrane

Questions 137–140

Match each description below with the appropriate lettered structure in the micrograph.

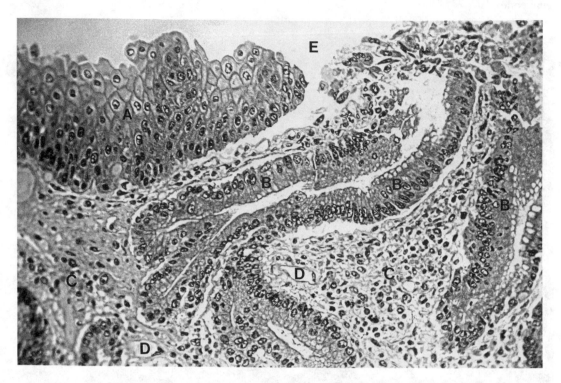

137. A loose areolar connective tissue often infiltrated with lymphocytes and plasma cells

138. A stratified squamous epithelium without keratinization

139. A simple columnar epithelium of mucus-secreting cells

140. This is a micrograph of the

(A) prostatic urethra
(B) cervix
(C) vagina
(D) gastroesophageal junction
(E) trachea

Questions 141–143

Match each description below with the appropriate lettered structure in this micrograph of a gastric gland.

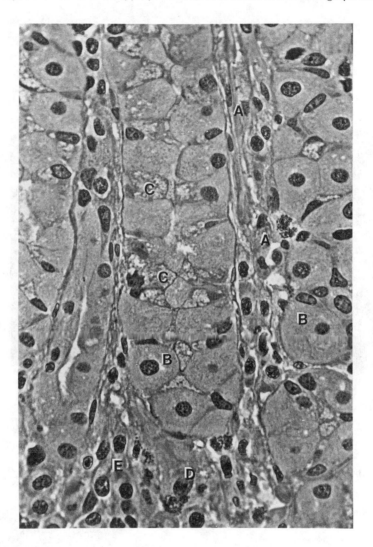

141. Secretes pepsinogen
142. Secretes HCl
143. Secretes gastric intrinsic factor

Questions 144–146

Match each description below with the appropriate lettered structure in this micrograph of liver parenchymal tissue.

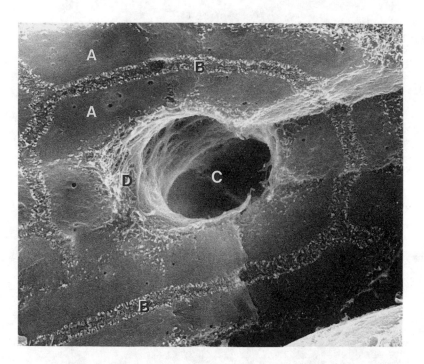

144. Receives blood from the portal vein and hepatic artery and drains blood into the central vein

145. A liver parenchymal cell

146. A bile canaliculus surrounded by tight junctions that form the blood-bile barrier

Questions 147–150

Match each description below with the most appropriate cardiovascular layer or layers.

(A) Tunica intima
(B) Tunica adventitia
(C) Both
(D) Neither

147. Homologous to the endocardium

148. Homologous to the epicardium

149. Layer that contains intrinsic blood vessels in the aorta

150. Present in muscular arteries

Questions 151–154

Match each cell characteristic below with the most appropriate leukocytes.

(A) Granulocytes
(B) Agranulocytes
(C) Both
(D) Neither

151. Phagocytic

152. Lack specific granules

153. Lack lysosomes

154. Have lobulated nuclei with 2–4 lobes

Questions 155–158

Match each cell characteristic below with the most appropriate adenohypophysial cells.

(A) Lactotropes
(B) Thyrotropes
(C) Both
(D) Neither

155. Present in the pars distalis

156. Present in the median eminence

157. Produces a low molecular weight polypeptide hormone

158. Produces a high molecular weight glycoprotein hormone

ANSWERS AND EXPLANATIONS

1. The answer is D. [*Chapter 17 II D 2 a–c*] The tongue has three types of papillae: filiform, fungiform, and circumvallate. Filiform, the most numerous papillae, are not associated with taste buds. Fungiform papillae are scattered among the filiform papillae on the dorsal surface of the tongue. Circumvallate papillae are present only at the root of the tongue. Both fungiform and circumvallate papillae have taste buds.

2. The answer is B. [*Chapter 21 V B 2*] The adrenal cortex exhibits striking zonation that includes the zona reticularis, zona fasciculata, and zona glomerulosa. Aldosterone is a mineralocorticoid synthesized in the zona glomerulosa. The zona reticularis and zona fasciculata both secrete cortisol under the stimulation of adrenocorticotropic hormone (ACTH). 17-α-Hydroxylase is a key enzyme in cortisol synthesis and occurs in both zones. Pregnenolone synthetase is an enzyme that is active in an early stage of steroid biosynthesis. Its product, pregnenolone, is a precursor for the synthesis of many steroids; thus, pregnenolone synthetase is widely distributed in the adrenal cortex.

3. The answer is A. [*Chapter 1 II A 1, E*] Fixation artifacts are artificial structures in specimens that result from specimen manipulation during fixation. Most techniques of preparing specimens for histologic examination produce some artifacts. Phase contrast microscopy specimens are live (unfixed) and therefore do not contain fixation artifacts. Staining creates an artifact of high contrast among cell structures. This type of artifact is useful for studying cell and tissue structure. The cell nucleus is not a fixation artifact.

4–5. The answers are: 4-C, 5-D. [*Chapter 8 II A 1 c, E*] This is a micrograph of pseudostratified epithelium. It may be mistaken for columnar epithelium, but close inspection reveals the curved basal nuclei typical of pseudostratified epithelium. Notice the prominent clear areas in the apices of the tall cells. These are the dilated cisternae of the Golgi apparatus. This secretory epithelium is present in seminal vesicles and produces some of the material in the ejaculate. The pseudostratified epithelium of the respiratory system is ciliated.

6. The answer is C. [*Chapter 11 VI D*] Azurophilic granules outnumber specific granules in the early stages of granulopoiesis but not in the late stages. Metamyelocytes have granules that are typical of a mature granulocyte but have an immature nuclear morphology. Myeloblasts and promyelocytes both lack specific granules.

7. The answer is B. [*Chapter 24 V B 6*] The prostate requires androgenic stimulation but does not actually store spermatozoa. The prostate has glands lined by a pseudostratified epithelium without cilia. Its secretions, which are rich in proteolytic enzyme, contribute a considerable amount to the volume of the ejaculate.

8. The answer is C. [*Chapter 21 III C 3*] Parathyroid hormone (PTH) is a low molecular weight hormone whose antagonistic effect on calcitonin causes a systemic increase in serum calcium. PTH promotes osteolysis and renal resorption of calcium and stimulates phosphate excretion.

9. The answer is A. [*Chapter 27 II B 2*] The ciliary body is located between the edge of the retina and the edge of the lens. It is covered by a double-layered epithelium that is continuous with the retina but lacks the photoreceptors characteristic of the more posterior part of the retina. The ciliary epithelium secretes the aqueous humor. Many long ciliary processes project from the ciliary body toward the lens. Ciliary muscles are the major components of the ciliary body. Cones are found in the posterior retina, not in the ciliary body.

10. The answer is E. [*Chapter 16 II A 3, B 2–3; VI C*] The nasal cavity has a pseudostratified ciliated columnar (PCC) epithelium. The olfactory mucosa also has a PCC epithelium that is tall and modified for olfaction. The PCC epithelium of the trachea contains six types of cells that help perform the diverse functions of this tubular structure. Ciliated cells are common in bronchioles; alveolar epithelium lacks ciliated cells but contains type I and type II cells.

11. The answer is C. [*Chapter 1 II B 1 a*] Acidophilic structures have a net positive charge and typically stain with eosin or orange G (negatively charged, acidic dyes). Toluidine blue is a positively charged, basic dye used to stain basophilic structures. No direct connection exists between acidophilia and how a specimen reacts to the periodic acid-Schiff (PAS) or the Feulgen reaction (a test for DNA). A structure stains acid-phosphatase positive because it has this enzymatic activity, not because it is acidophilic.

12. The answer is E. [*Chapter 9 I B*] Connective tissue contains extracellular fibers, collagen fibers, and amorphous ground substance. Many connective tissue cells are motile. Fibroblasts have significant intrinsic motile capacity. Many of the immigrant cells of connective tissues (e.g., neutrophils and macrophages) have substantial motile activity. A basal lamina is characteristic of an epithelium, not connective tissue.

13. The answer is D. [*Chapter 9 II B 2*] Each macrophage has numerous microfilaments as part of its motility system. Macrophages are active phagocytes; therefore, they have many lysosomes and surface projections and are active in pinocytosis. The macrophage nucleus is irregularly shaped.

14. The answer is B. [*Chapter 12 I B 2 e; VII B 3, 4*] Both T cells and B cells are small lymphocytes. In the light microscope, T cells and B cells appear to be similar, unless fluorescent microscopy is used to distinguish them on the basis of differences in surface antigens. T cells are derived from bone marrow stem cells and undergo functional maturation in the thymus, not the spleen. In the thymus, the thymic hormone thymosin stimulates T cell differentiation. T cells are abundant in the spleen as well as the thymus.

15. The answer is D. [*Chapter 19 II B 2*] Blood flows to the liver through branches of the portal veins, the hepatic sinusoids, and then the central veins. The central veins empty into sublobular veins; they do not nourish the connective tissue and biliary apparatus of the liver. This function is performed by branches of the hepatic artery.

16. The answer is A. [*Chapter 9 I B; III A*] Adipose tissue is a specialized connective tissue. Almost all types of connective tissue, including adipose tissue, contain at least some elastic fibers and amorphous ground substance. Adipose tissue also contains adipocytes, fibroblasts, and scattered collagen fibers. Basement membranes are present beneath most epithelial tissue; however, they are not present beneath adipose tissue.

17. The answer is C. [*Chapter 26 II C 3, 5*] Proximal convoluted tubules (PCTs) and distal convoluted tubules (DCTs) have numerous apical microvilli. Both tubules are lined by simple cuboidal epithelium that rests on a robust basement membrane. The PCT is very active in both ion and protein transport (by pinocytosis) and has many mitochondria to supply adenosine triphosphate (ATP) for these energy-dependent processes.

18. The answer is D. [*Chapter 25 VI C*] The dermis is a dense, irregular connective tissue under the epidermis. The dermis of thin skin is thick; the dermis of thick skin is thin. The dermis is highly vascular and its blood supply is important for thermoregulation. The dermal extracellular matrix contains collagen fibers, elastic fibers, hyaluronate, and chondroitin sulfate. Merkel cells exist only in the epidermis. They are innervated by myelinated nerve fibers and are sensory elements in skin.

19–20. The answers are: 19-A, 20-E. [*Chapter 19 III D 2*] This is a histologic section of the gallbladder. It is lined by a simple columnar epithelium that does not have goblet cells. The mucosa forms numerous folds and crypts.

Bronchi have hyaline cartilage and a pseudostratified ciliated columnar epithelium with goblet cells. The renal pelvis and ureters have transitional epithelium. Uterine tubes have simple columnar ciliated cells.

21. The answer is B. [*Chapter 10 II B*] The extracellular matrix of hyaline cartilage is a complex mixture of macromolecules. Glycoproteins and proteoglycan are abundant in the amorphous ground substance of hyaline cartilage. Cartilage proteoglycan is an immense macromolecule containing a protein backbone attached to three glycosaminoglycans: chondroitin sulfate, hyaluronate, and keratan sulfate. This matrix also contains an abundance of Type II collagen assembled into collagen fibers. Hyaline cartilage has very few elastic and reticular fibers. Elastic cartilage is rich in elastic fibers.

22. The answer is D. [*Chapter 27 IV A 2, B 4*] The auditory tube connects the middle ear cavity to the nasopharynx. It is lined near its nasopharyngeal orifice with pseudostratified epithelium similar to that found in the nasopharynx. It is surrounded in this location by a spiral of elastic cartilage, which helps keep it open. The auditory tube is associated with mucous glands and lymphoid nodules. Ceruminous glands are modified apocrine sweat glands in the external auditory meatus that secrete cerum (ear wax).

23. The answer is E. [*Chapter 24 III A 1*] Testosterone is the most important androgenic steroid. Testosterone synthesis is controlled by hypothalamic and adenohypophysial hormones. Testosterone inhibits the production of luteinizing hormone (LH) and releasing hormone. Testosterone is synthesized in Leydig cells from cholesterol esters stored in lipid droplets in Leydig cells. The secretions of the prostate, seminal vesicles, and bulbourethral glands are testosterone-dependent. Sertoli cells secrete an androgen binding protein that binds testosterone, thus maintaining the high concentration of testosterone required to maintain spermatogenesis.

24. The answer is A. [*Chapter 18 II D 4*] Chief cells are abundant in the deep portions of the glands in the fundic and corpic stomach. Chief cells secrete pepsinogen, a hydrolytic digestive enzyme, into the gastric lumen. Like all cells specialized to synthesize protein for export, chief cells have prominent nucleoli, cytoplasmic basophilia due to a well-developed rough endoplasmic reticulum (RER), a large Golgi apparatus, and zymogen granules. Pancreatic acinar cells also are specialized for the synthesis of digestive enzymes and have a similar ultrastructure.

25. The answer is B. [*Chapter 16 IV B; V A 2, 3*] The trachea and bronchi are lined by respiratory epithelium with ciliated cells and goblet cells. Both have an abundance of smooth muscle. The trachea has the dorsal trachealis smooth muscle and the bronchi have abundant smooth muscle between plates of hyaline cartilage. The chief difference between the trachea and bronchi is that the trachea has C-shaped rings of hyaline cartilage and the bronchi have plates of irregularly shaped hyaline cartilage. Small mucous glands are present in both the trachea and the bronchi.

26. The answer is A. [*Chapter 21 I A 4; II C 2*] Low thyroxine levels indirectly stimulate the secretion of thyrotropin releasing hormone (TRH). TRH secretion directly stimulates basophils in the adenohypophysis to secrete thyroid-stimulating hormone (TSH). TSH stimulates thyroid follicular epithelial cells to engulf and then degrade thyroglobulin intracellularly to release active thyroxine. Thyroxine is released into the systemic circulation and affects the basal metabolic rate. When blood thyroxine levels rise, TRH secretion, and thus TSH secretion, is inhibited. Thyroperoxidase is not part of the feedback loop. It is an enzyme on the apical cell surface of follicular epithelial cells that catalyses the oxidation of iodine so that it can be added to thyroglobulin.

27. The answer is C. [*Chapter 17 II C*] Enamel and dentin are hard components of the dental crown. Enamel is the hardest substance in the body. It covers the dental crown and is secreted by ameloblasts. Its peculiar hardness is due to hydroxyapatite accumulation and its lack of cellular processes. Dentin, which lies beneath the enamel, is bone-like in hardness and contains hydroxyapatite. Dentin also has odontoblast processes, which are similar to the osteocyte processes in bone.

28. The answer is D. [*Chapter 25 III A 2*] Eccrine sweat glands are found over most of the surface of the body, including the thick palmar skin. They have coiled ducts lined with stratified cuboidal epithelium. Eccrine sweat glands secrete a large volume of sweat when the body becomes overheated; evaporation of this sweat cools the body. Sweat is a hypotonic (not hypertonic) solution of water, sodium, chloride, urea, and other salts.

29. The answer is E. [*Chapter 9 II B 2; Chapter 12 VII B 1, 3*] The mononuclear phagocyte system (MPS) is a collection of monocyte-derived dedicated phagocytes in the human body. Components of the MPS include macrophages, monocytes, histiocytes, Kupffer cells, and osteoclasts. B lymphocytes differentiate into plasma cells, which secrete immunoglobulins. T lymphocytes have various modulatory or cytotoxic functions in the immune system. Neither B cells nor T cells are part of the MPS.

30. The answer is D. [*Chapter 1 II B 2*] A specimen that stains metachromatically with toluidine blue (i.e., stains purple) has a high net negative charge and could reasonably be expected to bind other basic dyes, such as hematoxylin and methylene blue. Acidic dyes, such as eosin, rarely stain a specimen metachromatically. DNA molecules do not have a highly negative net charge and therefore usually stain orthochromatically (blue). No direct connection exists between metachromasia and how a specimen reacts to the PAS reaction.

31. The answer is A. [*Chapter 11 III B*] Neutrophils and basophils are two types of granule-containing leukocytes called granulocytes. Neutrophils are the most common leukocytes in human blood; basophils are the rarest leukocytes. Histamine is a vasoconstrictor contained in basophilic granules. Neutrophilic granules do not contain histamine; however, they can be azurophilic. These azurophilic granules have a major role in the phagocytotic activity of neutrophils. Phagocytized bacteria release peptides and other substances that chemotactically attract neutrophils. Neutrophils have nuclei with 3–5 lobes.

32. The answer is B. [*Chapter 21 V C 1 d*] Cells in the adrenal cortex contain numerous large mitochondria with tubular cristae. In contrast, cells in the adrenal medulla have unremarkable mitochondria. Cells in the adrenal medulla also have abundant RER and a prominent Golgi apparatus located around the nucleus. The Golgi apparatus synthesizes prominent membrane-bound granules that are 200 nm in diameter. These granules store either epinephrine or norepinephrine.

33. The answer is E. [*Chapter 25 II A, B 1*] Cells in the stratum basale synthesize DNA and, therefore, show ^{3}H-thymidine incorporation. Lamellated granules are synthesized early during keratin formation and are located near the Golgi apparatus in cells in the stratum spinosum. Amorphous granules of basophilic material are prominent in the stratum granulosum. In the stratum corneum, lamellated granules are discharged into the spaces between cells.

34. The answer is D. [*Chapter 27 II A 2 a (5)*] The corneal epithelium, located on the anterior corneal surface, is a stratified squamous epithelium. It rests on Bowman's membrane, the second corneal layer. The stroma is the third corneal layer and comprises about 90% of the corneal thickness. The stroma is a mixture of fibroblasts, collagen fibers, and amorphous ground substance rich in glycosaminoglycans. The fourth corneal layer is a thick basement membrane called Descemet's membrane. The posterior corneal layer is the endothelium, which is a single layer of flattened cells (i.e., a simple squamous epithelium).

35. The answer is C. [*Chapter 21 II C 2 a*] Thyroglobulin is a glycoprotein. Thyroglobulin polypeptide chains are synthesized in the RER. Then, sugar residues are added in the RER and Golgi apparatus. Thyroglobulin is iodinated by extracellular peroxidases that are secreted into the follicular lumen along with thyroglobulin.

36. The answer is A. [*Chapter 18 V B, C*] G cells secrete gastrin and are present in the pyloric antrum. Gastrin is a polypeptide hormone that regulates hydrochloric acid secretion. I cells secrete a polypeptide called cholecystokinin (pancreozymin) and are present in most parts of the stomach. S cells secrete a polypeptide called secretin and are present in the distal stomach. Secretin stimulates alkaline secretions from the pancreas. EC cells secrete serotonin and are common throughout the gastrointestinal mucosa. A cells secrete glucagon and are scattered throughout the gastric mucosa.

37. The answer is B. [*Chapter 18 III C 1, 2, 3*] The epithelium of the small intestine is a columnar epithelium with goblet cells. The cells are joined by robust apical junctional complexes. Like all other epithelial layers, the epithelium of the small intestine rests on a basement membrane. Paneth cells probably contain apical granules of lysozyme. The glycocalyx is peculiar in that it appears to contain extracellular carbohydrate-hydrolyzing enzymes.

38. The answer is A. [*Chapter 11 VI C*] A basophilic erythroblast has much more cytoplasmic RNA than does a polychromatic erythroblast. The former uses cytoplasmic ribosomal RNA to synthesize hemoglobin. As the hemoglobin accumulates, the ribosomal RNA is replaced, and the tinctorial quality of the cytoplasm changes from basophilic to polychromatic and then to orthochromatic.

39. The answer is A. [*Chapter 11 II A–D*] Hemoglobin accounts for about 33% of the weight of a mature erythrocyte. Erythrocytes normally comprise about 40% of the total volume of blood. They lack mitochondria and are 7–8 μm in diameter. They circulate in the blood for approximately 120 days.

40–42. The answers are: 40-C, 41-E, 42-A. [*Chapter 8 I B 2; II E; Chapter 9 II A 1*] This is an example of loose (areolar) connective tissue. This connective tissue is common in the subcutaneous fascia and the mesenteries. The loose connective tissue in this micrograph is in the lamina propria, which is the connective tissue domain underlying moist epithelia in the gastrointestinal tract and elsewhere. In this sample, the loose connective tissue is underlying the pseudostratified ciliated columnar epithelium of the trachea.

43. The answer is C. [*Chapter 26 II C 2 a (2); Figure 26-5*] Two barriers prevent plasma proteins from entering the urinary filtrate. One is the glomerular basement membrane, which prevents passage of proteins with a molecular weight (MW) of 400,000 or more. The other is the podocyte slit diaphragm, which prevents passage of proteins with a MW of 160,000 or more. Low MW substances, such as a 40,000 MW tracer, pass both barriers and enter the urinary filtrate. Many of these proteins are resorbed in the proximal convoluted tubules in the renal cortex; however, a small amount of protein is lost in the urine.

44. The answer is D. [*Chapter 10 III A 1 f, B 5; V C*] The endosteum is an epithelioid layer of osteoblasts. It lines all inner cavities of gross bones; thus, all trabeculae of spongy bone and all thick layers of compact bone have an endosteum. The endosteum is a layer of osteoblasts between the marrow cavity and the bony tissue. Osteocytes are the cellular elements of bone. Osteoclasts are large, acidophilic, multinuclear cells that degrade bone. There are a few osteoclasts on the endosteal surface, but most of these cells are osteoblasts. The endosteum has the ability to form new bone and is involved in bone remodeling and repair.

45. The answer is C. [*Chapter 21 V A 2 a*] The adrenal cortex exhibits a striking morphologic zonation that reflects its functional zonation. All parenchymal cells in the adrenal cortex have round nuclei, an abundance of smooth endoplasmic reticulum (SER), and mitochondria with tubular cristae. These are characteristics associated with steroidogenic cells. Cells in the zona glomerulosa have fewer lipid droplets than cells in the zona fasciculata. They also have fewer lipofuscin granules than cells in the zona reticularis.

46. The answer is B. [*Chapter 5 II A, B; III A*] All of the cells in the list contain microtubules. Red blood cells have a cortical band of microtubules that help maintain their characteristic shape. Neutrophils have microtubules scattered throughout their cytoplasm. Tall columnar absorptive epithelial cells have microtubules aligned parallel to the long axis of the cell. Motor neuron axons contain many neurotubules (a type of microtubule) that are arranged parallel to the long axis of the axon. In cilia and flagella, microtubules are arranged in a characteristic 9 + 2 doublet configuration known as the axoneme. Since the spermatozoon is the only cell listed that has either a cilium or a flagellum, it is the only one with an axoneme.

47. The answer is A. [*Chapter 14 I B 2 b; II C 1; III A, B, C; V B 2 b*] Schwann cells are peripheral nervous system glia dedicated to myelination in the peripheral nervous system. Fibrous astrocytes and oligodendrogliocytes are both glial cells of the central nervous system (CNS). Ependymal cells also are CNS glial cells. They line the brain ventricles and central canal of the spinal cord. Purkinje cells are complex neurons in the cerebellum.

48. The answer is E. [*Chapter 14 V F 2, 3*] Cerebrospinal fluid (CSF) is a blood filtrate that is produced in the choroid plexus. It circulates in the central canal of the spinal cord and the ventricles of the brain and permeates through the subarachnoid space. The apical surfaces of the ependymal cells face the central canal and ventricles; consequently, they are bathed in CSF. CSF bathes the inner and outer surfaces of the CNS.

49. The answer is B. [*Chapter 13 IV B, C*] Cardiac myocytes are the predominant cell in the myocardium. They are striated muscle cells, and, like skeletal muscle cells, they have prominent sarcomeres with A bands, I bands, and Z lines. They exist as separate cellular entities with one nucleus in each cell and are therefore not anatomically syncytial. Their nuclei are located at the center of the muscle fiber. The action of cardiac myocytes is coordinated by numerous gap junctions, which establish ionic communication between adjacent cells and ensure coordinated beating of groups of cells.

50. The answer is B. [*Chapter 24 II A 4 a*] Testosterone is the chief androgenic steroid produced in the male reproductive system. It is synthesized by Leydig cells, which are present in the interstitial tissue between seminiferous tubules. Like other endocrine cells, Leydig cells are closely associated with fenestrated capillaries. The seminiferous tubules consist of a continuous layer of Sertoli cells and spermatogenic cells including spermatogonia, primary spermatocytes, secondary spermatocytes, spermatids, and spermatozoa.

51. The answer is E. [*Chapter 6 I C 1*] The nuclear envelope is the boundary between the nucleoplasm and the cytoplasm. It is a double unit membrane that contains nuclear pores. The pores are composed of octagonal subunits and are closed by a thin diaphragm. Nuclear pores allow macromolecules such as mRNA to pass from the nuclear genes into the cytoplasm, where they bind to ribosomes to direct protein synthesis. The outer unit membrane of the nuclear envelope is continuous in some locations with the membranes of the RER.

52. The answer is E. [*Chapter 9 II A 2*] Dense irregular connective tissue contains abundant, irregularly arranged cells and fibers. The dermis is an example of dense irregular connective tissue. Dense regular connective tissue contains many cells and fibers arranged in regular arrays. Tendons and ligaments are examples of dense regular connective tissue. Loose areolar connective tissues have fewer cells and fibers than dense connective tissues, and the cells and fibers are irregularly arranged. Mesenteries and the lamina propria are examples of loose areolar connective tissue.

53. The answer is D. [*Chapter 20 II B, C; III C 4*] The adenohypophysis consists of the pars distalis, pars intermedia, and pars tuberalis. These three regions contain basophils, acidophils, and chromophobes. Gonadotropes are one kind of basophil. Lactotropes are one kind of acidophil. Chromophobes are adenohypophysial cells that stain poorly.

The neurohypophysis consists of the pars nervosa (infundibular process), median eminence, and infundibular stem. Pituicytes are the chief parenchymal cells in the pars nervosa. Pituicytes are glial cells interspersed between the nerve processes that originate in the median eminence, course through the infundibular stem, and terminate in the pars nervosa.

54. The answer is B. [*Chapter 2 III A; IV B, C, E*] The plasma membrane is a phospholipoprotein bilayer. Cholesterol molecules become intercalated between the phospholipids in membranes, thus increasing membrane stability.

Phospholipids and proteins are amphipathic molecules because they have a hydrophilic portion which interacts strongly with water, and a hydrophobic component which interacts weakly with water. Hydrophilic components of the membrane dissolve in water in the extracellular compartment surrounding the cell or in the cell's cytoplasmic compartment. Hydrophobic components dissolve in each other within the plane of the bilayer.

Many membrane proteins are glycoproteins that have hydrophilic glycosylated components projecting into the extracellular compartment around the cell and hydrophobic components, which interact strongly with phospholipid hydrophobic components in the plane of the bilayer.

55–57. The answers are: 55-A, 56-C, 57-E. [*Chapter 8 III A 2 a*] This is an electron micrograph of microvilli in the small intestine. Microvilli contain numerous actin-rich microfilaments and are specializations of the apical surface of many epithelial cells involved in absorption. Microvilli increase the amount of cell surface area available for absorption. They are abundant in fluid-transporting epithelia, such as epithelia in the choroid plexus (CSF production), renal tubules (protein and ion transport), and small intestines (nutrient transport). Apical cells in the esophagus do not have microvilli.

58–59. The answers are: 58-B, 59-A. [*Chapter 18 II E 2*] This is a light micrograph of the muscularis externa in the stomach. It is smooth muscle. Notice that the tissue lacks striations (present in cardiac myocardium and Purkinje cells) and that the nuclei are located in the center of the cells. The density of cells per unit volume of tissue is too high for this to be lamina propria (loose connective tissue) of the gastric mucosa. Also, if this were lamina propria, lymphocytes and plasma cells would be visible. If this were dense irregular connective tissue (present in the dermis), more collagenous fibers would be visible.

60–62. The answers are: 60-B, 61-D, 62-C. [*Chapter 24 IV C*] This is a micrograph of the epididymis. The epididymis is a highly convoluted tubular organ that conveys sperm and fluid from the testis to the ductus (vas) deferens. Its luminal epithelium is a pseudostratified columnar epithelium with numerous tall apical stereocilia. These cells secrete poorly characterized substances, which are added to the seminal fluid, and remove other poorly characterized substances from the fluids that drain from the seminiferous tubules in the testis.

63–65. The answers are: 63-C, 64-B, 65-A. [*Chapter 10 II B*] This is a light micrograph of hyaline cartilage from a developing bone. In the adult, hyaline cartilage is found on the articular surfaces of bone, in tracheal cartilage, and at the ends of ribs. Hyaline cartilage consists of many chondrocytes and an extracellular matrix rich in chondroitin sulfate and Type II collagen.

66. The answer is B. [*Chapter 1 V C; Chapter 2 III B*] Freeze-fracture-etch is a technique that reveals the details of membrane structure. Fixation is not an essential part of this technique. Lightly fixed or unfixed specimens are immersed in glycerol, frozen, fractured in a vacuum, coated with metal in a vacuum, and then mounted for viewing in the transmission electron microscope. Specimens are coated with evaporated metals at 10^{-8} mm Hg, much less than one atmosphere. Glycerol infiltration and rapid freezing minimize the damaging effects of ice crystal formation. After the specimen is chemically digested, the thin metal coating becomes the specimen that is examined in the electron microscope.

67. The answer is D. [*Chapter 1 V A*] Autoradiography is used in a broad range of applications by varying the use of radiolabels and labelling techniques; however, cells must be metabolically active in order to incorporate the labelled precursor. Autoradiography can be used in conjunction with light or electron microscopy to study many types of cellular activity. For example, it can be used to reveal the sites of DNA synthesis, protein synthesis, and hormone receptors. Pulse-chase experiments allow scientists to study cell movement and the relationship between stem cells and their progeny.

68. The answer is A. [*Chapter 3 V A, B; IV B*] The Golgi apparatus consists of stacks of flattened lamella and vesicles. Its phospholipid bilayers are similar to plasma membranes; however, they have subtle biochemical and morphologic differences. The function of SER and RER is closely related to Golgi apparatus function, apparently contributing membranes to the forming face of the Golgi apparatus. Membrane-bound ribosomes exist on the RER, not the Golgi apparatus. The mRNA for protein synthesis is bound to RER ribosomes.

69. The answer is B. [*Chapter 3 VI B*] Zymogen granules exist in many secretory cells, such as pancreatic acinar cells. Condensing vacuoles that bud from the maturing face of the Golgi apparatus contain a dehydrated and condensed secretion product. These vacuoles are zymogen granule precursors and their product is less mature than the product in zymogen granules. Zymogen granules are membrane-bound vesicles that contain inactive enzymes. The enzymes are activated by peptidases in the gastrointestinal tract.

70. The answer is C. [*Chapter 3 VI A 3*] The signal hypothesis attempts to explain some aspects of the regulation of cellular secretory mechanisms. According to this hypothesis, secretory proteins are encoded in mRNA with a specific signal codon on the 3' end of the initiation codon (AUG). Signal codons are translated into signal peptides near the N-terminus, which interacts strongly with appropriate receptor proteins in RER membranes. Signal peptidases cleave signal peptides, allowing the N-terminus of the nascent polypeptide to float freely in the RER cisternal space where it begins to fold into its secondary and tertiary structure. During this conformational change, the newly formed polypeptide becomes too large to leave the RER lumen. The ribosomes release the polypeptide, and then a detachment factor releases the ribosomes from the RER membranes, sequestering the polypeptide in the RER lumen.

71. The answer is D. [*Chapter 4 II C*] Mitochondria are discrete, membrane bound cytoplasmic organelles that have many functions. They have their own genome and the ability to move and, therefore, are semi-autonomous organelles. Mitochondria synthesize DNA when they replicate and reproduce and contain enzymes that assist steroid biosynthesis. However, their primary function is ATP production, which is accomplished by oxidative phosphorylation involving TCA cycle enzymes and the electron transport chain. Enzymes that degrade nucleic acids, proteins, and polysaccharides are contained in lysosomes, not mitochondria.

72. The answer is C. [*Chapter 4 III A*] Lysosomes are specialized secretion products. They are discrete, membrane-bound cytoplasmic organelles that perform many functions. They contain an array of acid hydrolases, which are capable of degrading proteins and nucleic acids. They help degrade phagocytosed bacteria and can degrade effete intracellular organelles by fusing with them. Many lysosomal enzymes are glycoproteins; however, glycosylation occurs in the RER and Golgi apparatus during lysosome formation, not in the lysosomes themselves.

73. The answer is B. [*Chapter 5 II B*] Microtubules are cytoskeletal elements that help maintain cell shape and are involved in cell division. Skeletal muscle cells do not contain microtubules; however, they contain an abundance of microfilaments. Often, neurons are elongated cells containing neurotubules (a type of microtubule) that run parallel to the long axis of axons. The cilia of ciliated epithelia and the flagella of spermatozoa have microtubules in their axonemes. The stratum basale is a proliferative cell population, which contains numerous mitotic spindles. Mitotic spindles contain microtubules.

74–77. The answers are: 74-B, 75-E, 76-A, 77-D. [*Chapter 9 II B 1*] Fibroblasts are specialized to synthesize and secrete components of the extracellular matrix (ECM), such as collagen fibers, elastic fibers (*C*), and proteoglycans. Genes coding for the ECM polypeptides are located on genes in the nucleus (*A*). These genes are transcribed into mRNA in the nucleus. The mRNA is transported to the RER (*E*), binds to ribosomes studding the RER membranes, and is translated for polypeptide synthesis. Collagen fiber assembly and cross-linking occurs in the extracellular space (*D*). Mitochondria (*B*) synthesize the ATP used in the synthesis and secretion of extracellular matrix components.

78–80. The answers are: 78-B, 79-A, 80-C. [*Chapter 23 II B 2 a, C 2*] Granulosa cells secrete glycoproteins into the space around them in primary ovarian follicles. The glycoproteins coalesce into Call-Exner bodies, which probably fuse to form the antrum of the secondary follicle.

In the preovulatory follicle, granulosa cells secrete liquor folliculi, which accumulates in the antrum and causes its volume to increase. A mound of granulosa cells (the cumulus oophorus) surrounds the oocyte in the preovulatory follicle.

Theca interna cells secrete androgens that are converted to estrogens in granulosa cells. At ovulation, the mature follicle ruptures and releases the oocyte with an attached layer of granulosa cells called the corona radiata. At this time, the follicle becomes a corpus luteum, which is formed from granulosa cells and theca interna cells. Granulosa cells become granulosa lutein cells, cease glycoprotein secretion, and begin to secrete progesterone. Theca interna cells become theca lutein cells, which are believed to secrete steroids other than progesterone.

81–85. The answers are: 81-D, 82-C, 83-B, 84-A, 85-C. [*Chapter 19 IV B 2*] This is an electron micrograph of a pancreatic acinar cell; a classic example of a cell specialized for protein synthesis. This pancreatic acinar cell has ribosome-studded endoplasmic reticulum for protein synthesis (A), mitochondria for ATP production (B), granules of secretion product (D), and a Golgi apparatus with glycosyltransferase to add carbohydrate moieties to products destined for secretion (C).

86–90. The answers are: 86-A, 87-B, 88-D, 89-E, 90-A. [*Chapter 13 III C*] Thin filaments, which insert at the Z line (A), are the sole component of the I band (B). Thin and thick filaments do not overlap in the H band (D). The M line (E) is the middle of the sarcomere; the Z line is the end of the sarcomere. As the extent of the thick filament and thin filament overlap changes during the contraction and relaxation cycle, the I band changes length. The A band (C) is a region of overlap between thick and thin filaments.

91–93. The answers are: 91-C, 92-D, 93-A. [*Chapter 18 III C 1, F 3*] The ileal villus has a connective tissue core of lamina propria (A), which contains formed elements of the blood. The epithelium covering the villus is composed of tall columnar cells (E) with many microvilli in a brush border (D) and mucus-secreting goblet cells (C). This epithelium rests on a basement membrane (B).

94–98. The answers are: 94-B, 95-C, 96-A, 97-D, 98-E. [*Chapter 18 III C 1, F 2*] This is a transmission electron micrograph of the apical portion of a tall columnar intestinal epithelial cell in the jejunum. The nucleus (A) is bounded by a double unit membrane perforated by nuclear pores. Microvilli (B) are surface projections covered by a unit membrane and filled with actin-containing microfilaments. Microvilli increase the absorptive surface area of these cells. Mitochondria (C) are abundant in the apical cytoplasm. They produce ATP for active transport and motility of microvilli, among other things. The multivesicular body (D) present in this micrograph is part of the lysosomal system. The SER (E) contains membranes with enzymes for assembly of chylomicrons.

99–103. The answers are: 99-C, 100-D, 101-C, 102-B, 103-A. [*Chapter 18 III B, D, E*] The small intestine, like other gastrointestinal organs, has five layers in its wall. The mucosa lies next to the lumen and consists of an epithelium and a loose connective tissue domain under the epithelium called the lamina propria. The epithelium consists of tall columnar absorptive cells and goblet cells. The lamina propria contains fibroblasts, smooth muscle cells, lacteals, capillaries, and various wandering cells such as lymphocytes and plasma cells.

The muscularis mucosae (technically, part of the mucosa) is a thin layer of smooth muscle that forms the boundary between the mucosa and the submucosa. The muscularis mucosae is innervated by Meissner's plexus.

The submucosa is a connective tissue domain that joins the mucosa to the muscularis externa. The plicae circulares are permanent folds of mucosa and submucosa and are homologous to the rugae of the stomach. In the duodenum, the Brunner's glands extend into the submucosa.

The muscularis externa consists of an inner circular layer and an outer longitudinal layer of smooth muscle with the parasympathetic Auerbach's plexus between the two layers. The serosa/adventitia is external to the muscularis externa and consists of a thin layer of boundary connective tissue. In most parts of the small intestines, it is covered by the visceral component of the peritoneum (serosa). The proximal part of the duodenum is secondarily retroperitoneal and is therefore not covered by visceral peritoneum (adventitia).

104–108. The answers are: 104-B, 105-A, 106-C, 107-E, 108-D. [*Chapter 16 IV A 1*] The pseudostratified ciliated columnar epithelium in the trachea contains several cell types. The tall columnar ciliated cells have many apical cilia that beat proximally, away from the alveoli of the lungs. They carry debris, trapped in the mucus secreted by goblet cells, into the gastrointestinal tract and eventually out of the body.

Basal cells are short undifferentiated cells resting on the basement membrane. They have the ability to divide and differentiate into new cells to replace those lost from the epithelium.

Brush cells are tall columnar cells with an apical brush border but no cilia. They may be goblet cells that have lost their mucus or an intermediate step in basal cell differentiation into ciliated cells.

Granule cells have small granules in their cytoplasm. They are APUD (amine precursor uptake and decarboxylation) cells that secrete the granules to regulate mucus secretion, ciliary beating, smooth muscle contraction, and other functions in the respiratory epithelium.

109–113. The answers are: 109-D, 110-A, 111-B, 112-E, 113-C. [*Chapter 23 II A 1; III C; IV A 1; V A 1*] This is a diagram of organs in the female reproductive system. The vagina (A) has a stratified squamous unkeratinized epithelium. The vagina does not have glands; vaginal fluids are leakage from mucosal blood vessels. The cervix (B) has an epithelial transitional zone where the stratified squamous unkeratinized epithelium of the portion facing the vagina gives way to the simple columnar epithelium of the portion facing the uterus. The cervical canal has deep crypts that resemble glands. The uterus (C) and uterine tubes (E) are lined by simple columnar epithelium. Mucus-secreting cells predominate in the uterus, and ciliated cells predominate in the uterine tubes. The ovary (D) is coated with a simple cuboidal epithelium (germinal epithelium) that is continuous with the simple squamous epithelium (mesothelium) that lines the peritoneal cavity. The ovarian cortex consists of follicles containing gametogenic oocytes surrounded by stromal fibroblasts.

114–117. The answers are: 114-E, 115-B, 116-C, 117-D. [*Chapter 16 IV A 1*] The tracheal epithelium consists mainly of ciliated cells (A) and goblet cells (C). These ciliated cells move mucus and entrapped debris. The epithelium rests on a basement membrane (E). Beneath the basement membrane is a lamina propria with multicellular glands (B); beneath the lamina propria is cartilage (D).

118–121. The answers are: 118-B, 119-C, 120-E, 121-D. [*Chapter 10 II B 4*] Protein synthesis begins in the RER (B). Mitochondria (C) produce ATP. The secretion product vacuoles (E) contain materials destined for the extracellular matrix. Glucuronyltransferases are involved in glycosaminoglycan synthesis and are localized on Golgi membranes (D).

122–126. The answers are: 122-C, 123-D, 124-E, 125-B, 126-A. [*Chapter 27 IV C 4; Figure 27-8*] This is a light micrograph of the organ of Corti. It is contained within the scala media (D), which is also called the cochlear duct. The vestibular (Reissner's) membrane (A) is the boundary between the scala media and the scala vestibuli (E). The hair cells of the organ of Corti (B) are connected to the basilar membrane (C). When the basilar membrane vibrates, the tips of the hair cells are deformed, initiating an action potential.

127–131. The answers are: 127-A, 128-D, 129-C, 130-A, 131-B. [*Chapter 19 IV B 2 b*] Pancreatic acinar cells manufacture protein for export. Protein synthesis begins in the RER (A) and continues in the Golgi apparatus (C), where proteins are glycosylated. Next, condensing vacuoles (E) are converted into zymogen granules (D), just prior to their release from the cell. Protein synthesis is an energy-dependent process, driven by ATP from mitochondria (B).

132–136. The answers are: 132-D, 133-E, 134-B, 135-A, 136-C. [*Chapter 26 II C 2*] This is a low-power transmission electron micrograph of the renal cortex. In this micrograph, podocytes (E) and vascular endothelial cells (A) rest on a thick glomerular basement membrane (C). Urinary filtrate is produced as some blood constituents pass from the lumen of the capillary (B), through the capillary endothelial fenestrations, across the glomerular basement membrane, and between the podocyte foot processes at the filtration slits. After crossing these barriers, the filtrate enters the urinary space (D). An erythrocyte is shown in cross section in the vascular space just above the endothelial cell (A).

137–140. The answers are: 137-C, 138-A, 139-B, 140-D. [*Chapter 17 IV B; Chapter 18 II B, C, D*] This is a photomicrograph of the gastroesophageal junction. At this location, the stratified squamous unkeratinized epithelium of the esophagus (A) gives way to a simple columnar epithelium of the cardiac stomach (B). There are shallow gastric pits in the cardiac region consisting solely of mucus-secreting cells. Other visible structures include the esophageal lumen (E), a small blood vessel (D), and the loose areolar connective tissue of the mucosal lamina propria (C).

141–143. The answers are: 141-C, 142-B, 143-B. [*Chapter 18 II A, C 2, D*] This is a micrograph of the lower portion of the gastric pits, which are characteristic of the fundic/corpic stomach. They consist of a mixture of parietal cells (*B*), which secrete HCl and gastric intrinsic factor, and chief cells (*C*), which secrete pepsinogen. The other letters label the lamina propria (*A, D, E*).

144–146. The answers are: 144-C, 145-A, 146-B. [*Chapter 19 II B 2 a, D 1; III B*] This is a scanning electron micrograph of the liver. Liver parenchymal cells (*A*) secrete bile into the bile canaliculi (*B*) and blood proteins, such as serum albumin and transferrin, into the liver sinusoids (*C*). Liver sinusoids are modified fenestrated and discontinuous capillaries that receive blood from the portal vein and hepatic artery in the portal canals and carry it to the central veins.

147–150. The answers are: 147-A, 148-B, 149-B, 150-C. [*Chapter 15 I C 2 a, c*] The tunica adventitia and the epicardium are located at the abluminal portion of the cardiovascular system and are histologically similar. The tunica intima and the endocardium are located at the luminal portion of the cardiovascular system and are histologically similar. The tunica adventitia contains the vasa vasorum, which is the intrinsic blood supply of large vessels, such as the aorta. Muscular arteries have both an intima and an adventitia.

151–154. The answers are: 151-C, 152-B, 153-D, 154-A. [*Chapter 11 III A, B, F*] Neutrophils are one example of granulocytes and monocytes are one example of agranulocytes; both are phagocytes. All leukocytes have a nuclear envelope and mitochondria. Granulocytes contain specific granules and have several prominent nuclear lobes; agranulocytes do not. Granulocytes and agranulocytes both contain many lysosomes.

155–158. The answers are: 155-C, 156-D, 157-A, 158-B. [*Chapter 20 II C 1 c, 2 a, D 1*] Lactotropes are acidophils that secrete a low molecular weight polypeptide hormone called prolactin. Thyrotropes are basophils that secrete a high molecular weight glycoprotein hormone called thyroid-stimulating hormone (TSH). Both lactotropes and thyrotropes are present in the pars distalis of the adenohypophysis; neither are present in the median eminence of the neurohypophysis.

Index

Note: Page numbers in italics denote illustrations, those followed by *t* denote tables, those followed by Q denote questions, and those followed by E denote explanations.